Functions of Several Complex Variables and Their Singularities

Functions of Several Complex Variables and Their Singularities

Wolfgang Ebeling

Translated by Philip G. Spain

Graduate Studies in Mathematics

Volume 83

American Mathematical Society
Providence, Rhode Island

Originally published in the German language by
Friedr. Vieweg & Sohn Verlag, D-65189 Wiesbaden, Germany,
as "Wolfgang Ebeling: Funktionentheorie, Differentialtopologie und Singularitäten.
1. Auflage (1st edition)".

Translated by Philip G. Spain

2000 *Mathematics Subject Classification.* Primary 32–01;
Secondary 32S10, 32S55, 58K40, 58K60.

For additional information and updates on this book, visit
www.ams.org/bookpages/gsm-83

Library of Congress Cataloging-in-Publication Data

Ebeling, Wolfgang.
[Funktionentheorie, differentialtopologie und singularitäten. English]
Functions of several complex variables and their singularities / Wolfgang Ebeling ; translated by Philip Spain.
p. cm. — (Graduate studies in mathematics, ISSN 1065-7339 ; v. 83)
Includes bibliographical references and index.
ISBN 0-8218-3319-7 (alk. paper)
1. Functions of several complex variables. 2. Singularities (Mathematics) I. Title.

QA331.E27 2007
515′.94—dc22 2007060745

∞ The paper used in this book is acid-free and falls within the guidelines
established to ensure permanence and durability.
Visit the AMS home page at http://www.ams.org/

10 9 8 7 6 5 4 3 2 1 12 11 10 09 08 07

Contents

Foreword to the English translation

The German title of the book is "Funktionentheorie, Differentialtopologie und Singularitäten". The book is an introduction to the theory of functions of several complex variables and their singularities, with special emphasis on topological aspects. Its aim is to guide the reader from the fundamentals to more advanced topics of recent research. It originated from courses given by the author to German mathematics students at the University of Hanover.

I am very happy that the AMS has provided an English edition of my book. I am grateful to Edward Dunne, the editor of the book program, for his efforts. My particular thanks go to Philip Spain, who translated this book into English. He has done a very good job.

I have taken the opportunity to make some corrections and improvements in the text. I am grateful to Theo de Jong and Helmut Köditz for their comments and suggestions for improvement.

Hanover, January 2007 Wolfgang Ebeling

Introduction

The study of singularities of analytic functions can be considered as a subarea of the theory of functions of several complex variables and of algebraic/analytic geometry. It has in the meantime, together with the theory of singularities of differentiable mappings, developed into an independent subject, singularity theory. Through its connections with very many other mathematical areas and applications to natural and economic sciences and in technology (for example, under the heading 'catastrophe theory') this theory has aroused great interest. The particular appeal, but also its particular difficulty, lies in the fact that deep results and methods from various branches of mathematics come into play here.

The aim of this book is to present the foundations of the theory of functions of several complex variables and on this basis to develop the fundamental concepts of the theory of isolated singularities of holomorphic functions systematically. It is derived from lectures given by the author to mathematics students in their third and fourth year to introduce them to current research questions in the area of the theory of functions of several variables. The book has its genesis in this. As prerequisites we assume only an introductory knowledge of the theory of functions of a single complex variable and of algebra, such as students will normally acquire in their first two years of study. The first two chapters correspond to a continuation of the course on complex analysis and deal with Riemann surfaces and the theory of functions of several complex variables. They also present an introduction to local complex geometry. In the third chapter the results will be applied to deformation and classification of isolated singularities of holomorphic functions. These three chapters have grown from notes for the author's lectures on Riemann surfaces and the theory of functions of several complex variables delivered in Hanover in the winter semester of 1998/1999 and the summer semester of 1999. Parts of these notes go back to similar courses given in the winter semester of 1992/1993 and the summer semester of 1993.

The rest of the book deals with the topological study of these singularities begun in the now classical book of J. Milnor [**Mil68**]. Picard-Lefschetz theory is an important tool and can be viewed as a complex version of Morse theory. It is expounded at the beginning of the second volume of the extensive two-volume standard work of V. I. Arnold, S. M. Gusein-Zade and A. N. Varchenko [**AGV85, AGV88**]. These books assume considerable prior knowledge. We offer an introduction to this theory in the last two chapters of the present book. In the fourth chapter we first present the necessary foundations of algebraic and of differential topology. The fifth chapter introduces the topological study of singularities. It rests in part on [**AGV88**, Part I. The topological structure of isolated critical points of functions]. At the end of this chapter there is a survey of topical results, some presented without proof. The last two chapters are based on a course on singularities delivered by the author in Hanover in the winter semester of 1993/1994.

This book can be used for a course on functions of several complex variables, an introductory course on differential topology, or for a special course or seminar on an introduction to singularity theory. The first two chapters would be suitable for a further course on functions of several complex variables. The beginning of §1.1, §1.2, and the first four sections of Chapter 3 and Chapter 4 treat themes from differential topology and can be read independently of the rest of the book: they can therefore serve as the basis of an introductory course on differential topology. Chapter 3 and Chapter 5 can be used as reading for a seminar on *Introductory singularity theory*, with reference back to the results of the previous chapters according to the state of knowledge of the participants.

Naturally the themes discussed here are only a small choice from a great variety of possibilities. This choice has been shaped by the author's own predilections and by his work. Nevertheless the author hopes that his book presents a good foundation for the study of the more advanced literature indicated in the bibliography.

I thank Sigrid Guttner and Robert Wetke most sincerely for their careful preparation of the majority of the LaTeX files. Robert Wetke also deserves special thanks for the preparation of the computer diagrams. I am most grateful to Dr. Michael Lönne and Dr. Jörg Zintl for their help in proofreading.

Hanover, January 2001 Wolfgang Ebeling

List of figures

List of tables

Chapter 1

Riemann surfaces

1.1. Riemann surfaces

Let us start from the concept of a topological manifold.

Definition. An *n-dimensional topological manifold* M is a Hausdorff space with a countable basis for its topology with the property that every point $a \in M$ has an open neighborhood U which is homeomorphic to an open neighborhood V of $\mathbb{R}^n$.

A homeomorphism $\varphi : U \to V$, $U \subset M$ open, $V \subset \mathbb{R}^n$ open, is called a *chart*.

A family $\mathfrak{A} = \{\varphi_i : U_i \to V_i\}_{i \in I}$ of charts is called an *atlas* of M if $\bigcup_{i \in I} U_i = M$.

Let $\varphi_1 : U_1 \to V_1$, $\varphi_2 : U_2 \to V_2$ be two charts with $U_1 \cap U_2 \neq \emptyset$. Then

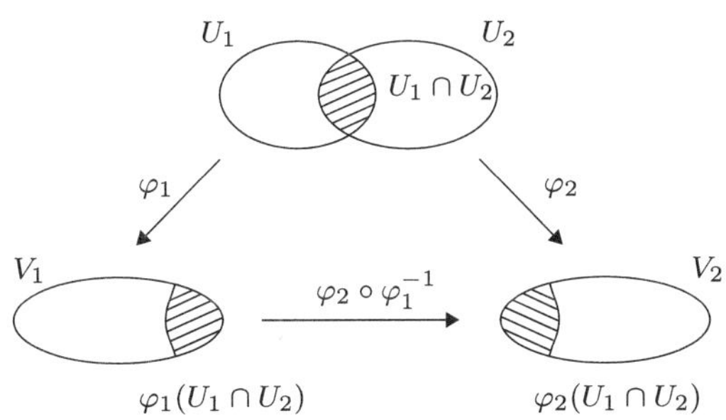

Figure 1.1. Transition function

the map

$$\varphi_2 \circ \varphi_1^{-1} : \varphi_1(U_1 \cap U_2) \longrightarrow \varphi_2(U_1 \cap U_2)$$

is called the corresponding *transition function* (see Figure 1.1).

A 2-dimensional topological manifold is also called a (topological) *surface.*

Definition. Let X be a surface. In what follows we identify $\mathbb{R}^2$ in the natural way with $\mathbb{C}$.

An atlas of X is called *complex* (or *holomorphic*) if all its transition functions are holomorphic.

Two complex atlases $\mathfrak{U}$ and $\mathfrak{V}$ are called *equivalent* if and only if $\mathfrak{U} \cup \mathfrak{V}$ is also a complex atlas.

A *complex structure* on X is an equivalence class of complex atlases on X.

An (abstract) *Riemann surface* is a connected surface with a complex structure.

Remark 1.1. Each complex structure contains a uniquely determined maximal atlas $\mathfrak{A}^*$: If $\mathfrak{A}$ is an arbitrary atlas in the corresponding equivalence class, then

$$\mathfrak{A}^* := \left\{ \varphi : U \longrightarrow V \text{ chart} \;\middle|\; \begin{array}{l} \text{the transition functions of } \varphi \\ \text{to all charts of } \mathfrak{A} \text{ are holomorphic} \end{array} \right\}$$

is a maximal atlas.

Convention. If X is a Riemann surface, then a chart of X is always a chart of the maximal atlas of the complex structure.

Example 1.1. The complex plane $\mathbb{C}$ is a Riemann surface. The set $\{\mathrm{id} : \mathbb{C} \to \mathbb{C}\}$ forms a complex atlas.

Example 1.2. Every region $G \subset \mathbb{C}$ is a Riemann surface: the set $\{\mathrm{id}|_G : G \to G\}$ is an atlas. More generally, if X is a Riemann surface and $Y \subset X$ is an open and connected subset of X, then Y is also a Riemann surface. The charts $\varphi : U \to V$ of X, where $U \subset Y$, form another atlas.

Example 1.3. The Riemann sphere $\hat{\mathbb{C}} = \mathbb{C} \cup \{\infty\}$ is a Riemann surface. We put

$$U_1 := \hat{\mathbb{C}} \setminus \{\infty\} = \mathbb{C}, \quad \varphi_1 = \mathrm{id} : U_1 \to \mathbb{C},$$

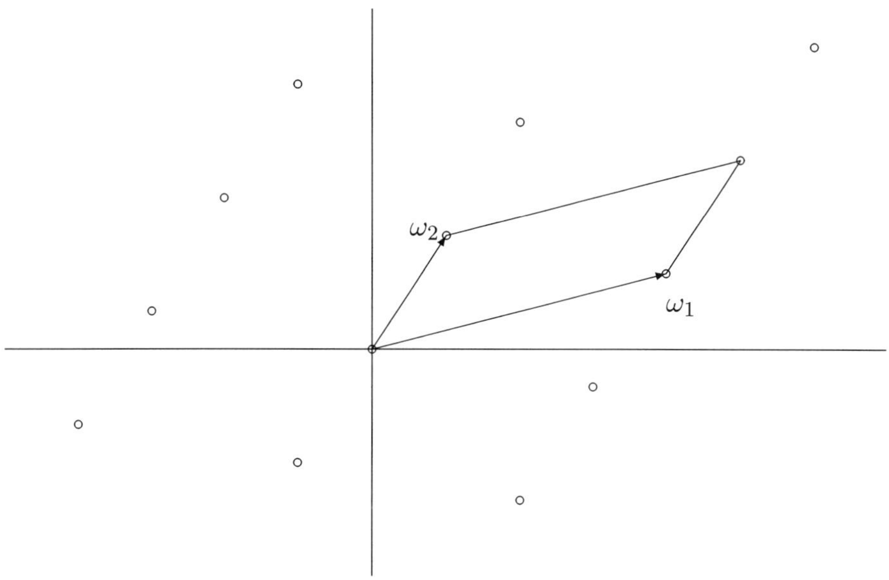

Figure 1.2. Lattice L and parallelogram P

$$U_2 := \hat{\mathbb{C}} \setminus \{0\} = \mathbb{C}^* \cup \{\infty\}, \quad \varphi_2 : \begin{array}{l} U_2 \to \mathbb{C} \\ z \mapsto \begin{cases} \frac{1}{z} & \text{for } z \in \mathbb{C}^*, \\ 0 & \text{for } z = \infty. \end{cases} \end{array}$$

Here $\mathbb{C}^* = \mathbb{C} \setminus \{0\}$. Then $\{\varphi_i : U_i \to \mathbb{C} \,|\, i = 1, 2\}$ is an atlas of $\hat{\mathbb{C}}$. This atlas is complex: we have $\varphi_1(U_1 \cap U_2) = \varphi_2(U_1 \cap U_2) = \mathbb{C}^*$ and $\varphi_2 \circ \varphi_1^{-1} : \mathbb{C}^* \to \mathbb{C}^*, z \mapsto 1/z$, is biholomorphic.

Example 1.4. Let $\omega_1, \omega_2 \in \mathbb{C}$ be two linearly independent elements over $\mathbb{R}$. We define

$$L := \{m_1\omega_1 + m_2\omega_2 \,|\, m_1, m_2 \in \mathbb{Z}\}.$$

Such a subset $L \subset \mathbb{C}$ is called a *lattice* in $\mathbb{C}$. A lattice L is a subgroup of $\mathbb{C}$ that as an abelian group is isomorphic to $\mathbb{Z} \times \mathbb{Z}$. A lattice L defines an equivalence relation on $\mathbb{C}$: two complex numbers $z, z' \in \mathbb{C}$ are called equivalent modulo L (in symbols $z \equiv z' \,(\operatorname{mod} L)$) if and only if $z - z' \in L$. The set of all equivalence classes will be denoted by $\mathbb{C}/L$. Each equivalence class has a representative in the parallelogram

$$P := \{\lambda_1\omega_1 + \lambda_2\omega_2 \,|\, 0 \le \lambda_1 \le 1, 0 \le \lambda_2 \le 1\}.$$

Moreover P satisfies: no two interior points are equivalent, and no boundary point is equivalent to any interior point. Furthermore, P is compact. Such a domain is called a *fundamental domain* and P is also called the *fundamental parallelogram*. One obtains $\mathbb{C}/L$ $(= \mathbb{R}^2/L)$ from P by identifying the opposite sides:

$$\begin{aligned} \alpha &\equiv \alpha + \omega_2 \pmod{L}, & \alpha = \lambda_1\omega_1\,(0 \le \lambda_1 \le 1), \\ \beta &\equiv \beta + \omega_1 \pmod{L}, & \beta = \lambda_2\omega_2\,(0 \le \lambda_2 \le 1). \end{aligned}$$

Graphically, this means that one first identifies two opposite sides to obtain a cylinder, then identifies the two ends: one obtains a *torus*.

We can next introduce a topology $\mathbb{C}/L$ as follows. Let

$$\pi : \mathbb{C} \to \mathbb{C}/L$$

be the quotient map. We declare a subset $U \subset \mathbb{C}/L$ to be open if $\pi^{-1}(U)$ is open in $\mathbb{C}$. With this topology $\mathbb{C}/L$ is a Hausdorff space and the quotient map $\pi : \mathbb{C} \to \mathbb{C}/L$ is continuous. Since $\mathbb{C}$ is connected, $\mathbb{C}/L$ is also connected. Since $\mathbb{C}/L$ is obtained from P by the identification of opposite sides, and P is compact, $\mathbb{C}/L$ is also compact.

We next introduce a complex structure on $\mathbb{C}/L$. Let $a \in \mathbb{C}/L$. We choose a preimage point $z \in \pi^{-1}(a)$ and a small open disc V around z containing only pairwise inequivalent points modulo L. By definition of the topology on $\mathbb{C}/L$ the set $U := \pi(V) \subset \mathbb{C}/L$ is then open, and $\pi|_V : V \to U$ is a homeomorphism. Its inverse map $\varphi : U \to V$ is a chart on $\mathbb{C}/L$ around a.

Let $\mathfrak{U}$ be the set of all such charts. Let $\varphi_1 : U_1 \to V_1$ and $\varphi_2 : U_2 \to V_2$ be two charts from $\mathfrak{U}$. Then the transition function $\varphi_2 \circ \varphi_1^{-1} : \varphi_1(U_1 \cap U_2) \to \varphi_2(U_1 \cap U_2)$ is the translation by a lattice vector, hence is holomorphic.

Therefore $\mathbb{C}/L$ is a compact Riemann surface when endowed with the complex structure defined by the atlas $\mathfrak{U}$.

We next want to define holomorphic maps between Riemann surfaces.

Definition. Let X be a Riemann surface and $Y \subset X$ an open subset. A function $f : Y \to \mathbb{C}$ called *holomorphic* if for each chart $\varphi : U \to V$ on X the function $f \circ \varphi^{-1} : \varphi(U \cap Y) \to \mathbb{C}$ is holomorphic. We denote by $\mathcal{O}(Y)$ the set of all holomorphic functions on Y.

Remark 1.2. (i) $\mathcal{O}(Y)$ is a $\mathbb{C}$-algebra.

(ii) The condition need be verified only for some family of charts that covers Y.

(iii) A chart $\varphi : U \to V$ is holomorphic. The coordinate on X defined through φ is called a *local coordinate* or a *local uniformizing parameter* of X, and (U, φ) is called a *coordinate neighborhood* of $a \in U$. One often writes z instead of φ.

Definition. Let X be a Riemann surface, $a \in X$, and let $f : X \setminus \{a\} \to \mathbb{C}$ be a function. The point a is called an *isolated singularity of* f if there is an open neighborhood U of a for which f is holomorphic in $U \setminus \{a\}$.

Let a be an isolated singularity of f. The singularity a is called *removable* (resp. a *pole of order* n, resp. an *essential singularity*) if there is a chart $\varphi : U \to V$ with $a \in U$ for which $f \circ \varphi^{-1}$ has a removable singularity (resp. a pole of order n, resp. an essential singularity) at $z_0 = \varphi(a)$.

Remark 1.3. All these definitions are independent of the coordinates. For example, let a be a pole of f of order n with respect to $\varphi : U \to V$ with $\varphi(a) = z_0$. Then

$$f \circ \varphi^{-1}(z) = \frac{a_{-n}}{(z - z_0)^n} + \dots \qquad \text{(Laurent expansion).}$$

Let $\varphi' : U' \to V'$ be another chart, $\varphi'(a) = z_0'$. Then

$$\varphi \circ \varphi'^{-1}(z') = z = z_0 + b(z' - z_0') + \dots$$

and $b \neq 0$, so $\varphi \circ \varphi'^{-1}$ is biholomorphic. Thus

$$\begin{aligned} f \circ \varphi'^{-1}(z') &= f \circ \varphi^{-1} \circ (\varphi \circ \varphi'^{-1})(z') \\ &= \frac{a_{-n}}{(b(z' - z_0') + \dots)^n} + \dots \\ &= \frac{a_{-n}}{b^n (z' - z_0')^n} + \dots . \end{aligned}$$

Note, however, that the coefficient a_{-n} of $(z - z_0)^{-n}$ depends on the chart. There is no particular value to identify as a residue!

The following proposition is an immediate consequence of Riemann's removable singularity theorem in the complex plane.

Proposition 1.1 (Riemann's removable singularity theorem)**.** *Let X be a Riemann surface, $U \subset X$ an open subset, and $a \in U$. Furthermore, let $f : U \setminus \{a\} \to \mathbb{C}$ be a holomorphic function. If a is a removable singularity of f, then f can be extended in a unique way to a holomorphic function $f : U \to \mathbb{C}$.*

Remark 1.4. In particular, a point a must be a removable singularity if f is bounded in a neighborhood of a.

Definition. Let X, Y be Riemann surfaces. A map $f : X \to Y$ is called *holomorphic* if f is continuous and if the map

$$\psi \circ f \circ \varphi^{-1} : V \to V'$$

is holomorphic for each pair of charts $\varphi : U \to V$ of X and $\psi : U' \to V'$ of Y with $f(U) \subset U'$.

A map $f : X \to Y$ is called *biholomorphic* if f is bijective and f and f^{-1} are holomorphic.

Two Riemann surfaces are called *isomorphic* or *conformally equivalent* if there is a biholomorphic map $f : X \to Y$.

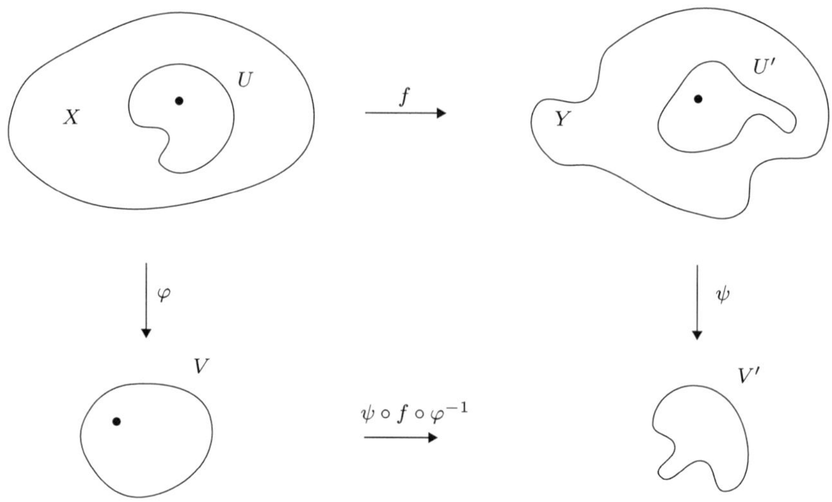

Figure 1.3. Definition of a holomorphic map

Remark 1.5. Holomorphic maps $f : X \to Y = \mathbb{C}$ are holomorphic functions.

Remark 1.6. If X, Y, Z are Riemann surfaces and $f : X \to Y$, $g : Y \to Z$ are holomorphic, then $g \circ f : X \to Z$ is also holomorphic.

Remark 1.7. A map $f : X \to Y$ is holomorphic if and only if f is continuous and for each open set $V \subset Y$ and each holomorphic function $\varphi \in \mathcal{O}(V)$ the function $\varphi \circ f : f^{-1}(V) \to \mathbb{C}$ is holomorphic, i.e. $\varphi \circ f \in \mathcal{O}(f^{-1}(V))$. A holomorphic map $f : X \to Y$ thus induces a map

$$\begin{aligned} f^* : \mathcal{O}(V) &\to \mathcal{O}(f^{-1}(V)) \\ \varphi &\mapsto \varphi \circ f. \end{aligned}$$

This map f^* is a ring homomorphism. If $g : Y \to Z$ is holomorphic, $W \subset Z$ is open, $V := g^{-1}(W)$, and $U := f^{-1}(V)$, then for $(g \circ f)^* : \mathcal{O}(W) \to \mathcal{O}(U)$ we have

$$(g \circ f)^* = f^* \circ g^*.$$

Proposition 1.2 (Identity theorem). *Let X, Y be Riemann surfaces and let $f_1, f_2 : X \to Y$ be two holomorphic maps that agree on a subset $A \subset X$ having an accumulation point $a \in X$. Then $f_1 \equiv f_2$.*

Proof. Let

$$G := \{x \in X \mid \text{there is a neighborhood } W \text{ of } x \text{ with } f_1|_W = f_2|_W\}.$$

By definition G is open. We shall show that G is also closed.

Let b be an accumulation point of G. From the continuity of f_1 and f_2 it follows that $f_1(b) = f_2(b)$. There are therefore charts $\varphi : U \to V$ on X and $\psi : U' \to V'$ on Y with $b \in U$ and $f_i(U) \subset U'$, $i = 1, 2$. Without loss of generality we may assume that U is connected. The map

$$g_i := \psi \circ f_i \circ \varphi^{-1} : V \to V' \subset \mathbb{C}$$

is holomorphic for $i = 1, 2$. From the identity theorem for holomorphic functions of one complex variable (cf. e.g. [**Ahl78**]), and since $U \cap G \neq \emptyset$, it follows that $g_1 \equiv g_2$. Hence

$$f_1|_U = f_2|_U,$$

so $b \in G$ and therefore G is closed.

Since X is connected, it follows that $G = \emptyset$ or $G = X$. But from the identity theorem for holomorphic functions of one complex variable we have $a \in G$. Thus $G = X$, i.e. $f_1 \equiv f_2$ on the whole of X. □

Definition. Let X be a Riemann surface. A *meromorphic function* on X is a holomorphic function $f : X \setminus Z \to \mathbb{C}$ where

(i) Z is a discrete subset of X;

(ii) the points of Z are poles of f.

Let $\mathcal{M}(X)$ be the set of all meromorphic functions on X.

We shall now interpret meromorphic functions as holomorphic maps into the Riemann sphere.

Proposition 1.3. *Let X be a Riemann surface. If f is a meromorphic function on X, and if one defines $f(a) := \infty$ for a pole a of f, one then has a holomorphic map $f : X \to \hat{\mathbb{C}}$.*

Conversely, if $f : X \to \hat{\mathbb{C}}$ is a holomorphic map, then either f is constant, and equal to ∞, or $f^{-1}(\infty)$ is a discrete subset of X and f is meromorphic on X.

Proof. a) First let f be meromorphic on X with pole set Z. On defining $f(a) := \infty$ for $a \in Z$, one obtains a continuous map $f : X \to \hat{\mathbb{C}}$. Let $\varphi : U \to V$ be a chart of X, $\psi : U' \to V'$ a chart of $\hat{\mathbb{C}}$ with $f(U) \subset U'$. We need to show that

$$g := \psi \circ f \circ \varphi^{-1} : V \to V'$$

is holomorphic. Since f is holomorphic on $X \setminus Z$, then g is holomorphic on $V \setminus \varphi(U \cap Z)$. But since $g : V \to V' \subset \mathbb{C}$ is continuous, it follows from the Riemann removable singularity theorem (Proposition 1.1) that g is holomorphic on all of V.

b) The converse follows from the identity theorem (Proposition 1.2). □

Proposition 1.4. *The set $\mathcal{M}(X)$ of meromorphic functions on the Riemann surface X is a field with $\mathbb{C} \subset \mathcal{O}(X) \subset \mathcal{M}(X)$.*

Proof. It follows from Proposition 1.3 that the identity theorem (Proposition 1.2) holds for meromorphic functions on X too. Hence an $f \in \mathcal{M}(X)$, $f \not\equiv 0$, has only isolated zeros. Therefore $1/f$ is meromorphic on X. □

Proposition 1.5 (Local form of holomorphic maps). *Let X, Y be Riemann surfaces and $f : X \to Y$ a holomorphic map that is not constant, with $a \in X$ and $b := f(a)$.*

There are then a natural number $k \geq 1$, a chart $\varphi : U \to V$ of X around a with $\varphi(a) = 0$, and a chart $\psi : U' \to V'$ of Y with $\psi(b) = 0$ and $f(U) \subset U'$, such that the map $F := \psi \circ f \circ \varphi^{-1} : V \to V'$ can be expressed as

$$F(z) = z^k$$

in terms of the coordinate z in V.

Proof. Let $\varphi_1 : U_1 \to V_1$, $\psi : U' \to V'$ be charts of X resp. Y with $a \in U_1$, $\varphi_1(a) = 0$, $b \in U'$, $\psi(b) = 0$ and $f(U_1) \subset U'$. Consider the function

$$f_1 := \psi \circ f \circ \varphi_1^{-1} : V_1 \to V' \subset \mathbb{C}.$$

By the identity theorem f_1 is not constant. Since $f_1(0) = 0$, there is a $k \in \mathbb{N}$, $k \geq 1$, such that

$$f_1(z) = z^k g(z),$$

for a holomorphic function $g : V_1 \to \mathbb{C}$ with $g(0) \neq 0$. Choose $\varepsilon > 0$ such that for all $z \in \mathbb{C}$ with $|z| < \varepsilon$ we have $z \in V_1$ and $|g(z) - g(0)| < |g(0)|$. In the ε-neighborhood of 0 there is a holomorphic function h with $h(z)^k = g(z)$ (i.e., a branch of $\sqrt[k]{g(z)}$ is defined).

There are open neighborhoods V_2, V of 0 such that the map

$$\begin{aligned} \alpha : V_2 &\to V \\ z &\mapsto zh(z) \end{aligned}$$

is holomorphic. Let $U := \varphi_1^{-1}(V_2)$, $\varphi = \alpha \circ \varphi_1|_U : U \to V$. By construction of the map $F := \psi \circ f \circ \varphi^{-1}$ we have $F(z) = z^k$.

As in Remark 1.3 one can show that the number k is independent of the choice of charts $\varphi_1 : U_1 \to V_1$ and $\psi : U' \to V'$. □

It follows from this proposition that there is a neighborhood U of a such that each $y \in f(U)$ with $y \neq b$ has exactly k preimages under f.

Definition. We call the number k in Proposition 1.5 the *order* or *multiplicity* of f at the point a. Symbolically, $k = \text{ord}_a f$.

Corollary 1.1. *Let X, Y be Riemann surfaces and $f : X \to Y$ a holomorphic map that is not constant. Then f is open, i.e., the image under f of each open set in X is open in Y.*

Proof. It follows immediately from Proposition 1.5 that if U is an open neighborhood of $a \in X$, then so is $f(U)$ an open neighborhood of $f(a)$. □

Corollary 1.2. *Let X, Y be Riemann surfaces and $f : X \to Y$ an injective holomorphic map. Then $f : X \to f(X)$ is biholomorphic.*

Proof. If f injective, then $\mathrm{ord}_a f = 1$ for all $a \in X$ and in suitable local coordinates f has the form $f(z) = z$. It follows that $f^{-1} : f(X) \to X$ is holomorphic. □

Corollary 1.3. *Let X be a Riemann surface and $f : X \to \mathbb{C}$ a holomorphic function that is not constant. Then the real function $|f|$ has no local maximum on X.*

Proof. If $|f|$ has a local maximum at $a \in X$, then there is an open neighborhood U of a with $|f(x)| \leq |f(a)|$ for all $x \in U$. But then $|f(a)|$ is a boundary point of $|f|(U)$ in contradiction to the fact that $f(U)$ is open. □

Corollary 1.4. *Let X be a compact Riemann surface and $f : X \to \mathbb{C}$ a holomorphic function. Then f is constant.*

Proof. Since X is compact and $|f|$ is continuous, $|f|$ must attain a maximum on X. The conclusion follows from Corollary 1.3. □

1.2. Homotopy of paths, fundamental groups

We now need to prepare several topological tools concerning the homotopy of paths. One can find them in most books on algebraic topology, e.g. in [**GH81**].

A *path* in a topological space X is a continuous map $\gamma : I = [0,1] \to X$.

Definition. Let X be a topological space and $x_0, x_1 \in X$. Let $\gamma_1, \gamma_2 : I \to X$ be two paths with the same start and end point, i.e., $\gamma_1(0) = \gamma_2(0) = x_0, \gamma_1(1) = \gamma_2(1) = x_1$. The paths γ_1 and γ_2 are said to be *homotopic*, in symbols $\gamma_1 \simeq \gamma_2$, when there is a continuous map $F : I \times I \to X$ with

(i) $F(s, 0) = \gamma_1(s)$ for $s \in I$,

(ii) $F(s, 1) = \gamma_2(s)$ for $s \in I$,

(iii) $F(0, t) = x_0$ for $t \in I$,

(iv) $F(1, t) = x_1$ for $t \in I$.

The map F is called a *homotopy* between γ_1 and γ_2.

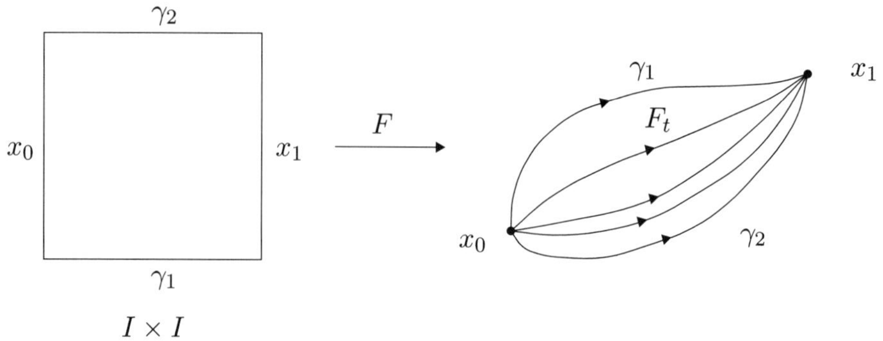

Figure 1.4. A homotopy F between γ_1 and γ_2

For fixed $t \in I$ we obtain a path

$$\begin{aligned} F_t \ : \ & I \to X \\ & s \mapsto F_t(s) := F(s,t). \end{aligned}$$

F_t is a path from x_0 to x_1 for each $t \in I$. The family $(F_t)_{0 \le t \le 1}$ can be viewed as a deformation of γ_1 to γ_2 (cf. Figure 1.4).

One can verify immediately that

(1) $\gamma_1 \simeq \gamma_1$,

(2) $\gamma_1 \simeq \gamma_2 \Rightarrow \gamma_2 \simeq \gamma_1$,

(3) $\gamma_1 \simeq \gamma_2, \gamma_2 \simeq \gamma_3 \Rightarrow \gamma_1 \simeq \gamma_3$.

If, furthermore, one defines the *product* of two paths γ_1 and γ_2 with $\gamma_1(1) = \gamma_2(0)$ by

$$\begin{aligned} \gamma_1\gamma_2 \ &: \ I \longrightarrow X \\ \gamma_1\gamma_2(s) &= \begin{cases} \gamma_1(2s), & 0 \le s \le \frac{1}{2}, \\ \gamma_2(2s-1), & \frac{1}{2} \le s \le 1, \end{cases} \end{aligned}$$

then

(4) $\gamma_1 \simeq \gamma_1', \gamma_2 \simeq \gamma_2' \Rightarrow \gamma_1\gamma_2 \simeq \gamma_1'\gamma_2'$.

All these claims are easy to verify. In the case of (4) one proceeds as follows: If $F : I \times I \to X$ is a homotopy between γ_1 and γ_1' and $G : I \times I \to X$ is a homotopy between γ_2 and γ_2', then a homotopy is defined between $\gamma_1\gamma_2$ and $\gamma_1'\gamma_2'$ by

$$\begin{aligned} FG \ &: I \times I \longrightarrow X \\ FG(s,t) &= \begin{cases} F(2s,t), & 0 \le s \le \frac{1}{2}, \\ G(2s-1,t), & \frac{1}{2} \le s \le 1. \end{cases} \end{aligned}$$

We remark that $F(1,t) = \gamma_1(1) = \gamma_2(0) = G(0,t)$. This can be represented symbolically as in Figure 1.5. In this way we have defined an *equivalence*

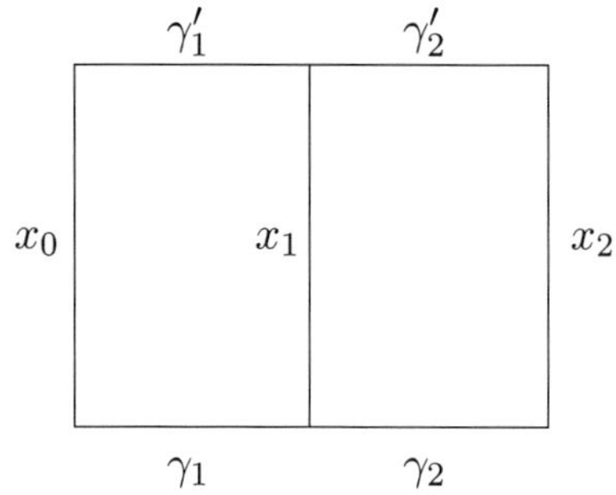

Figure 1.5. The homotopy FG

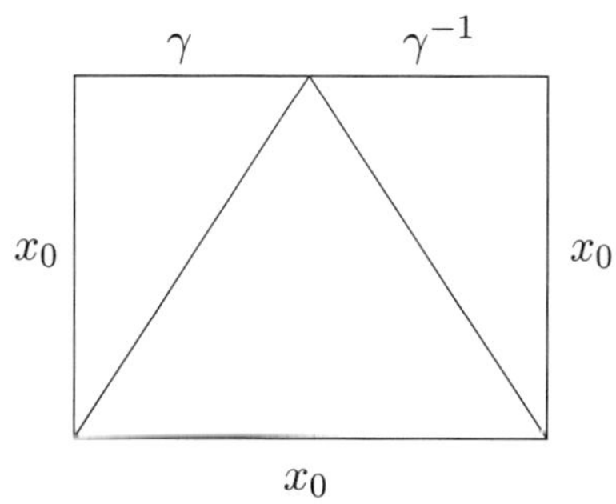

Figure 1.6. The homotopy F between the constant path x_0 and $\gamma\gamma^{-1}$

relation on the set of all paths that is compatible with the product of paths.

Next we consider a fixed point $x_0 \in X$ and closed paths $\gamma : I \to X$ with start and end point x_0.

Proposition 1.6. *Let $\pi_1(X, x_0)$ be the set of homotopy classes of closed paths with start and end point x_0. Under this product of paths $\pi_1(X, x_0)$ is a group whose neutral element is given by the constant path x_0, and in which the inverse element of the class $[\gamma]$ is given by $[\gamma^{-1}]$ where $\gamma^{-1}(s) = \gamma(1-s)$.*

Proof. All the group properties are easily verified. We show here that $\gamma\gamma^{-1} \simeq x_0$. A homotopy between the constant path x_0 and $\gamma\gamma^{-1}$ can be given concretely as follows:

$$F(s,t) = \begin{cases} \gamma(2s), & 0 \le 2s \le t, \\ \gamma(t), & t \le 2s \le 2-t, \\ \gamma^{-1}(2s-1), & 2-t \le 2s \le 2. \end{cases}$$

This can be shown diagrammatically as in Figure 1.6. The map F is obviously continuous on each triangle. Since F is well defined, i.e., the various definitions agree on the intersections of these triangles, F must be continuous on all $I \times I$. It is clear that F provides the desired homotopy. □

Definition. The group $\pi_1(X, x_0)$ is called the *fundamental group of X with respect to the base point x_0*.

Notation. We denote the homotopy class of the closed path γ by $[\gamma]$. For the composition in $\pi_1(X, x_0)$ we therefore have by definition that $[\gamma_1][\gamma_2] = [\gamma_1\gamma_2]$.

The next obvious question is the extent to which the fundamental group $\pi_1(X, x_0)$ depends on the *base point* x_0.

Proposition 1.7. *If α is a path from x_0 to x_1, then a group isomorphism is defined by $\alpha_* : \pi_1(X, x_0) \to \pi_1(X, x_1)$, $[\gamma] \mapsto [\alpha^{-1}\gamma\alpha]$.*

Proof. Clearly α_* is well defined and is a group homomorphism. Its inverse is given by $(\alpha^{-1})_*$. □

We recall that a topological space X is said to be *path connected* when every two points x_0, x_1 in X can be joined by a path.

Corollary 1.5. *The fundamental group of a path connected topological space X is independent of the base point, up to isomorphism.*

Definition. A path connected topological space X is called *simply connected* when its fundamental group is trivial, i.e., $\pi_1(X, x_0) = 0$ for any, and so for every, base point x_0.

Proposition 1.8. *Let X be a path connected simply connected topological space and let $x_0, x_1 \in X$. Then any two paths $\gamma_1, \gamma_2 : I \to X$ from x_0 to x_1 are homotopic.*

Proof. Let x_0 (resp. x_1) denote the constant path x_0 (resp. x_1). Since $\pi_1(X, x_0) = 0$, it follows that $\gamma_1\gamma_2^{-1} \simeq x_0$. From property (4) it follows that

$$\gamma_1\gamma_2^{-1}\gamma_2 \simeq x_0\gamma_2.$$

Now $[\gamma_2^{-1}\gamma_2] \in \pi_1(X, x_1) = 0$, hence $\gamma_2^{-1}\gamma_2 \simeq x_1$. It follows that

$$\gamma_1 x_1 \simeq x_0\gamma_2.$$

The map $F : I \times I \to X$ with

$$F(s,t) := \begin{cases} \gamma_1((2-t)s), & 0 \le s \le \frac{1}{2} + \frac{1}{2}t, \\ x_1, & \frac{1}{2} + \frac{1}{2}t \le s \le 1, \end{cases}$$

is a homotopy between $\gamma_1 x_1$ and γ_1. Correspondingly one has a homotopy between $x_0\gamma_2$ and γ_2. Therefore $\gamma_1 \simeq \gamma_2$. □

Example 1.5. a) Each star shaped subset $X \subset \mathbb{R}^n$ is simply connected. We recall that a subset $X \subset \mathbb{R}^n$ is called *star shaped with respect to a point* $x_0 \in X$ when for each point $x \in X$ the (joining) interval $\{\lambda x_0 + (1-\lambda)x \mid 0 \leq \lambda \leq 1\}$ is a subset of X. A subset $X \subset \mathbb{R}^n$ is called *star shaped* when it is star shaped with respect to some point $x_0 \in X$. Let $X \subset \mathbb{R}^n$ be star shaped with respect to x_0 and $\gamma : I \to X$ a closed path with start and end point x_0. Then

$$\begin{aligned} F \ : \ & I \times I \to X \\ & (s,t) \mapsto tx_0 + (1-t)\gamma(s) \end{aligned}$$

specifies a homotopy between γ and x_0. Thus $\pi_1(X, x_0) = 0$. In particular, it follows that $\mathbb{C}$ is simply connected.

b) The Riemann sphere $\hat{\mathbb{C}}$ is simply connected. To prove this, one considers $U_1 := \hat{\mathbb{C}} \setminus \{\infty\}$ and $U_2 := \hat{\mathbb{C}} \setminus \{0\}$. Since U_1 and U_2 are homeomorphic to $\mathbb{C}$, we see that U_1 and U_2 are simply connected. Now let $\gamma : I \to \hat{\mathbb{C}}$ be a closed path with start and end point 0. Since I is compact and γ is continuous, there is a division

$$0 = t_0 < t_1 < \ldots < t_{2n} < t_{2n+1} = 1$$

of the interval I such that $\gamma([t_{2k}, t_{2k+1}]) \subset U_1$ for $k = 0, \ldots, n$, $\gamma([t_{2k-1}, t_{2k}]) \subset U_2$ for $k = 1, \ldots, n$ and $\gamma(t_l) \neq \infty$ for $l = 0, \ldots, 2n+1$. Let $\gamma_l := \gamma|_{[t_{l-1}, t_l]}$, $l = 1, \ldots, 2n+1$. By Proposition 1.8 one can find γ'_{2k} homotopic to the paths γ_{2k}, $k = 1, \ldots, n$, avoiding the point ∞. The path

$$\gamma' := \gamma_1 \gamma'_2 \gamma_3 \cdots \gamma'_{2n} \gamma_{2n+1}$$

is then a path homotopic to γ with $\gamma'(I) \subset U_1$. Since U_1 is simply connected, γ', and therefore γ, is homotopic to the constant path 0.

1.3. Coverings

We now introduce the concept of a covering. The book [**GH81**] is again a standard reference.

Definition. Let $p : Y \to X$ be a continuous map of topological spaces. The map $p : Y \to X$ is a *covering* if every point $x \in X$ has a neighborhood U with the property:

(*) $p^{-1}(U)$ is a disjoint union of open sets S_k in Y such that $p|_{S_k} : S_k \to U$ is a homeomorphism.

The S_k are then called the *sheets* over U.

Example 1.6. a) The map $\exp : \mathbb{C} \to \mathbb{C}^*$, $z \mapsto \exp(z)$, is a covering. For a $z_0 \in \mathbb{C}^*$ the set $\exp^{-1}(z_0)$ consists of the various values of the logarithm of z_0:

$$\log z_0 = \ln|z_0| + i(\arg z_0 + 2\pi k), \quad k \in \mathbb{Z}, \quad -\pi < \arg z_0 \leq \pi.$$

Let Δ be a small disc around z_0 with $0 \notin \Delta$. Then $\exp^{-1}(\Delta)$ is the disjoint union of the sheets

$$S_k := \{\ln|z| + i(\arg z + 2\pi i k) \mid z \in \Delta, -\pi < \arg z \leq \pi\}, \quad k \in \mathbb{Z}.$$

b) The map $p_k : \mathbb{C}^* \to \mathbb{C}^*$, $z \mapsto z^k$ $(k \geq 1)$, is a covering.

c) Let $L \subset \mathbb{C}$ be a lattice and $\pi : \mathbb{C} \to \mathbb{C}/L$ the quotient map. Then π is a covering.

Definition. A continuous map $p : Y \to X$ of topological spaces is called a *local homeomorphism* when each point $y \in Y$ has an open neighborhood V that is mapped homeomorphically onto an open neighborhood $U = p(V)$ of $p(y)$ by p.

A covering $p : Y \to X$ is a local homeomorphism, but the converse is not true in general.

Given a local homeomorphism, one can now ask about the "lifting" of paths, homotopies or maps from Z to X.

Definition. If Z is another topological space and $f : Z \to X$ is a continuous map, then a *lifting* of f with respect to $p : Y \to X$ is a continuous map $\tilde{f} : Z \to Y$ with $p \circ \tilde{f} = f$, i.e., the diagram

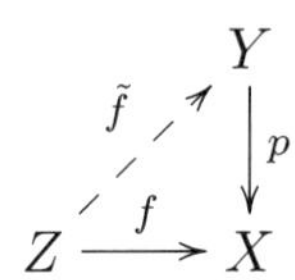

commutes.

Proposition 1.9. *Let X, Y be Hausdorff spaces, $p : Y \to X$ a local homeomorphism, Z a connected topological space, $f : Z \to X$ a continuous map, and $x_0 \in X$, $y_0 \in Y$, $z_0 \in Z$ with $p(y_0) = x_0$ and $f(z_0) = x_0$. Then there is at most one lifting $\tilde{f} : Z \to Y$ of f with $\tilde{f}(z_0) = y_0$.*

Proof. Let $\tilde{f}' : Z \to Y$ be another such lifting. We put

$$A = \{z \in Z \mid \tilde{f}(z) = \tilde{f}'(z)\}.$$

Then A is not empty, for $\tilde{f}(z_0) = \tilde{f}'(z_0) = y_0$ so $z_0 \in A$. Since

$$B = \{z \in Z \mid \tilde{f}(z) \neq \tilde{f}'(z)\},$$

Z is the disjoint union of A and B. We next show that A and B are open. Since Z is connected, it follows that $A = Z$.

Let $z_1 \in A$ and $y := \tilde{f}(z_1) = \tilde{f}'(z_1)$. Since p is a local homeomorphism, there is an open neighborhood V of y and an open neighborhood U of $p(y) = f(z_1)$ such that $p|_V : V \to U$ is a homeomorphism. Since $\tilde{f}$, $\tilde{f}'$ are continuous, there is a neighborhood W of z_1 with $\tilde{f}(W) \subset V$ and $\tilde{f}'(W) \subset V$. Since $p \circ \tilde{f} = p \circ \tilde{f}' = f$, we have $\tilde{f}|_W = \tilde{f}'|_W = (p|_V)^{-1} \circ f|_W$. It follows that $W \subset A$. So A is open.

Now let $z_1 \in B$. Then $\tilde{f}(z_1) \neq \tilde{f}'(z_1)$. Since Y is a Hausdorff space, there are an open neighborhood V_1 of $\tilde{f}(z_1)$ and an open neighborhood V_1' of $\tilde{f}'(z_1)$ with $V_1 \cap V_1' = \emptyset$. By the continuity of $\tilde{f}$ and $\tilde{f}'$ there is an open neighborhood W of z_1 with $\tilde{f}(W) \subset V_1$ and $\tilde{f}'(W) \subset V_1'$. Since V_1 and V_1' are disjoint, it follows that $W \subset B$. Thus B is also open. □

Proposition 1.10. *Let $y_0 \in Y$, $x_0 \in X$, $p : Y \to X$ be a covering with $p(y_0) = x_0$ and γ a path in X with $\gamma(0) = x_0$. Then there is a unique lifting $\tilde{\gamma}$ of γ with $\tilde{\gamma}(0) = y_0$.*

Proof. The uniqueness of $\tilde{\gamma}$ follows immediately from the proposition above. To show the existence of $\tilde{\gamma}$, we subdivide the interval $I = [0,1]$ into subintervals $[t_k, t_{k+1}]$ with $0 = t_0 < t_1 < \cdots < t_n = 1$, such that $\gamma([t_k, t_{k+1}])$ is contained in a set U_k for which the property $(*)$ holds. We consider U_0 and the sheet S_0 over U_0 with $y_0 \in S_0$. Then there is a unique lifting $\tilde{\gamma}_0$ of $\gamma|_{[t_0,t_1]}$ with $\tilde{\gamma}_0(0) = y_0$. Suppose that we have now a lifting $\gamma_i : [0, t_{i+1}] \to Y$ of $\gamma|_{[0,t_{i+1}]}$ with $\tilde{\gamma}_i(0) = y_0$. Then there is a lifting $\gamma_{i+1}' : [t_{i+1}, t_{i+2}] \to Y$ with $\gamma_{i+1}'(t_{i+1}) = \tilde{\gamma}_i(t_{i+1})$. By concatenating $\tilde{\gamma}_i$ and γ_{i+1}', we obtain a lifting $\tilde{\gamma}_{i+1}$ of $\gamma|_{[0,t_{i+2}]}$ with $\tilde{\gamma}_{i+1}(0) = y_0$. □

Proposition 1.11. *Let X, Y be Hausdorff spaces, $p : Y \to X$ a local homeomorphism, $x_0, x_1 \in X$, and $y_0 \in Y$ with $p(y_0) = x_0$. Let $F : I \times I \to X$ be a continuous map with $F(0,t) = x_0$, $F(1,t) = x_1$ for all $t \in I$. We put*

$$\gamma_t(s) := F(s,t).$$

Suppose that each path γ_t has a lifting $\tilde{\gamma}_t : I \to Y$ with $\tilde{\gamma}_t(0) = y_0$ (i.e., $p \circ \tilde{\gamma}_t = \gamma_t$).

Then $\tilde{\gamma}_0$ and $\tilde{\gamma}_1$ have the same end point and F has a unique lifting to a homotopy $\tilde{F} : I \times I \to Y$ between $\tilde{\gamma}_0$ and $\tilde{\gamma}_1$.

Proof. We define

$$\begin{aligned} \tilde{F} \; : \; & I \times I \to Y \\ & (s,t) \mapsto \tilde{\gamma}_t(s). \end{aligned}$$

We have to show that $\tilde{F}$ is continuous. Since $F = p \circ \tilde{F}$ and $F(1,t) = x_1$ for all $t \in I$, we have $\tilde{F}(1,t) \in p^{-1}(x_1)$ for all $t \in I$. Since $p^{-1}(x_1)$ is discrete and $\{1\} \times I$ is connected, it follows that there is a y_1 with $p(y_1) = x_1$ and that $\tilde{F}(1,t) = y_1$ for all $t \in I$. From this it follows that $\tilde{\gamma}_0(1) = \tilde{\gamma}_1(1) = y_1$ and that $\tilde{F}$ is a homotopy between $\tilde{\gamma}_0$ and $\tilde{\gamma}_1$.

We prove the continuity of $\tilde{F}$ indirectly. Suppose there is a point $(s_0, t_0) \in I \times I$ at which $\tilde{F}$ is not continuous. Let

$$\sigma := \inf\{s \in I \mid \tilde{F} \text{ is not continuous at } (s, t_0)\}.$$

Let $x := F(\sigma, t_0)$, $y := \tilde{F}(\sigma, t_0) = \tilde{\gamma}_{t_0}(\sigma)$. There are an open neighborhood V of y and an open neighborhood U of x such that $p|_V : V \to U$ is a homeomorphism. We distinguish two cases:

Case 1: $\sigma = 0$. Since $F(0,t) = x_0$ for all $t \in I$ and F is continuous, there is an $\varepsilon_0 > 0$ such that $F([0,\varepsilon_0] \times I) \subset U$. By Proposition 1.9 we have

$$\tilde{\gamma}_t|_{[0,\varepsilon_0]} = (p|_V)^{-1} \circ \gamma_t|_{[0,\varepsilon_0]} \quad \text{for all } t \in I,$$

i.e., $\tilde{F} = (p|_V)^{-1} \circ F$ on $[0,\varepsilon_0] \times I$. From this it follows that $\tilde{F}$ is continuous on $[0, \varepsilon_0) \times I$, contradicting the definition of σ.

Case 2: $\sigma > 0$. Since F is continuous, there is an $\varepsilon > 0$ such that $F(U_\varepsilon(\sigma) \times U_\varepsilon(t_0)) \subset U$, where

$$U_\varepsilon(\tau) = \{t \in I \mid |t - \tau| < \varepsilon\}.$$

In particular $\gamma_{t_0}(U_\varepsilon(\sigma)) \subset U$ by Proposition 1.9, and therefore

$$\tilde{\gamma}_{t_0}|_{U_\varepsilon(\sigma)} = (p|_V)^{-1} \circ \gamma_{t_0}|_{U_\varepsilon(\sigma)}.$$

Let $s_1 \in U_\varepsilon(\sigma)$ with $s_1 < \sigma$. Then $\tilde{F}(s_1, t_0) = \tilde{\gamma}_{t_0}(s_1) \in V$. Also, $\tilde{F}$ is continuous in (s_1, t_0), by the definition of σ. From this there is a $\delta > 0$, $\delta \le \varepsilon$, such that $\tilde{F}(s_1, t) = \tilde{\gamma}_t(s_1) \in V$ for all $t \in U_\delta(t_0)$. From the uniqueness of the lifting it again follows that

$$\tilde{\gamma}_t|_{U_\varepsilon(\sigma)} = (p|_V)^{-1} \circ \gamma_t|_{U_\varepsilon(\sigma)} \quad \text{for all } t \in U_\delta(t_0),$$

i.e., $\tilde{F} = (p|_V)^{-1} \circ F$ on $U_\varepsilon(\sigma) \times U_\delta(t_0)$. Therefore $\tilde{F}$ is continuous in a neighborhood of (σ, t_0). This contradicts the definition of σ. □

For a covering $p : Y \to X$ one can now ask generally about the existence of *liftings of continuous maps*, i.e., we pose the question of whether given a continuous map $f : Z \to X$ and given points $x_0 \in X$, $y_0 \in Y$, $z_0 \in Z$ with $f(z_0) = p(y_0) = x_0$ there is a continuous map $\tilde{f} : Z \to Y$ with $\tilde{f}(z_0) = y_0$ such that the diagram

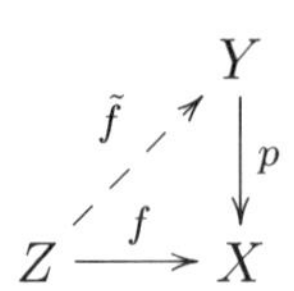

commutes.

Definition. A topological space X is called *locally path connected* if for each point x and each neighborhood U of x there is a path connected neighborhood V of x with $V \subset U$.

Proposition 1.12. *Let X, Y, Z be locally path connected topological spaces. Let $p : Y \to X$ be a covering. If Z is simply connected, then every continuous map $f : Z \to X$ is liftable, i.e., given points $x_0 \in X$, $y_0 \in Y$, $z_0 \in Z$ with $f(z_0) = p(y_0) = x_0$, there is a continuous map $\tilde{f} : Z \to Y$ with $\tilde{f}(z_0) = y_0$ and $p \circ \tilde{f} = f$.*

Proof. Let $z_1 \in Z$ be arbitrary and $\gamma : I \to Z$ a path from z_0 to z_1. Let $\tilde{\gamma} : I \to Y$ be the lifting of $f \circ \gamma : I \to X$ with $\tilde{\gamma}(0) = y_0$. The end point $\tilde{\gamma}(1)$ depends only on z_1 and not on the choice of γ: If $\gamma_1 : I \to Z$ is another path from z_0 to z_1, then by Proposition 1.8, $\gamma \simeq \gamma_1$, since Z is simply connected. Thus $f \circ \gamma \simeq f \circ \gamma_1$. By Proposition 1.11 the liftings of $f \circ \gamma$ and $f \circ \gamma_1$ have the same end point. Thus $\tilde{f}(z_1) = \tilde{\gamma}(1)$ defines a map $\tilde{f} : Z \to Y$ uniquely.

By construction $p \circ \tilde{f} = f$. We have still to show that $\tilde{f}$ is continuous at any arbitrary $z \in Z$. Let U_z be a path connected neighborhood of z such that $f(U_z)$ lies in an open set $V \subset X$ having the property $(*)$. Let S_z be the sheet of Y over V in which $\tilde{f}(z)$ lies. Then $\tilde{f}(U_z) \subset S_z$. For $z' \in U_z$ we can specify $\tilde{f}(z')$ as follows: Let $\gamma_{z'} : I \to Z$ be a path from z to z'. Then $\tilde{f}(z')$ is the end point of that same lifting $\tilde{\gamma}_{z'}$ of $f \circ \gamma_{z'}$ that has $\tilde{f}(z)$ as start point. Since $f \circ \gamma_{z'}$ takes values in U_z, the path $\tilde{\gamma}_{z'}$ takes its values in S_z. Thus $\tilde{f}(z') = \tilde{\gamma}_{z'}(1) \in S_z$. The map $p_z := p|_{S_z} : S_z \to V$ is a homeomorphism and $\tilde{f} = p_z^{-1} \circ f$ on U_z. It follows that $\tilde{f}$ is continuous on U_z and so at z. □

Definition. A *universal covering* of X is a covering $p : Y \to X$ with a simply connected space Y.

Remark 1.8. If $p : Y \to X$ is a universal covering, $q : Z \to X$ is an arbitrary covering, and X is locally path connected, then by Proposition 1.12 there is a commutative diagram

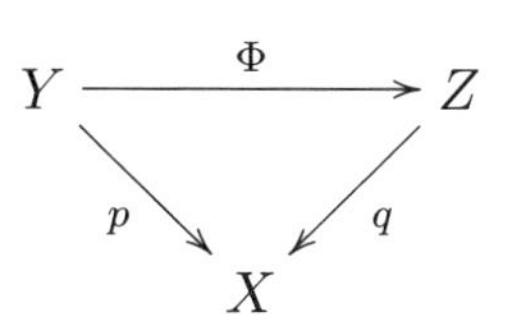

In this sense the universal covering is the "largest" covering of X.

Definition. Two coverings $p : Y \to X$ and $q : Z \to X$ are called *equivalent* if there is a homeomorphism $\Phi : Y \to Z$ such that the diagram

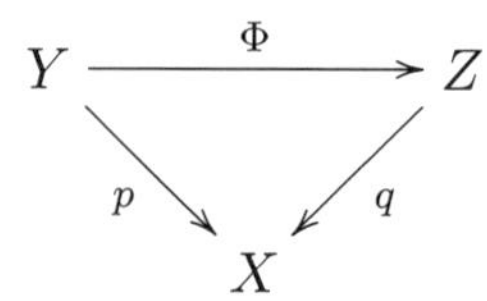

commutes.

Corollary 1.6. *Let X be locally path connected. Then any two universal coverings of X are equivalent.*

Proof. For two universal coverings $p : Y \to X$ and $q : Z \to X$ there is a diagram

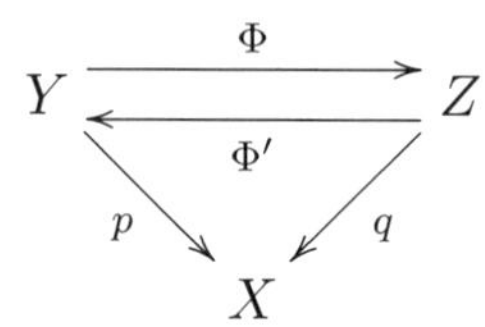

Since $\Phi'\Phi(y_0) = y_0$, it follows in view of the uniqueness of the lifting that $\Phi'\Phi = \mathrm{id}_Y$ and similarly that $\Phi\Phi' = \mathrm{id}_Z$. □

Example 1.7. The covering $\exp : \mathbb{C} \to \mathbb{C}^*$ is the universal covering of $\mathbb{C}^*$.

The question of the existence of a universal covering can now be answered.

Definition. A space X is called *semi-locally simply connected* if every point $x \in X$ has a neighborhood U such that every closed path lying in U is nullhomotopic in X.

A space X that has a universal covering necessarily has this property. Topological manifolds are examples of such spaces.

Theorem 1.1. *A connected locally path connected semi-locally simply connected space X always has a universal covering $p : \tilde{X} \to X$.*

Proof. We choose a fixed point $x_0 \in X$. Consider all pairs (x, α) where $x \in X$ and α is a homotopy class of paths from x_0 to x. Let $\tilde{X}$ be the set of all such pairs. We define a map $p : \tilde{X} \to X$ by $p((x, \alpha)) = x$.

We must now endow $\tilde{X}$ with a suitable topology. Let $(x_1, \alpha) \in \tilde{X}$ and let U be an open neighborhood of x_1 such that every closed path lying in U is nullhomotopic in X. For $x \in U$ let γ_x be a path from x_1 to x running through U: its homotopy class $[\gamma_x]$ depends only on x, by Proposition 1.8. We put

$$\tilde{U}(\alpha) := \{(x, \beta) \in \tilde{X} \mid x \in U,\ \beta = \alpha[\gamma_x]\}.$$

We call a subset $W \subset \tilde{X}$ open if W is the union of sets of the form $\tilde{U}(\alpha)$.

a) We show that a topology is defined on $\tilde{X}$ in this way. It is enough to show that each intersection $\tilde{U}_1(\alpha_1) \cap \tilde{U}_2(\alpha_2)$ can be rewritten as the union of sets of the form $\tilde{U}(\alpha)$. Let $(x, \gamma) \in \tilde{U}_1(\alpha_1) \cap \tilde{U}_2(\alpha_2)$. Then $x \in U_1 \cap U_2$ and there exists an open path connected neighborhood $V \subset U_1 \cap U_2$ of x. Then

$$(x, \gamma) \in \tilde{V}(\gamma) \subset \tilde{U}_1(\alpha_1) \cap \tilde{U}_2(\alpha_2).$$

b) The map p is continuous and open, for sets of the form $\tilde{U}(\alpha)$ are mapped homeomorphically via p onto the corresponding open sets $U \subset X$.

c) We show that p is a covering. Let U be an open neighborhood of $x_1 \in X$ such that every closed path lying in U is nullhomotopic in X. We show that U has the covering property $(*)$ with respect to p. Let $(x, \beta) \in p^{-1}(U)$. Then $(x, \beta) = (x, \alpha[\gamma_x])$ where γ_x is a path in U from x_1 to x and $\alpha = \beta[\gamma_x]^{-1}$. Thus $(x, \beta) \in \tilde{U}(\alpha)$. Hence $p^{-1}(U)$ is the union of the sets $\tilde{U}(\alpha)$ over all homotopy classes α of paths in X from x_0 to x_1. This union is disjoint: for if $(x, \beta) \in \tilde{U}(\alpha_1) \cap \tilde{U}(\alpha_2)$, then $\beta = \alpha_1[\gamma_x] = \alpha_2[\gamma_x]$, thus $\alpha_1 = \alpha_2$. So finally p maps each set $\tilde{U}(\alpha)$ homeomorphically onto U.

d) We show that $\tilde{X}$ is path connected. Let $y_0 = (x_0, 0)$, where 0 is the homotopy class of the constant paths with start point x_0. Furthermore, let $y = (x, \alpha) \in \tilde{X}$. Let γ be a path from x_0 to x representing the homotopy class α. For $\tau \in I = [0, 1]$ we put $\gamma_\tau(t) = \gamma(\tau t)$ and $\alpha_\tau = [\gamma_\tau]$. (The path γ_τ is up to reparametrization the path $\gamma|_{[0,\tau]}$.) Then we consider the map

$$\begin{aligned} \hat{\gamma} : I &\rightarrow \tilde{X} \\ t &\mapsto (\gamma(t), \alpha_t) . \end{aligned}$$

By construction $\hat{\gamma}(0) = y_0$ and $\hat{\gamma}(1) = y$. We show that $\hat{\gamma}$ is continuous. Towards this, let $\tau \in I$ and $\tilde{V}$ be an open neighborhood of $\hat{\gamma}(\tau)$. Without loss of generality $\tilde{V} = \tilde{U}(\alpha_\tau)$, where U is an open neighborhood of $\gamma(\tau)$. Since γ is continuous, there is an interval $I^* \subset I$ which is open in I with $\tau \in I^*$ and $\gamma(I^*) \subset U$. Then $\hat{\gamma}(I^*) \subset \tilde{U}(\alpha_\tau)$. Next, for $t > \tau$, $t \in I^*$, we have $\alpha_t = [\gamma_\tau \gamma|_{[\tau,t]}]$ and from $\gamma([\tau, t]) \subset U$ it follows that $(\gamma(t), \alpha_t) \in \tilde{U}(\alpha_\tau)$. For $t < \tau$, $t \in I^*$, one concludes similarly. Thus $\hat{\gamma}$ is a lifting of γ and is a path in $\tilde{X}$ from y_0 to y.

e) We show finally that $\tilde{X}$ is simply connected. Let $\hat{\gamma}$ be a path in $\tilde{X}$ from $y_0 = (x_0, 0)$ to y_0 and $\gamma = p \circ \hat{\gamma}$. Then, by the uniqueness of the lifting (Proposition 1.9), $\hat{\gamma}$ is the uniquely determined lifting of γ with $\hat{\gamma}(0) = y_0$. By d) $\hat{\gamma}(t) = (\gamma(t), \alpha_t)$ with $\alpha = [\gamma]$. From $\hat{\gamma}(1) = y_0$ it follows that $[\gamma] = \alpha_1 = 0$, i.e., γ is nullhomotopic in X. By Proposition 1.11, $\hat{\gamma}$ is then nullhomotopic in $\tilde{X}$.

We have demonstrated Theorem 1.1 completely. □

Definition. Let $p : Y \to X$ be a covering and $x \in X$. We call the subset $p^{-1}(x)$ of Y the *fiber* of p over x. Let $q : Z \to X$ be another covering. A continuous map $\Phi : Y \to Z$ is called *fiber preserving* if the diagram

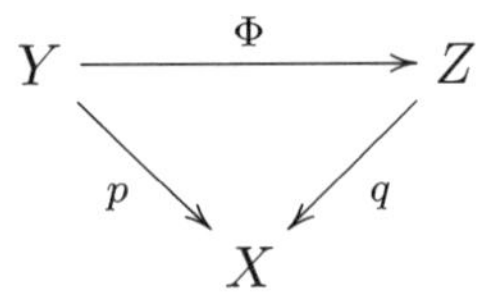

commutes, i.e. if $q \circ \Phi = p$. A fiber preserving homeomorphism $\Phi : Y \to Y$ is also called a *covering transformation* of the covering $p : Y \to X$.

Clearly the covering transformations form a group.

Proposition 1.13. *Let $p : \tilde{X} \to X$ be the universal covering, $x_0 \in X$. There is then a natural isomorphism between the group G of covering transformations and the fundamental group $\pi_1(X, x_0)$.*

Proof. Choose $y_0 \in \tilde{X}$ with $p(y_0) = x_0$. We define a map

$$\chi : G \to \pi_1(X, x_0)$$

as follows: Let Φ be a covering transformation, $\tilde{\gamma} : I \to \tilde{X}$ a path from y_0 to $\Phi(y_0)$. Since $\tilde{X}$ is simply connected, all paths from y_0 to $\Phi(y_0)$ are homotopic to each other. Since $p(\Phi(y_0)) = x_0$, it follows that $\gamma = p \circ \tilde{\gamma}$ is a closed path in X with start point x_0. We define

$$\chi(\Phi) := [\gamma] \in \pi_1(X, x_0).$$

Since the class $[\gamma] \in \pi_1(X, x_0)$ depends only on the point $\Phi(y_0)$, the map $\chi : G \to \pi_1(X, x_0)$ is well defined.

a) χ is a homomorphism: Given $\Phi_1, \Phi_2 \in G$, let $\tilde{\gamma}_1$, resp. $\tilde{\gamma}_2$, be paths in $\tilde{X}$ from y_0 to $\Phi_1(y_0)$, resp. $\Phi_2(y_0)$. Then $\Phi_1 \circ \tilde{\gamma}_2$ is a path from $\Phi_1(y_0)$ to $\Phi_1\Phi_2(y_0)$ with $p \circ (\Phi_1 \circ \tilde{\gamma}_2) = p \circ \tilde{\gamma}_2$, $\tilde{\gamma}_1(\Phi_1 \circ \tilde{\gamma}_2)$ a path from y_0 to $\Phi_1\Phi_2(y_0)$. Therefore,

$$\begin{aligned} \chi(\Phi_1\Phi_2) &= [p(\tilde{\gamma}_1(\Phi_1 \circ \tilde{\gamma}_2))] \\ &= [p(\tilde{\gamma}_1)p(\Phi_1 \circ \tilde{\gamma}_2)] \\ &= [(p \circ \tilde{\gamma}_1)(p \circ \tilde{\gamma}_2)] \\ &= [p \circ \tilde{\gamma}_1][p \circ \tilde{\gamma}_2] \\ &= \chi(\Phi_1)\chi(\Phi_2). \end{aligned}$$

b) χ is surjective: Let $[\gamma] \in \pi_1(X, x_0)$, $\tilde{\gamma}$ be the uniquely determined lifting of γ with start point y_0. Then by Corollary 1.6 there is a $\Phi \in G$ with $\Phi(y_0) = \tilde{\gamma}(1)$. From the definition of χ it follows that $\chi(\Phi) = [\gamma]$.

c) χ is injective: $\chi(\Phi) = [p \circ \tilde{\gamma}] = 0$ means that the path $p \circ \tilde{\gamma}$ is homotopic to the constant path x_0. It follows from Proposition 1.11 that the path $\tilde{\gamma}$

in $\tilde{X}$ is then homotopic to the constant path y_0. So $\Phi(y_0) = \tilde{\gamma}(1) = y_0$. Since $\tilde{X}$ is connected, it follows from Proposition 1.9 that $\Phi = \mathrm{id}_{\tilde{X}}$. □

Example 1.8. Consider the universal covering $\exp : \mathbb{C} \to \mathbb{C}^*$. The translations of $\mathbb{C}$ by $2\pi ik$, $k \in \mathbb{Z}$, are covering transformations. A map $\Phi : \mathbb{C} \to \mathbb{C}$ with $\exp \circ \Phi = \exp$ is however necessarily of the form $\Phi(z) = z + 2\pi ik(z)$ for some function $k : \mathbb{C} \to \mathbb{Z}$. If Φ is continuous, then k is also continuous and therefore constant. Hence the covering transformations are precisely the translations by $2\pi ik$, $k \in \mathbb{Z}$. Thus $G = 2\pi i\mathbb{Z}$ and so $\pi_1(\mathbb{C}^*, z_0)$ is isomorphic to $\mathbb{Z}$.

Let $p : \tilde{X} \to X$ again be the universal covering and G the group of covering transformations. Then we have a map

$$\begin{array}{rcl} \tilde{X} \times G & \to & \tilde{X} \\ (y, \Phi) & \mapsto & \Phi(y). \end{array}$$

In this way the group G acts on $\tilde{X}$.

Definition. One says that a group G *acts* on a set X (from the right) if there is a map

$$\begin{array}{rcl} X \times G & \to & X \\ (x, g) & \mapsto & xg \end{array}$$

with the following properties:

(i) $(xg)g' = x(gg')$ for all $x \in X$ and $g, g' \in G$,

(ii) $x1 = x$ for all $x \in X$.

The *orbit* of an element $x \in X$ is the set

$$xG := \{xg \,|\, g \in G\}.$$

A *fixed point* of an element $g \in G$ is an $x \in X$ with $xg = x$. One says that G acts *fixed point free* on X if no $g \neq 1$ has a fixed point.

Now let $x_0 \in X$ and $y_0, y_1 \in \tilde{X}$ with $p(y_0) = p(y_1) = x_0$. By Corollary 1.6 there is a $g \in G$ with $g(y_0) = y_1$. By Proposition 1.9, g is uniquely determined. Therefore G acts fixed point free on $\tilde{X}$. Next we consider on $\tilde{X}$ the equivalence relation

$$y \sim y' :\Leftrightarrow y' \in yG.$$

This means that y and y' are equivalent precisely when they lie in the same orbit. The equivalence classes are precisely the orbits of G. Let $\tilde{X}/G$ be the quotient set. We endow $\tilde{X}/G$ with the quotient topology. This means that a subset $U \subset \tilde{X}/G$ is open precisely when $\pi^{-1}(U)$ is open in $\tilde{X}$ where $\pi : \tilde{X} \to \tilde{X}/G$ is the quotient map. With this terminology:

Proposition 1.14. *Let $p : \tilde{X} \to X$ be the universal covering with covering transformation group G. Then the quotient space $\tilde{X}/G$ is homeomorphic to X.*

Proof. Since G transforms the fibers of p to themselves, and since for two points $y_0, y_1 \in \tilde{X}$ in the same fiber there is just one $g \in G$ with $g(y_0) = y_1$, the orbits of G are just the fibers of p. We thus obtain a natural map

$$\begin{aligned} q \ : \ & \tilde{X}/G \to X \\ & yG \mapsto p(y). \end{aligned}$$

This map is a homeomorphism: First let $V \subset X$ be open. Since

$$\pi^{-1}(q^{-1}(V)) = p^{-1}(V),$$

$\pi^{-1}(q^{-1}(V))$ is open and then $q^{-1}(V)$ is also open by the definition of the quotient topology. If conversely $U \subset \tilde{X}/G$ is open, then so is $\pi^{-1}(U)$ open, and so too is

$$q(U) = q\pi\pi^{-1}(U) = p(\pi^{-1}(U)).$$

This proves Proposition 1.14. □

Example 1.9. Let $L \subset \mathbb{C}$ be a lattice. Then L acts on $\mathbb{C}$ by

$$\begin{aligned} \mathbb{C} \times L &\to \mathbb{C} \\ (z, \omega) &\mapsto z + \omega. \end{aligned}$$

Now

$$z \sim z' \Leftrightarrow z' = z + \omega \text{ for a } \omega \in L \Leftrightarrow z' - z \in L \Leftrightarrow z \equiv z' \pmod{L}.$$

Thus two complex numbers $z, z' \in \mathbb{C}$ are equivalent in the above sense exactly when they are equivalent modulo L in the sense of Example 1.4. The quotient space $\mathbb{C}/L$ coincides with the torus constructed in Example 1.4. The map $\pi : \mathbb{C} \to \mathbb{C}/L$ is the universal covering of $\mathbb{C}/L$ since $\mathbb{C}$ is simply connected. The covering transformations of this covering are exactly the translations $\mathbb{C} \to \mathbb{C}$, $z \mapsto z + \omega$, for $\omega \in L$. Therefore G is isomorphic to L and the action of G on $\mathbb{C}$ agrees with the above action of L on $\mathbb{C}$.

Next we consider local homeomorphisms $p : Y \to X$ where X is a Riemann surface.

Proposition 1.15. *Let X be a Riemann surface, Y a connected Hausdorff space and $p : Y \to X$ a local homeomorphism. Then there is exactly one complex structure on Y for which p is holomorphic.*

Remark 1.9. By Corollary 1.2, p is even locally biholomorphic.

Proof. For every point $y \in Y$ there exist an open neighborhood U of y and an open neighborhood U' of $p(y)$ such that $p|_U : U \to U'$ is a homeomorphism. Without loss of generality we may assume that U' is the chart domain of a chart $\varphi' : U' \to V$ of X. Then the map

$$\varphi = \varphi' \circ p|_U : U \to V \subset \mathbb{C}$$

is a chart of Y. If $\psi = \psi' \circ p|_{U_1} : U_1 \to V_1$ is another such chart, then the transition function satisfies

$$(\psi' \circ p|_{U_1}) \circ (\varphi' \circ p|_U)^{-1} = \psi' \circ (\varphi')^{-1}$$

on $(\varphi' \circ p|_U)(U \cap U_1)$. Hence all the transition functions are biholomorphic. These charts therefore form a complex atlas of Y. It is immediately clear that with respect to this atlas p is holomorphic.

As to uniqueness, if one has a maximal complex atlas of Y such that p is holomorphic, then the maps $\varphi' \circ p|_U$ are holomorphic, and so are charts of this atlas. □

Proposition 1.16. *Let X, Y, Z be Riemann surfaces, $p : Y \to X$ a holomorphic local homeomorphism and $f : Z \to X$ holomorphic. Then every lifting $\tilde{f} : Z \to Y$ of f is also holomorphic.*

Proof. Let $z_0 \in Z$, $y_0 := \tilde{f}(z_0)$, $x_0 := p(y_0) = f(z_0)$. Then there are an open neighborhood V of y_0 and an open neighborhood U of x_0 such that $p|_V : V \to U$ is biholomorphic. Let W be an open neighborhood of z_0 with $f(W) \subset U$. Then

$$\tilde{f}|_W = (p|_V)^{-1} \circ (f|_W).$$

Since $(p|_V)^{-1}$ and $f|_W$ are holomorphic, so too is $\tilde{f}|_W$ holomorphic. □

Corollary 1.7. *Let X, Y, Z be Riemann surfaces and let $p : Y \to X$ and $q : Z \to X$ be holomorphic coverings. Then every fiber preserving continuous map $\Phi : Y \to Z$ is holomorphic.*

Proof. It follows from Proposition 1.16 that Φ is a lifting of p with respect to q. □

Corollary 1.8. *Let Y, X be Riemann surfaces and $p : Y \to X$ a holomorphic covering. Then the covering transformations are biholomorphic maps.*

Proof. This follows immediately from the preceding corollary. □

Example 1.10. We can now classify all coverings of $\mathbb{C}^*$. Let $p : Y \to \mathbb{C}^*$ be such a covering. By Proposition 1.15 we can introduce a complex structure on Y so that p is holomorphic. The covering $\exp : \mathbb{C} \to \mathbb{C}^*$ is a holomorphic covering of Riemann surfaces. By Proposition 1.12 there is a map $f : \mathbb{C} \to Y$ such that the following diagram commutes:

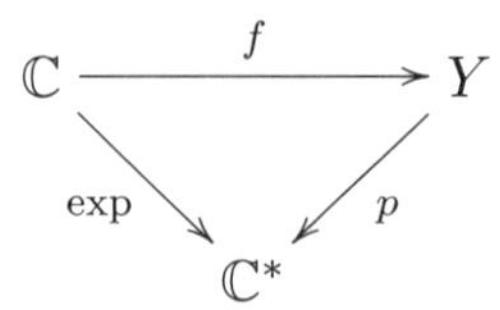

By Corollary 1.7, f is holomorphic. Since $\exp : \mathbb{C} \to \mathbb{C}^*$ is the universal covering, $f : \mathbb{C} \to Y$ is also a covering and is the universal covering of Y. Let G be the covering transformation group of $\exp : \mathbb{C} \to \mathbb{C}^*$ and let H be that of $f : \mathbb{C} \to Y$. Since each fiber of f lies in a fiber of exp, we see that H is a subgroup of G. Since $G = 2\pi i\mathbb{Z}$, either $H = 0$ or $H = 2\pi ik\mathbb{Z}$ for a natural number $k \geq 1$. In the first case, by Proposition 1.13, Y is simply connected and f is therefore biholomorphic. So assume $H = 2\pi ik\mathbb{Z}$. By Proposition 1.14, Y is then homeomorphic and thus biholomorphically equivalent to $\mathbb{C}/2\pi ik\mathbb{Z}$ and thus to $\mathbb{C}^*$. Up to biholomorphic equivalence we can now write f as $f(w) = \exp(w/k)$ and p as $p(z) = z^k$. Up to biholomorphic equivalence

$$\exp : \mathbb{C} \to \mathbb{C}^* \quad \text{and} \quad p_k : \mathbb{C}^* \to \mathbb{C}^*, z \mapsto z^k$$

are the unique coverings of $\mathbb{C}^*$. There is a corresponding result for the coverings of a punctured disc.

1.4. Analytic continuation

Let us now consider the analytic continuation of holomorphic functions along paths. Analytic continuation can be most easily conveniently formulated through the concept of a sheaf of germs of holomorphic functions. So we must next explain what germs of holomorphic functions are.

Let X be a Riemann surface, $a \in X$ a fixed point, U, V open neighborhoods of a.

Definition. Two functions $f : U \to \mathbb{C}$, $g : V \to \mathbb{C}$ are called *equivalent at the point* a (symbolically $f \sim_a g$) when there exists an open neighborhood W of a such that $W \subset U \cap V$ and $f|_W = g|_W$. An equivalence class of such functions is called a *function germ at* a.

If U is an open neighborhood of a and $f : U \to \mathbb{C}$ is a function, then f belongs to an equivalence class. This class is called the *germ of the function* f and is denoted by $\bar{f}$.

Two functions f, g equivalent at a assume the same value $f(a) = g(a)$ at a; this value is called the *value* of the function germ $\bar{f}$ at a. Note that the germ of a function at a depends not on its value at the point a but on its behavior on all of an open neighborhood of a:

Example 1.11. Let $X = \mathbb{C}$, $a = 0$. The functions $f(z) = z$ and $g(z) = z^2$ are not equivalent at 0 even though $f(0) = g(0) = 0$.

Next we consider holomorphic functions in particular. For a region $U \subset X$ we put

$$\mathcal{O}(U) = \{f : U \to \mathbb{C} \text{ holomorphic}\}.$$

By Remark 1.2, $\mathcal{O}(U)$ is a $\mathbb{C}$-algebra, so in particular it is a commutative ring with unit. If $a \in U$, we write

$$\mathcal{O}_a$$

for the set of all germs of holomorphic functions at a. On defining the sum and product of germs representativewise, $\mathcal{O}_a$ inherits the structure of a commutative ring with unity. The ring $\mathcal{O}_a$ is called the *ring of holomorphic function germs at a*.

Let U be an open neighborhood of a. We have a map

$$\rho_a : \mathcal{O}(U) \to \mathcal{O}_a,$$

taking a holomorphic function $f : U \to \mathbb{C}$ to its function germ $\rho_a(f) := \bar{f}$ at a. This map is a ring homomorphism.

By $\mathbb{C}\{z - w\}$ we denote the ring of all power series

$$\sum_{\nu=0}^{\infty} a_\nu (z - w)^\nu, \quad a_\nu \in \mathbb{C},$$

that converge on an open disc $\Delta_r(w) = \{z \in \mathbb{C} \mid |z - w| < r\}$ of radius $r > 0$ around $w \in \mathbb{C}$, where the radius r can depend on the power series.

Proposition 1.17. *Let $X \subset \mathbb{C}$ be a region, $a \in X$. The ring $\mathcal{O}_a$ of holomorphic function germs at a is isomorphic to the ring $\mathbb{C}\{z - a\}$ of convergent power series at a.*

Proof. We define a map

$$\Phi : \mathcal{O}_a \to \mathbb{C}\{z - a\}$$

by associating its power series expansion at a to a holomorphic function germ represented by a holomorphic function $f : U \to \mathbb{C}$, U an open neighborhood of a.

This map is well defined: Two representatives of the same germ agree in an open neighborhood of a and therefore by the identity theorem have the same power series expansion at a. It is clear that Φ is bijective and a ring homomorphism. ∎

Now let X again be a Riemann surface and $\mathcal{O}_X$ the disjoint union of the $\mathcal{O}_x$ for $x \in X$:

$$\mathcal{O}_X := \bigcup_{x \in X} \mathcal{O}_x.$$

Let

$$p : \mathcal{O}_X \to X, \quad p(\bar{f}) = x \text{ for } \bar{f} \in \mathcal{O}_x,$$

be the canonical projection. We next want to endow the set $\mathcal{O}_X$ with a topology. For $U \subset X$ open and $f : U \to \mathbb{C}$ holomorphic let

$$\sigma(U, f) := \{\rho_x(f) \mid x \in U\} \subset \mathcal{O}_X.$$

On $\mathcal{O}_X$ we consider the topology generated by all the sets $\sigma(U, f)$, i.e., we call a subset $V \subset \mathcal{O}_X$ open when it is the union of sets of the form $\sigma(U, f)$. This defines a topology on $\mathcal{O}_X$: To establish this, it is enough to show that each intersection $\sigma(U_1, f_1) \cap \sigma(U_2, f_2)$ is again the union of appropriate sets $\sigma(W, h)$. Let $\varphi \in \sigma(U_1, f_1) \cap \sigma(U_2, f_2)$, $p(\varphi) = x$. Then $x \in U_1 \cap U_2$ and

$$\varphi = \rho_x(f_1) = \rho_x(f_2).$$

There is therefore an open neighborhood $W \subset U_1 \cap U_2$ of x with $f_1|_W = f_2|_W =: h$. It follows that $\varphi \in \sigma(W, h) \subset \sigma(U_1, f_1) \cap \sigma(U_2, f_2)$.

Proposition 1.18. *The map $p : \mathcal{O}_X \to X$ is a local homeomorphism.*

Proof. If $\varphi \in \mathcal{O}_X$ and $f : U \to \mathbb{C}$ is a representative of φ, then $\sigma(U, f)$ is an open neighborhood of φ. The map $p|_{\sigma(U,f)} : \sigma(U, f) \to U$ is bijective and, by definition of the topology on $\mathcal{O}_X$, continuous and open, so it is a homeomorphism. □

Definition. We call $\mathcal{O}_X$ with the projection $p : \mathcal{O}_X \to X$ the *sheaf of germs of holomorphic functions* on X.

Proposition 1.19. *$\mathcal{O}_X$ is a Hausdorff space.*

Proof. Let $\bar{f}, \bar{g} \in \mathcal{O}_X$, $\bar{f} \neq \bar{g}$, with representatives $f : U \to \mathbb{C}$ and $g : V \to \mathbb{C}$.

If $p(\bar{f}) \neq p(\bar{g})$, we can assume $U \cap V = \emptyset$. Then $\sigma(U, f)$, resp. $\sigma(V, g)$, are disjoint neighborhoods of $\bar{f}$, resp. $\bar{g}$.

The other case is $p(\bar{f}) = p(\bar{g})$. Then without loss of generality we can assume that $U = V$ and that U is a region. We claim that then $\sigma(U, f)$, resp. $\sigma(U, g)$, are disjoint neighborhoods of $\bar{f}$, resp. $\bar{g}$. For if $\sigma(U, f) \cap \sigma(U, g) \neq \emptyset$, we would then have $\rho_x(f) = \rho_x(g)$ for some $x \in U$. But it follows from the identity theorem that then $f \equiv g$, and so $\bar{f} = \bar{g}$. □

Next we consider the analytic continuation of function germs.

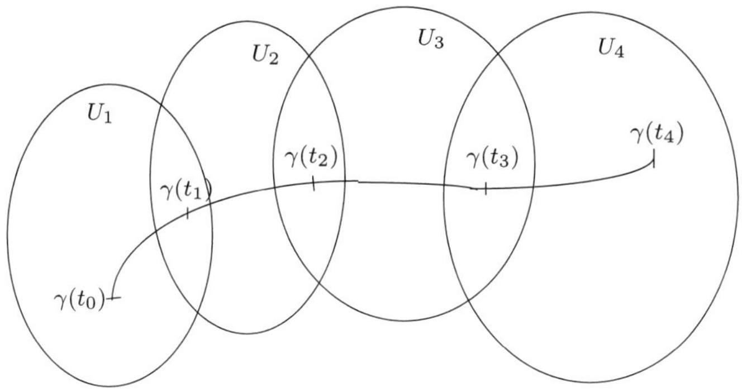

Figure 1.7. Analytic continuation along a path

Definition. Let X be a Riemann surface, $\gamma : [0,1] \to X$ a path with $\gamma(0) = x_0$ and $\gamma(1) = x_1$, $\bar{f} \in \mathcal{O}_{x_0}$, $\bar{g} \in \mathcal{O}_{x_1}$. One says that $\bar{g}$ *arises from* $\bar{f}$ *by analytic continuation along* γ if and only if there are a subdivision

$$0 = t_0 < t_1 < \ldots < t_{n-1} < t_n = 1$$

of the interval $[0,1]$, regions $U_j \subset X$ with $\gamma([t_{j-1}, t_j]) \subset U_j$, and holomorphic functions $f_j : U_j \to \mathbb{C}$ $(j = 1, \ldots, n)$ such that

(i) $\rho_{x_0}(f_1) = \bar{f}$, $\rho_{x_1}(f_n) = \bar{g}$,

(ii) $f_j = f_{j+1}$ on the connected component V_j of $\gamma(t_j)$ in $U_j \cap U_{j+1}$.

Example 1.12. a) Let $X = \mathbb{C}$ and let $\bar{f}$ be a function germ given by the power series

$$f(z) = \sum_{n=1}^{\infty} \frac{(-1)^{n-1}}{n}(z-1)^n$$

at $x_0 = 1$. Each branch of the logarithm arises from this by analytic continuation along a suitable path. One obtains e.g. the branch that has the value $5\pi i$ at $x_1 = -1$ by analytic continuation along a path that starts at $x_0 = 1$ and winds two and a half times counterclockwise around the origin.

b) Local primitive functions are continuable along any path: Let $G \subset \mathbb{C}$ be a region, $f : G \to \mathbb{C}$ holomorphic, $\varphi \in \mathcal{O}_a$ the germ of a local primitive function of f in $a \in G$, and $\gamma : [0,1] \to G$ a path in G with $\gamma(0) = a$. Then one can find a subdivision

$$0 = t_0 < t_1 < \ldots < t_n = 1$$

of the interval $[0,1]$ such that for each $j = 1, \ldots, n$ the set $\gamma([t_{j-1}, t_j])$ lies in an open subset $U_j \subset G$ on which a local primitive function F_j of f exists.

We choose the functions F_j so that $\rho_a(F_1) = \varphi$ and $F_{j+1}(\gamma(t_j)) = F_j(\gamma(t_j))$ for $j = 1, \ldots, n-1$. Then

$$F_j|_{U_j \cap U_{j+1}} = F_{j+1}|_{U_j \cap U_{j+1}},$$

so $\rho_{\gamma(1)}(F_n)$ arises from φ by analytic continuation along γ.

In this way we can also define the integral of a holomorphic function along a continuous (not necessarily piecewise continuously differentiable) path γ:

$$\int_\gamma f(z)dz := F_n(\gamma(1)) - F_1(\gamma(0)).$$

In particular one can define in this way the winding number of an arbitrary (continuous) closed path in $\mathbb{C} \setminus \{z_0\}$ around $z_0 \in \mathbb{C}$: one takes $f(z) = 1/(z - z_0)$.

By means of the map $p : \mathcal{O}_X \to X$ one can interpret analytic continuation along a path as follows: Such a continuation can be regarded as a path in the sheaf $\mathcal{O}_X$. More precisely:

Proposition 1.20. *Let X be a Riemann surface, $\gamma : [0,1] \to X$ a path with $\gamma(0) = x_0$ and $\gamma(1) = x_1$. Then the function germ $\bar{g} \in \mathcal{O}_{x_1}$ arises by analytic continuation along γ from $\bar{f} \in \mathcal{O}_{x_0}$ if and only if there is a lifting $\tilde{\gamma} : [0,1] \to \mathcal{O}_X$ of γ with $\tilde{\gamma}(0) = \bar{f}$ and $\tilde{\gamma}(1) = \bar{g}$.*

Proof. a) First let $\bar{g}$ arise from $\bar{f}$ by analytic continuation along the path γ. Let $0 = t_0 < t_1 < \ldots < t_n = 1$ be a subdivision of $[0,1]$, $U_j \subset X$ regions with $\gamma([t_{j-1}, t_j]) \subset U_j$, and let $f_j : U_j \to \mathbb{C}$ $(j = 1, \ldots, n)$ be holomorphic functions such that properties (i) and (ii) of the definition of analytic continuation hold. Then we can specify a path $\tilde{\gamma} : [0,1] \to \mathcal{O}_X$ by putting $\tilde{\gamma}(t) = \rho_{\gamma(t)}(f_j)$ for $t_{j-1} \le t \le t_j$: By the definition of the topology on $\mathcal{O}_X$ the path $\tilde{\gamma}$ is continuous. We have $\tilde{\gamma}(0) = \bar{f}$ and $\tilde{\gamma}(1) = \bar{g}$. Thus $\tilde{\gamma}$ is a lifting of γ with the desired properties.

b) Now let $\tilde{\gamma} : [0,1] \to \mathcal{O}_X$ be a lifting of γ with $\tilde{\gamma}(0) = \bar{f}$ and $\tilde{\gamma}(1) = \bar{g}$. There is a subdivision $0 = t_0 < t_1 < \ldots < t_n = 1$ of the compact interval $[0,1]$ such that each subpath $\tilde{\gamma}|_{[t_{j-1}, t_j]}$ lies in an open set of the form $\sigma(U_j, f_j)$ $(j = 1, \ldots, n)$. The t_j, U_j, f_j have the properties required in the definition of analytic continuation. □

Proposition 1.20 now enables us to apply the facts from §1.3. From the uniqueness of lifting of local homeomorphisms (Proposition 1.9) it follows that the analytic continuation of a function germ along a path is unique when it exists. Moreover:

Proposition 1.21 (Monodromy theorem). *Let X be a Riemann surface, $\gamma_0, \gamma_1 : [0,1] \to X$ two homotopic paths from x_0 to x_1, $H : [0,1] \times [0,1] \to X$*

a homotopy between γ_0 and γ_1 and $\bar{f} \in \mathcal{O}_{x_0}$ a function germ that can be continued analytically along each path $H_t : [0,1] \to X, H_t(s) = H(s,t)$. Then the analytic continuations of $\bar{f}$ along γ_0 and γ_1 yield the same function germ $\bar{g} \in \mathcal{O}_{x_1}$.

Proof. By Proposition 1.18, $p : \mathcal{O}_X \to X$ is a local homeomorphism. Proposition 1.19 asserts that $\mathcal{O}_X$ is a Hausdorff space. The result therefore follows from Proposition 1.11. □

1.5. Branched meromorphic continuation

In contrast to real functions several important complex functions fail to have unique inverse functions. So, for example, $\sqrt{z}$ cannot be considered as a function in the proper sense for a complex variable z. Given a $z \neq 0$, we cannot in a unique way assign a function value $w = f(z) = \sqrt{z}$ so that f becomes a holomorphic function on $\mathbb{C}$. The expression $\sqrt{z}$ can only stand for a so-called *multivalued* function: to a $z \in \mathbb{C}$ are associated the solutions of the equation $w^2 - z = 0$. For $z \neq 0$ one has two distinct values. As such a multivalued function, $\sqrt{z}$ is defined also on $\hat{\mathbb{C}}$ if we put $\sqrt{\infty} = \infty$. In this section we shall examine the following problem: Do there exist a Riemann surface X and a holomorphic map $p : X \to \hat{\mathbb{C}}$ such that $\sqrt{z}$ can be interpreted as a single valued meromorphic function on X?

This problem leads us back to analytic continuation. We follow the presentation in [**FL88**].

Definition. Let $G \subset \mathbb{C}$ be a region and $f : G \to \mathbb{C}$ holomorphic. A *branched meromorphic continuation* of f is a Riemann surface X together with

a) a holomorphic map $p : X \to \hat{\mathbb{C}}$,
b) a holomorphic map $j : G \to X$ with $p \circ j = \mathrm{id}_G$,
c) a holomorphic map $\hat{f} : X \to \hat{\mathbb{C}}$ with $\hat{f} \circ j = f$.

Example 1.13. Let $G = \{z \in \mathbb{C} \,|\, \mathrm{Re}\, z > 0\}$ and let $f : G \to \mathbb{C}$ be defined by $f(z) = \sqrt{z} = \sqrt{|z|}e^{i \arg z/2}$, $-\pi < \arg z < \pi$, i.e., f is a branch of the square root. Let $p_2 : \hat{\mathbb{C}} \to \hat{\mathbb{C}}$ be defined by $p_2(z) = z^2$ for $z \in \mathbb{C}$ and $p_2(\infty) = \infty$, and $j : G \to \hat{\mathbb{C}}$ by $j(z) = \sqrt{|z|}e^{i \arg z/2}$ for $z \in G$ and $\hat{f} = \mathrm{id}_{\hat{\mathbb{C}}} : \hat{\mathbb{C}} \to \hat{\mathbb{C}}$. Then $(\hat{\mathbb{C}}, p_2, j, \hat{f})$ is a branched meromorphic continuation of f.

Definition. A branched meromorphic continuation $(X, p, j, \hat{f})$ of f is called *complete* when for every branched meromorphic continuation $(X_1, p_1, j_1, \hat{f}_1)$ of f there is a holomorphic map $\varphi : X_1 \to X$ with $\varphi \circ j_1 = j$ and $p \circ \varphi = p_1$.

Proposition 1.22. *The complete branched meromorphic continuation of a holomorphic function is uniquely determined up to isomorphism.*

Proof. Let $(X_1, p_1, j_1, \hat{f}_1)$ and $(X_2, p_2, j_2, \hat{f}_2)$ be two complete branched meromorphic continuations of $f : G \to \mathbb{C}$. There are then fiber preserving holomorphic maps $\varphi : X_1 \to X_2$ and $\psi : X_2 \to X_1$ with $\varphi \circ j_1 = j_2$ and $\psi \circ j_2 = j_1$. But this just means that $\psi \circ \varphi$ on $j_1(G)$ and $\varphi \circ \psi$ on $j_2(G)$ are the identity. By the open mapping theorem $j_1(G)$ is open in X_1 and $j_2(G)$ is open in X_2. It follows from the identity theorem that then $\psi \circ \varphi = \mathrm{id}_{X_1}$ and $\varphi \circ \psi = \mathrm{id}_{X_2}$. Hence $\varphi : X_1 \to X_2$ is biholomorphic. □

Theorem 1.2. *Let $G \subset \mathbb{C}$ be a region and $f : G \to \mathbb{C}$ holomorphic. Then there exists a complete branched meromorphic continuation of f.*

Definition. The corresponding Riemann surface X together with the holomorphic map $p : X \to \hat{\mathbb{C}}$ is also called the *Riemann surface of f*.

Proof. a) Let $a \in G$ and let $\bar{f}$ be the function germ of f at a. Let $\tilde{X}$ be the path connected component of $\bar{f}$ in $\mathcal{O}_{\hat{\mathbb{C}}}$ and $\tilde{p} : \tilde{X} \to \hat{\mathbb{C}}$ the restriction projection $p : \mathcal{O}_{\hat{\mathbb{C}}} \to \hat{\mathbb{C}}$ on $\tilde{X}$. By Proposition 1.18, $p : \mathcal{O}_{\hat{\mathbb{C}}} \to \hat{\mathbb{C}}$ is a local homeomorphism. By Proposition 1.15 there is a complex structure on $\tilde{X}$ such that $\tilde{X}$ is a Riemann surface and the map $\tilde{p} : \tilde{X} \to \hat{\mathbb{C}}$ is holomorphic.

Let $j : G \to \tilde{X}$ be defined by $j(z) = \rho_z(f)$. Then j is a lifting of the inclusion $\iota : G \hookrightarrow \hat{\mathbb{C}}$ with respect to the map $\tilde{p} : \tilde{X} \to \hat{\mathbb{C}}$. By Proposition 1.16, j is holomorphic. The map j takes G onto $\sigma(G, f)$, and $\tilde{p} \circ j = \mathrm{id}_G$. Finally we specify $\tilde{f} : \tilde{X} \to \hat{\mathbb{C}}$ by

$$\tilde{f}(\varphi) = \varphi(\tilde{p}(\varphi)) \qquad \text{for } \varphi \in \tilde{X}.$$

For each set $\sigma(V, g) \subset \tilde{X}$ we then have $\tilde{f} = g \circ \tilde{p}$ on $\sigma(V, g)$. Thus $\tilde{f}$ is holomorphic and $\tilde{f} \circ j = f$ on G. (The quadruple $(\tilde{X}, \tilde{p}, j, \tilde{f})$ is also called a *complete analytic continuation* of f.)

b) We next continue the holomorphic map $\tilde{p} : \tilde{X} \to \hat{\mathbb{C}}$ to a holomorphic map $p : X \to \hat{\mathbb{C}}$ with branch points. To this end we must "insert" branch points and poles into $\tilde{X}$.

Let V be an open set in $\tilde{X}$ for which $\tilde{p}|_V : V \to \tilde{p}(V)$ is a k-sheeted covering of a punctured disc $\tilde{p}(V) = \Delta_r^*(a_V)$ with center $a_V \in \hat{\mathbb{C}}$, i.e., $\Delta_r^*(a_V) = \{z \in \mathbb{C} \,|\, 0 < |z - a_V| < r\}$ for $a_V \in \mathbb{C}$ and $\Delta_r^*(\infty) = \{z \in \mathbb{C} \,|\, |z| > 1/r\}$.

Claim. V is a path connected component of $\tilde{p}^{-1}(\Delta_r^*(a_V))$.

Let $x_0 \in V$ and let x_1 be a point in $\tilde{p}^{-1}(\Delta_r^*(a_V))$ joined to x_0 by a path $\tilde{\gamma}$ in $\tilde{p}^{-1}(\Delta_r^*(a_V))$. Then $\tilde{\gamma}$ is the lifting of a path γ in $\Delta_r^*(a_V)$ which joins the points $\tilde{p}(x_0)$ and $\tilde{p}(x_1)$. By Proposition 1.10 there is a unique lifting $\tilde{\gamma}_1$ of the path γ with respect to the covering $\tilde{p}|_V : V \to \Delta_r^*(a_V)$ with $\tilde{\gamma}_1(0) = x_0$. We put $J := \{t \in [0,1] \,|\, \tilde{\gamma}_1(t) = \tilde{\gamma}(t)\}$. Then J is closed, for $\tilde{X}$

is a Hausdorff space. Thus J is of the form $[0, t_0]$ for a $t_0 \in [0,1]$. Suppose $t_0 < 1$. Let U be a neighborhood of $\tilde{\gamma}(t_0)$ in $\tilde{X}$ for which $\tilde{p}|_U : U \to \tilde{p}(U)$ is a homeomorphism. Then there is an $\varepsilon > 0$ such that for the open interval $J' = (t_0 - \varepsilon, t_0 + \varepsilon)$ we have $\tilde{\gamma}(J'), \tilde{\gamma}_1(J') \subset U$. It then follows that

$$\tilde{\gamma}(t_0 + \frac{\varepsilon}{2}) = (\tilde{p}|_U)^{-1}(\gamma(t_0 + \frac{\varepsilon}{2})) = \tilde{\gamma}_1(t_0 + \frac{\varepsilon}{2}),$$

a contradiction. Hence $t_0 = 1$ and thus $x_1 = \tilde{\gamma}(1) = \tilde{\gamma}_1(1) \in V$. This proves our Claim.

By Example 1.10 there is a biholomorphic map $\psi = \psi_V : V \to \Delta^*_\rho(0)$, $\rho = r^{1/k}$, such that the following diagram commutes:

$$\begin{array}{ccc} V & \xrightarrow{\ \psi\ } & \Delta^*_\rho(0) \\ \Big\downarrow{\scriptstyle \tilde{p}|_V} & & \Big\downarrow{\scriptstyle p_k} \\ \Delta^*_r(a_V) & \xrightarrow{\ \varphi\ } & \Delta^*_r(0) \end{array}$$

Here $p_k : \Delta^*_\rho(0) \to \Delta^*_r(0)$ is the map $z \mapsto z^k$ and $\varphi : \Delta^*_r(a_V) \to \Delta^*_r(0)$ the map $z \mapsto z - a_V$ for $a_V \in \mathbb{C}$ and $z \mapsto 1/z$ for $a_V = \infty$. Now the map $p_k : \Delta^*_\rho(0) \to \Delta^*_r(0)$ can be continued to a map $p_k : \Delta_\rho(0) \to \Delta_r(0)$ with $p_k(0) = 0^k = 0$. This map is branched at 0.

We next distinguish two cases: Over a_V there may lie a point $\tilde{a}_V \in \tilde{X}$, or not. If there is such a point, then we must have $k = 1$ and for $a_V \in \mathbb{C}$ the function $\tilde{f} \circ \psi^{-1}$ must have a removable singularity at 0. If there is no such point over a_V, we insert into V one of the points $\hat{a}_V$ corresponding to 0 in $\Delta^*_\rho(0)$ provided that $\tilde{f}$ can be meromorphically continued to this point.

We call V *admissible* if $\tilde{f} \circ \psi^{-1}$ has a removable singularity or a pole at 0, unless $k = 1$, $a_V \in \mathbb{C}$ and $\tilde{f} \circ \psi^{-1}$ has a removable singularity at 0. We call two admissible sets V_1, V_2 equivalent, symbolically $V_1 \sim V_2$, if $a_{V_1} = a_{V_2}$ and $V_1 \subset V_2$ or $V_2 \subset V_1$. The equivalence class of V will be denoted by $\hat{a}_V$.

We next put

$$X := \tilde{X} \cup \{\hat{a}_V \mid V \text{ admissible}\}.$$

The maps $\tilde{p} : \tilde{X} \to \hat{\mathbb{C}}$ and $\tilde{f} : \tilde{X} \to \hat{\mathbb{C}}$ are continued to maps $p : X \to \hat{\mathbb{C}}$ and $f : X \to \hat{\mathbb{C}}$ by stipulating

$$\begin{aligned} p(\hat{a}_V) &:= a_V, \\ \hat{f}(\hat{a}_V) &:= \lim_{z \to 0} \tilde{f} \circ \psi_V^{-1}(z). \end{aligned}$$

It is clear that these definitions are independent of the choices of the representative V of $\hat{a}_V$ and of the map ψ_V.

c) We now show that the complex structure of $\tilde{X}$ can be continued to X so that X becomes a Riemann surface and $p : X \to \hat{\mathbb{C}}$ and $\hat{f} : X \to \hat{\mathbb{C}}$ become holomorphic maps.

To do this, we define a topology on X as follows: Let V be an admissible set with corresponding point $\hat{a}_V$. We continue $\psi_V : V \to \Delta^*_\rho(0)$ to $\hat{\psi}_V : V \cup \{\hat{a}_V\} \to \Delta_\rho(0)$ by $\hat{\psi}_V(\hat{a}_V) = 0$. We call a subset $U \subset X$ open if $U \cap \tilde{X}$ is open and if U together with each point $\hat{a}_V$ also contains a subset $\hat{\psi}_V^{-1}(\Delta_{\rho'}(0))$ with $0 < \rho' \leq \rho$. This defines a topology on X with respect to which $p : X \to \hat{\mathbb{C}}$ is continuous. Since $\tilde{X}$ is path connected, so too is X path connected.

We next show that X is a Hausdorff space: Let $x_1, x_2 \in X$, $x_1 \neq x_2$. If $p(x_1) \neq p(x_2)$ or $x_1, x_2 \in \tilde{X}$, then x_1 and x_2 have disjoint neighborhoods. Now let $x_1 = \hat{a}_{V_1}$, $x_2 = \hat{a}_{V_2}$, $p(x_1) = p(x_2) = a$. We can then assume that the representatives V_1 and V_2 have been chosen so that $\tilde{p}|_{V_1} : V_1 \to \Delta^*_r(a)$ and $\tilde{p}|_{V_2} : V_2 \to \Delta^*_r(a)$ are coverings over the same punctured disc $\Delta^*_r(a)$. By the Claim of b), V_1 and V_2 are then path connected components of $\tilde{p}^{-1}(\Delta^*_r(a))$. Therefore V_1 and V_2 are either disjoint or the same. However, they can be identical only if $x_1 = x_2$. Finally, let $x_1 = \hat{a}_V$ and $x_2 \in \tilde{X}$ with $p(x_1) = p(x_2) = a$. Then there are a neighborhood W of x_2, a disc $\Delta_r(a)$ around a, and a representative V of x_1 such that $\tilde{p}|_W : W \to \Delta_r(a)$ is a homeomorphism and $\tilde{p}|_V : V \to \Delta^*_r(a)$ is a k-sheeted covering. It follows again from the Claim in b) that V and $W \setminus \{x_2\}$ are path connected components of $\tilde{p}^{-1}(\Delta^*_r(a))$. But they cannot be the same, for otherwise we would have $k = 1$ and $\tilde{f} \circ \psi_V^{-1}$ would have a removable singularity at 0, so V would not be admissible. Hence $V \cup \{\hat{a}_V\}$ and W are disjoint neighborhoods of x_1 and x_2.

We now define a complex structure on X. The map $\hat{\psi}_V : V \cup \{\hat{a}_V\} \to \Delta_\rho(0)$ for an admissible V is a chart around $\hat{a}_V$. One obtains a complex atlas of X on adding to a complex atlas of $\tilde{X}$ the charts $\hat{\psi}_V : V \cup \{\hat{a}_V\} \to \Delta_\rho(0)$ for all admissible V. Since by definition the maps $\hat{f} \circ \hat{\psi}_V^{-1} : \Delta_\rho(0) \to \hat{\mathbb{C}}$ and $p \circ \hat{\psi}_V^{-1} : \Delta_\rho(0) \to \Delta_r(a_V)$ are holomorphic, so too $\hat{f} : X \to \hat{\mathbb{C}}$ and $p : X \to \hat{\mathbb{C}}$ are holomorphic.

d) We must now further show that the branched meromorphic continuation $(X, p, j, \hat{f})$ of f is complete. To achieve this, let $(X_1, p_1, j_1, \hat{f}_1)$ be an arbitrary branched meromorphic continuation of f. We choose a fixed point $z_0 \in G$. Let $\tilde{X}_1$ be the Riemann surface obtained from X_1 by removing the branch points of p_1 and the poles of $\hat{f}_1$. First let $x_1 \in \tilde{X}_1$. Let $\gamma_1 : [0,1] \to X_1$ be a path from $j_1(z_0)$ to x_1. Then there are a subdivision

$$0 = t_0 < t_1 < \ldots < t_{n-1} < t_n = 1$$

of the interval $[0,1]$ and regions $V_j \subset X_1$ with $\gamma_1([t_{j-1},t_j]) \subset V_j$ such that $p_1|_{V_j} : V_j \to p_1(V_j)$ is injective. Let $U_j = p_1(V_j)$, $\gamma = p_1 \circ \gamma_1$, $g = \hat{f}_1 \circ (p_1|_{V_n})^{-1}$ and $y_1 = p_1(x_1)$. Then the germ $\bar{g}$ of g in y_1 arises by analytic continuation along γ from the germ $\bar{f}$ of f in z_0. By Proposition 1.20 there is a path $\tilde{\gamma} : [0,1] \to \mathcal{O}_{\hat{\mathbb{C}}}$ with $\tilde{\gamma}(0) = \bar{f}$, $\tilde{\gamma}(1) = \bar{g}$ and $\tilde{p} \circ \tilde{\gamma} = \gamma$. Since $\tilde{X}$ is the path connected component of $\bar{f}$ in $\mathcal{O}_{\hat{\mathbb{C}}}$, it follows that $\bar{g} \in \tilde{X}$ and $\tilde{\gamma}$ is a path in $\tilde{X}$. We now put

$$\varphi(x_1) := \bar{g}.$$

Now we consider the case when x_1 is a branch point of p_1. Let V_1 be a neighborhood of x_1 such that for $V_1^* = V_1 \setminus \{x_1\}$ we have that $p_1|_{V_1^*} : V_1^* \to p_1(V_1^*) = \Delta_r^*(a)$ is a k-sheeted covering. Let V be the path connected component of $\varphi(V_1^* \cap \tilde{X}_1)$ in $\tilde{p}^{-1}(\Delta_r^*(a))$. By decreasing r (if necessary) we can assume that $\tilde{p}|_V : V \to \Delta_r^*(a)$ is a covering. If V is admissible, we put

$$\varphi(x_1) := \hat{a}_V.$$

Otherwise there is a point $x \in \tilde{X} \cap \bar{V}$ over a. Then we put

$$\varphi(x_1) := x.$$

One defines φ correspondingly if x_1 is a pole of $\hat{f}_1$. It is easy to show that the map $\varphi : X_1 \to X$ defined in this way is holomorphic and has the properties $\varphi \circ j_1 = j$ and $p \circ \varphi = p_1$ as demanded in the definition of completeness.

This completes the proof of Theorem 1.2. □

1.6. The Riemann surface of an algebraic function

We considered the multivalued function $t = t(z) = \sqrt{z}$ in §1.5. This was a special case of the (so-called) algebraic functions that we now want to study.

Definition. An *algebraic function* is a multivalued function $t = t(z)$ that satisfies an algebraic equation

$$t^n c_1(z) t^{n-1} + \ldots + c_n(z) = 0$$

where the coefficients $c_j(z)$ are convergent power series in z.

In this section we shall study the Riemann surfaces of algebraic functions. First we consider polynomials

$$P(z,t) = t^n + c_1(z)t^{n-1} + \ldots + c_n(z)$$

in the complex variable t with coefficients $c_j = c_j(z)$ in the ring $\mathbb{C}\{z\}$ of convergent power series in the variable z. Such a polynomial is also an element of $\mathbb{C}\{z\}[t]$. The quotient field of $\mathbb{C}\{z\}$ is the field $K = \mathbb{C}\{\{z\}\}$ of all Laurent series with finite principal part

$$\varphi(z) = \sum_{\nu=k}^{\infty} a_\nu z^\nu, \quad k \in \mathbb{Z},\ a_\nu \in \mathbb{C},$$

that converge in a punctured disc $\Delta_r^*(0) = \{z \in \mathbb{C} \mid 0 < |z| < r\}$ of radius $r > 0$ around 0. The radius r may depend on the series $\varphi(z)$. There is a natural inclusion $\mathbb{C}\{z\} \subset K = \mathbb{C}\{\{z\}\}$. Thus we can interpret the polynomial $P(z,t)$ also as an element of $K[t]$.

Let now K be an arbitrary field and let

$$\begin{aligned} f(x) &= a_0x^n + a_1x^{n-1} + \ldots + a_n, \\ g(x) &= b_0x^m + b_1x^{m-1} + \ldots + b_m \end{aligned}$$

be two general polynomials from $K[x]$.

Definition. The *resultant* $R(f,g)$ of f and g is the determinant

$$\left|\begin{array}{ccccccc} a_0 & a_1 & \cdots & a_n & & & \\ & a_0 & a_1 & \cdots & & a_n & \\ & & \cdots & \cdots & \cdots & & \\ & & & a_0 & a_1 & \cdots & a_n \\ b_0 & b_1 & \cdots & b_m & & & \\ & b_0 & b_1 & \cdots & & b_m & \\ & & \cdots & \cdots & \cdots & & \\ & & & b_0 & b_1 & \cdots & b_m \end{array}\right| \begin{array}{l} \left.\vphantom{\begin{array}{c}a\\a\\a\\a\end{array}}\right\} m \text{ rows} \\ \left.\vphantom{\begin{array}{c}a\\a\\a\\a\end{array}}\right\} n \text{ rows} \end{array}$$

Proposition 1.23. *Let*

$$\begin{aligned} f(x) &= a_0x^n + a_1x^{n-1} + \ldots + a_n, \\ g(x) &= b_0x^m + b_1x^{m-1} + \ldots + b_m \end{aligned}$$

be polynomials from $K[x]$ with $a_0b_0 \neq 0$. Then $f(x)$ and $g(x)$ have a common factor of degree ≥ 1 precisely when $R(f,g) = 0$.

Proof. Suppose that $f(x)$ and $g(x)$ have a common factor $p(x)$ of degree ≥ 1. Then

$$\begin{aligned} f(x) &= h(x) \cdot p(x), \\ g(x) &= k(x) \cdot p(x), \end{aligned}$$

where $h(x) \neq 0$ and $k(x) \neq 0$ are polynomials with degree $h(x) < n$ and degree $k(x) < m$. Then

$$k(x)f(x) - h(x)g(x) = 0. \tag{1.1}$$

Conversely, if $k(x), h(x) \neq 0$ are polynomials with

$$\text{degree}\, h(x) < \text{degree}\, f(x)$$

and

$$\operatorname{degree} k(x) < \operatorname{degree} g(x)$$

that satisfy this equation, there must be a common factor of degree ≥ 1 of $f(x)$ and $g(x)$. For then, since the ring $K[x]$ is a unique factorization domain, $k(x)f(x)$ and $h(x)g(x)$ have the same decomposition into irreducible factors in $K[x]$. Thus all irreducible factors of $f(x)$ must also divide the product $h(x)g(x)$. Since degree $h(x) <$ degree $f(x)$, they cannot all divide the polynomial $h(x)$. Therefore at least one irreducible factor of $f(x)$ must divide the polynomial $g(x)$. Therefore the existence of solutions $h(x), k(x) \neq 0$ of the equation (1.1) with degree $h(x) <$ degree $f(x)$ and degree $k(x) <$ degree $g(x)$ is equivalent to the existence of a common factor of degree ≥ 1 of $f(x)$ and $g(x)$. Let now

$$\begin{aligned} h(x) &= c_0 x^{n-1} + c_1 x^{n-2} + \ldots + c_{n-1}, \\ k(x) &= d_0 x^{m-1} + d_1 x^{m-2} + \ldots + d_{m-1}. \end{aligned}$$

Then the equation (1.1) is equivalent to the following linear system of equations for the coefficients c_j and d_l:

$$\begin{aligned} a_0 d_0 - b_0 c_0 &= 0, \\ a_1 d_0 + a_0 d_1 - b_1 c_0 - b_0 c_1 &= 0, \\ \vdots \quad &\vdots \quad \vdots \\ a_{n-1} d_0 + a_{n-2} d_1 + \cdots + a_0 d_{n-1} - b_{n-1} c_0 - b_{n-2} c_1 - \cdots - b_0 c_{n-1} &= 0, \\ a_n d_0 + a_{n-1} d_1 + \cdots + a_0 d_n - b_n c_0 - b_{n-1} c_1 - \cdots - b_1 c_{n-1} &= 0, \\ a_n d_1 + a_{n-1} d_2 + \cdots + a_0 d_{n+1} - b_{n+1} c_0 - b_n c_1 - \cdots - b_2 c_{n-1} &= 0, \\ \vdots \quad &\vdots \quad \vdots \\ a_n d_{m-2} + a_{n-1} d_{m-1} - b_m c_{n-2} - b_{m-1} c_{n-1} &= 0, \\ a_n d_{m-1} - b_m c_{n-1} &= 0. \end{aligned}$$

The existence of a nontrivial solution of this system of equations is however equivalent to the vanishing of the determinant of the coefficient matrix. If one multiplies the last n columns by the factor -1 and transposes the matrix, then the determinant of this matrix is the resultant $R(f,g)$. □

Definition. Let $f(z)$ be a polynomial in $\mathbb{C}[z]$. Then the *discriminant* of $f(z)$ is defined as

$$\Delta(f) = R(f, f').$$

Corollary 1.9. *The discriminant of a nonconstant polynomial $f(z) \in \mathbb{C}[z]$ vanishes if and only if $f(z)$ has a multiple zero.*

Proof. A multiple zero of $f(z)$ is also a zero of the derivative $f'(z)$. Thus $f(z)$ has a multiple zero precisely when $f(z)$ and $f'(z)$ have a common factor of degree ≥ 1. By Proposition 1.23 this condition is, however, equivalent to $\Delta(f) = R(f, f') = 0$. □

We now apply Proposition 1.23 to polynomials from $\mathbb{C}\{z\}[t]$.

Proposition 1.24. *Let $P(z,t)$ and $Q(z,t)$ be polynomials in $\mathbb{C}\{z\}[t]$ without common factor of degree ≥ 1. Let $R > 0$ be the minimum of the radii of convergence of the coefficients of these polynomials. Then the set of all $z \in \Delta_R(0)$ for which the equations*

$$P(z,t) = 0, \quad Q(z,t) = 0$$

have a common solution $t \in \mathbb{C}$ is a discrete subset of $\Delta_R(0)$.

Proof. As remarked above, we can interpret $P(z,t)$ and $Q(z,t)$ as elements of $K[t]$, where $K = \mathbb{C}\{\{z\}\}$. From the Gauss lemma (cf. e.g. [**Lan02**, V, §6]) it follows that $P(z,t)$ and $Q(z,t)$ have no common factor of degree ≥ 1 in $K[t]$. By Proposition 1.23 their resultant $R = R(P,Q)$ in the field K differs from 0. By definition the resultant R is a polynomial in the coefficients of $P(z,t)$ and $Q(z,t)$, which are elements of $\mathbb{C}\{z\}$. Hence $R = R(z)$ is a nontrivial convergent power series in z.

Now let $z \in \Delta_R(0)$ be chosen so that equations $P(z,t) = 0$, $Q(z,t) = 0$ have a common solution $t \in \mathbb{C}$. Then the corresponding polynomials $f(t) = P(z,t)$, $g(t) = Q(z,t)$ in $\mathbb{C}[t]$ have a common zero, hence a common linear factor. By Proposition 1.23 we must have $R(z) = 0$. But the zeros of a nontrivial convergent power series form a discrete subset of $\mathbb{C}$. This proves Proposition 1.24. □

Next we consider a polynomial

$$P(z,t) = t^n + c_1(z)t^{n-1} + \ldots + c_n(z) \in \mathbb{C}\{z\}[t]$$

that is irreducible in the ring $\mathbb{C}\{z\}[t]$. Let $R > 0$ be the minimum of the radii of convergence of the power series $c_1(z), \ldots, c_n(z)$. We call a point $z \in \Delta_R(0)$ *critical* if $P(z,t)$ has a multiple zero. Let $C \subset \Delta_R(0)$ be the set of all critical points. Now $P(z,t)$ is irreducible and the derivative $(\partial P/\partial t)(z,t)$ has degree $n-1$, so $P(z,t)$ and $(\partial P/\partial t)(z,t)$ have no common factor of degree ≥ 1. By Proposition 1.24 the set of all $z \in \Delta_R(0)$ for which $P(z,t)$ and $(\partial P/\partial t)(z,t)$ have a common zero t is a discrete subset of $\Delta_R(0)$. Therefore C is discrete.

Now let $a \in \Delta_R(0) \setminus C$. Then the equation $P(a,t) = 0$ has exactly n distinct solutions. The following proposition shows that these solutions vary holomorphically in z for z near a.

Proposition 1.25. *Let $R > 0$, $\Delta_R(a) := \{z \in \mathbb{C} \mid |z-a| < R\}$ and $c_1, \ldots, c_n$ be holomorphic functions in $\Delta_R(a)$. Let $t_0 \in \mathbb{C}$ be a simple zero of the polynomial*

$$t^n + c_1(a)t^{n-1} + \ldots + c_n(a).$$

Then there exist an r with $0 < r \leq R$ and a holomorphic function f : $\Delta_r(a) \to \mathbb{C}$ with $f(a) = t_0$ and

$$f^n + c_1 f^{n-1} + \ldots + c_n = 0 \quad \text{on } \Delta_r(a).$$

Proof. For $z \in \Delta_R(a)$ and $t \in \mathbb{C}$ put

$$P(z,t) = t^n + c_1(z)t^{n-1} + \ldots + c_n(z).$$

By assumption there is an $\varepsilon > 0$ such that the function $t \mapsto P(a,t)$ on $\overline{\Delta_\varepsilon(t_0)} = \{t \in \mathbb{C} \,|\, |t - t_0| \leq \varepsilon\}$ has the unique zero t_0. By the continuity of P there is an r with $0 < r \leq R$ such that P has no zero on

$$\{(z,t) \in \mathbb{C}^2 \,|\, |z-a| < r, |t - t_0| = \varepsilon\}.$$

Let $\gamma : I \to \mathbb{C}$ be a parameterization of $\partial\overline{\Delta_\varepsilon(t_0)}$. For a fixed $z \in \Delta_r(a)$

$$\nu(z) = \frac{1}{2\pi i} \int_\gamma \frac{\frac{\partial P}{\partial t}(z,t)}{P(z,t)} dt$$

gives, by the argument principle, the number of zeros of the function $t \mapsto P(z,t)$ in $\overline{\Delta_\varepsilon(t_0)}$. Since $\nu(a) = 1$ and $\nu(z)$ is a continuous function of z, we must have $\nu(z) = 1$ for all $z \in \Delta_r(a)$. For $z \subset \Delta_r(a)$ let $f(z)$ be the unique zero of $t \mapsto P(z,t)$ in $\overline{\Delta_\varepsilon(t_0)}$. By the residue theorem

$$f(z) = \frac{1}{2\pi i} \int_\gamma \frac{t\frac{\partial P}{\partial t}(z,t)}{P(z,t)} dt.$$

Now the integrand depends holomorphically on z so $f : \Delta_r(a) \to \mathbb{C}$ is holomorphic. By construction $P(z, f(z)) = 0$ for all $z \in \Delta_r(a)$. □

Theorem 1.3. *Let*

$$P(z,t) = t^n + c_1(z)t^{n-1} + \ldots + c_n(z) \in \mathbb{C}\{z\}[t]$$

be an irreducible polynomial. Furthermore, let $R > 0$ be the minimum of the radii of convergence of the power series $c_1(z), \ldots, c_n(z)$, $a_0 \in \Delta_R(0)\setminus C$, $0 < r \leq R$, and let $f : \Delta_r(a_0) \to \mathbb{C}$ be a holomorphic function with $P(z, f(z)) = 0$ for all $z \in \Delta_r(a_0)$. Let $(\hat{X}, \hat{p}, j, \hat{f})$ be the Riemann surface of f. We put $X := \hat{p}^{-1}(\Delta_R(0))$, $p := \hat{p}|_X$, $\tilde{X} := X \setminus p^{-1}(C)$.

Then $p|_{\tilde{X}} : \tilde{X} \to \Delta_R(0) \setminus C$ is an n-sheeted covering, and for each point $a \in \Delta_R(0) \setminus C$, if $p^{-1}(a) = \{x_1, \ldots, x_n\}$, then $\hat{f}(x_1), \ldots, \hat{f}(x_n)$ are the distinct solutions of $P(a,t) = 0$.

Proof. Let $A = \Delta_R(0)\setminus C$ and $\tilde{Y} \subset \mathcal{O}_A$ be the set of all germs $\bar{g} \in \mathcal{O}_a$, $a \in A$, of functions $g : U \to \mathbb{C}$, $U \subset A$ open neighborhood of a, with $P(z, g(z)) = 0$ for all $z \in U$. Let $\tilde{p} : \tilde{Y} \to A$ be the canonical projection.

By Proposition 1.25 for each $a \in A$ there are an open neighborhood $U \subset A$ and holomorphic functions $g_1, \ldots, g_n \in \mathcal{O}(U)$ with

$$P(z,t) = (t - g_1(z)) \cdots (t - g_n(z)) \quad \text{for all } z \in U.$$

Then

$$\tilde{p}^{-1}(U) = \bigcup_{j=1}^{n} \sigma(U, g_j).$$

The sets $\sigma(U, g_j)$ are disjoint and are mapped homeomorphically onto U by $\tilde{p}$. Thus $\tilde{p} : \tilde{Y} \to A$ is a covering.

It is now enough to show that $\tilde{Y}$ is path connected. For if $\bar{f}$ is the germ of f at a, then $\bar{f} \in \tilde{Y}$ and by the construction of $\hat{X}$ as in the proof of Theorem 1.2 the path connected component of $\bar{f}$ in $\mathcal{O}_{\hat{\mathbb{C}}}$ is contained in $\hat{X}$. Then $\tilde{Y} \subset X$ and the theorem follows from this.

We now show that $\tilde{Y}$ is path connected. To do this, it is enough to show that for an $a \in A$ the germs $\bar{g}_1, \ldots, \bar{g}_n \in \mathcal{O}_a$ lie in the same path connected component of $\tilde{Y}$: for an arbitrary $y \in \tilde{Y}$ is a function germ at a point $b \in A$ that can be joined to a by a path γ in A, and, by analytic continuation along γ, there arises from y one of the germs $\bar{g}_1, \ldots, \bar{g}_n \in \mathcal{O}_a$, say $\bar{g}_j$. By Proposition 1.20 this means that y and $\bar{g}_j$ are joined by a path.

Now let $a \in A$ and let $\bar{g}_1, \ldots, \bar{g}_n$ be the points of $\tilde{p}^{-1}(a)$. Suppose that $\bar{g}_1, \ldots, \bar{g}_n$ do not lie in the same path connected component of $\tilde{Y}$. By renumbering (if necessary) we may assume that the path connected component of $\bar{g}_1$ contains the germs $\bar{g}_1, \ldots, \bar{g}_k$, but not $\bar{g}_{k+1}, \ldots, \bar{g}_n$, where $k < n$. The equation

$$\prod_{j=1}^{k} (t - g_j(z)) = t^k - \sigma_1(z) t^{k-1} + \ldots + (-1)^k \sigma_k(z),$$

must hold, where $\sigma_1, \ldots, \sigma_k$ are the elementary symmetric functions in $g_1, \ldots, g_k$, i.e.,

$$\begin{aligned}
\sigma_1(g_1, \ldots, g_k) &= g_1 + g_2 + \ldots + g_k, \\
\sigma_2(g_1, \ldots, g_k) &= g_1 g_2 + \ldots + g_1 g_k + g_2 g_3 + \ldots + g_{k-1} g_k = \sum_{j<l} g_j g_l, \\
&\vdots \\
\sigma_k(g_1, \ldots, g_k) &= g_1 g_2 \cdots g_k.
\end{aligned}$$

These are polynomials in $g_1, \ldots, g_k$. They can therefore be continued analytically along all paths in A. Since the analytic continuation of the germs $\bar{g}_1, \ldots, \bar{g}_n$ along closed paths leads to a permutation of these germs, the $\sigma_1, \ldots, \sigma_k$ are single valued holomorphic functions on A.

Now let $b \in C \subset \Delta_R(0)$. We consider the behavior as $z \to b$. The solutions t of $P(z,t) = 0$ are bounded as $z \to b$, since the power series $c_j(z)$ $(1 \leq j \leq n)$, being convergent in $\Delta_R(0)$, are bounded as $z \to b$; and for $|t| \geq 1$ it follows from $P(z,t) = 0$ that

$$\begin{aligned} |t| &= \left| c_1(z) + \frac{c_2(z)}{t} + \ldots + \frac{c_n(z)}{t^{n-1}} \right| \\ &\leq |c_1(z)| + |c_2(z)| + \ldots + |c_n(z)|. \end{aligned}$$

From this it follows that the solutions $g_1, \ldots, g_k$, and with them also the elementary symmetric functions $\sigma_1, \ldots, \sigma_k$ in $g_1, \ldots, g_k$, are holomorphic in a neighborhood of 0. Hence

$$Q(z,t) = \prod_{j=1}^{k} (t - g_j(z))$$

is an element of $\mathbb{C}\{z\}[t]$. As a polynomial in $\mathbb{C}\{z\}[t]$ it has degree k. Since $P(z,t)$ is irreducible by hypothesis, and of degree $n > k$, then $P(z,t)$ and $Q(z,t)$ cannot have a common factor of degree ≥ 1. By Proposition 1.24 the set of all $z \in \Delta_R(0)$ for which $P(z,t) = 0$, $Q(z,t) = 0$ have a common solution t is discrete. Now

$$P(z, g_1(z)) = Q(z, g_1(z)) = 0$$

for all z sufficiently near to a. This is a contradiction. Thus Theorem 1.3 is proved. □

Example 1.14. Let $\lambda > 0$, $G = \{z \in \mathbb{C} \mid |z| < \sqrt{\lambda}\}$ and let $f : G \to \mathbb{C}$ be defined by $f(z) = \sqrt{\lambda - z^2}$ with $f(0) = \sqrt{\lambda}$. The Riemann surface (X, p) of f can be described as follows, by Theorem 1.2: $X = \hat{\mathbb{C}}$ and $p : \hat{\mathbb{C}} \to \hat{\mathbb{C}}$. Let $Y = \hat{\mathbb{C}} \setminus \{-\sqrt{\lambda}, \sqrt{\lambda}\}$. Then $\tilde{X} = p^{-1}(Y)$ and $p|_{\tilde{X}} : X \to Y$ is a 2-sheeted covering. Exactly one point lies over the points $-\sqrt{\lambda}$ and $\sqrt{\lambda}$, respectively. Since $wf(1/w) = \sqrt{(\sqrt{\lambda}w - 1)(\sqrt{\lambda}w + 1)}$, two points lie over ∞. Let $X_1 = p^{-1}(\mathbb{C})$. We next want to describe the map $p|_{X_1} : X_1 \to \mathbb{C}$ topologically. To do this, we take two copies of the complex plane and cut them along the interval from $-\sqrt{\lambda}$ to $\sqrt{\lambda}$. We paste the two "edges" of the "cuts" crosswise (respectively + with −; cf. Figure 1.8). The concepts of "cut" and "paste" can be defined strictly mathematically. The "glueing together" of two spaces is equivalent e.g. to passing to a suitable quotient space of their disjoint union with the quotient topology. The very graphic terminology of cutting and pasting was already introduced by B. Riemann, long before the development of set theoretic topology with whose help we can now define these concepts mathematically. The resulting surface is homeomorphic to a cylinder. If we are now to add in the two points lying over ∞, we must

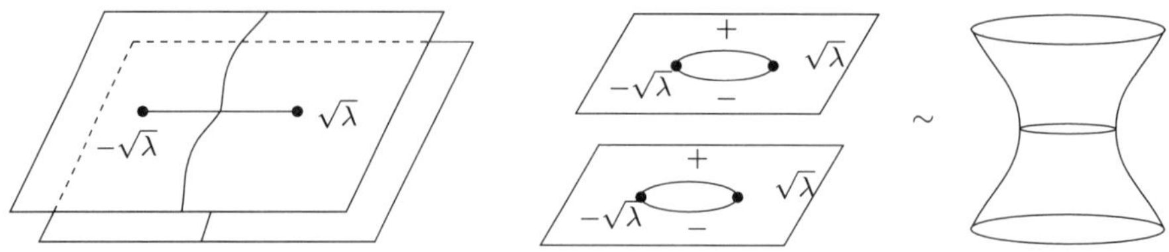

Figure 1.8. The Riemann surface of the function $\sqrt{\lambda - z^2}$

paste, respectively, a point to the upper and to the lower boundary of the cylinders, so obtaining the Riemann sphere $\hat{\mathbb{C}}$.

1.7. Puiseux expansion

As an application of Theorem 1.3 we demonstrate the following proposition.

Theorem 1.4 (Puiseux). *Let*

$$P(z,t) = t^n + c_1(z)t^{n-1} + \ldots + c_n(z) \in \mathbb{C}\{z\}[t]$$

be an irreducible polynomial with $c_j(0) = 0$ *for* $j = 1, \ldots, n$. *Then there is a convergent power series*

$$\varphi(\zeta) = \sum_{\nu=0}^{\infty} a_\nu \zeta^\nu \in \mathbb{C}\{\zeta\}$$

such that in the ring of convergent power series $\mathbb{C}\{\zeta\}$

$$P(\zeta^n, \varphi(\zeta)) = 0.$$

Definition. If one puts $z = \zeta^n$ and rewrites the series $\varphi(\zeta)$ as a series in the variable z, one obtains a series

$$\varphi(\zeta) = \varphi(\sqrt[n]{z}) = \sum_{\nu=0}^{\infty} a_\nu z^{\frac{\nu}{n}}$$

with fractional exponents. This series is called the *Puiseux series* corresponding to the polynomial $P(z,t)$. The series

$$t = \varphi(\sqrt[n]{z}) = \sum_{\nu=0}^{\infty} a_\nu z^{\frac{\nu}{n}}$$

then solves the equation

$$P(z,t) = 0.$$

Proof. Let $R > 0$ be the minimum of the radii of convergence of the power series $c_1(z), \ldots, c_n(z)$. Let $r > 0$ be chosen so small that in the disc $\Delta_r(0)$ the origin is the only critical point of $P(z,t)$. Furthermore, let $a \in \Delta_r(0) \setminus \{0\}$ and $r' > 0$ be such that $\Delta_{r'}(a) \subset \Delta_R(0)$ and $f : \Delta_{r'}(a) \to \mathbb{C}$ is a holomorphic function with $P(z, f(z)) = 0$ for all $z \in \Delta_{r'}(a)$. Let $(\hat{X}, \hat{p}, j, \hat{f})$ be the Riemann surface of f which exists, by Theorem 1.2, and let $Y = \hat{p}^{-1}(\Delta_r^*(0))$. By Theorem 1.3, $\hat{p}|_Y : Y \to \Delta_r^*(0)$ is then an n-sheeted covering. By the classification of the coverings of the punctured disc $\Delta_r^*(0)$ (Example 1.10) there is a biholomorphic map $\psi : \Delta_\rho^*(0) \to Y$, $\rho = \sqrt[n]{r}$, such that

$$\hat{p}(\psi(\zeta)) = \zeta^n \qquad \text{for all } \zeta \in \Delta_\rho^*(0).$$

By putting $\psi(0) = \hat{p}^{-1}(0) \in \hat{X}$ the map ψ can be continued to a holomorphic map $\psi : \Delta_\rho(0) \to \hat{X}$ with $\hat{p}(\psi(\zeta)) = \zeta^n$ for all $\zeta \in \Delta_\rho(0)$. We put

$$g(\zeta) = \hat{f}(\psi(\zeta)) \qquad \text{for all } \zeta \in \Delta_\rho(0).$$

Then g is a holomorphic function on $\Delta_\rho(0)$. Let $\varphi(\zeta)$ be the power series expansion of g at the origin. By the construction and Theorem 1.3

$$P(\zeta^n, g(\zeta)) = 0 \qquad \text{for all } \zeta \in \Delta_\rho(0).$$

Hence in $\mathbb{C}\{z\}$ we have

$$P(\zeta^n, \varphi(\zeta)) = 0,$$

as was to be shown. □

1.8. The Riemann sphere

We now assemble a few facts about the Riemann sphere $\hat{\mathbb{C}}$ which we shall need in §3.9.

Definition. Given four points $\tau_0, \tau_1, \tau_2, \tau_3 \in \hat{\mathbb{C}}$, of which the last three are pairwise distinct, the *cross ratio*

$$\mathrm{DV}(\tau_0, \tau_1, \tau_2, \tau_3) := \frac{\tau_0 - \tau_1}{\tau_0 - \tau_3} : \frac{\tau_2 - \tau_1}{\tau_2 - \tau_3}$$

is defined.

The biholomorphic maps of $\hat{\mathbb{C}}$ to itself are the *fractional linear transformations*

$$T(z) = \frac{az + b}{cz + d}, \qquad a, b, c, d \in \mathbb{C}, \qquad ad - bc \neq 0$$

(cf. e.g. [**Ahl78**, Chap. 3, Sec. 3]).

We next want to show that the cross ratio is an invariant under fractional linear transformations. To do this, we note that a fractional linear transformation $T \neq \mathrm{id}$ can have at most two fixed points, i.e., points $z_0 \in \hat{\mathbb{C}}$ with $T(z_0) = z_0$: If $T(z) = az + b$, $T \neq \mathrm{id}$, then ∞, and, if $a \neq 1$, the point $b/(1 - a)$, are the only fixed points. For a general T with

$$T(z) = \frac{az + b}{cz + d}$$

(with $c \neq 0$) the fixed points are simply the solutions of the quadratic equation

$$cz^2 + (d-a)z = b.$$

It follows that a fractional linear transformation is fully specified by giving the images of the three distinct points $\tau_1, \tau_2, \tau_3 \in \hat{\mathbb{C}}$. Then $T_1(z_0) = T_2(z_0)$, so z_0 is a fixed point of the map $T_2^{-1} \circ T_1$.

Next we consider the map

$$\begin{array}{rccc} T: & \hat{\mathbb{C}} & \longrightarrow & \hat{\mathbb{C}} \\ & z & \longmapsto & \mathrm{DV}(z, \tau_1, \tau_2, \tau_3) \end{array}$$

for pairwise distinct $\tau_1, \tau_2, \tau_3 \in \hat{\mathbb{C}}$. This is a fractional linear transformation: this is clear for $\tau_1, \tau_2, \tau_3 \in \mathbb{C}$, and

$$\begin{aligned} \mathrm{DV}(z, \infty, \tau_2, \tau_3) &= \frac{\tau_2 - \tau_3}{z - \tau_3}, \\ \mathrm{DV}(z, \tau_1, \infty, \tau_3) &= \frac{z - \tau_1}{z - \tau_3}, \\ \mathrm{DV}(z, \tau_1, \tau_2, \infty) &= \frac{z - \tau_1}{\tau_2 - \tau_1}. \end{aligned}$$

The fractional linear transformation $z \mapsto \mathrm{DV}(z, \tau_1, \tau_2, \tau_3)$ maps $\tau_1, \tau_2, \tau_3 \in \hat{\mathbb{C}}$ to $0, 1, \infty$, so it is the uniquely determined fractional linear transformation that maps these three points to $0, 1, \infty$.

Proposition 1.26. *Let $\tau_0, \tau_1, \tau_2, \tau_3$ be points in $\hat{\mathbb{C}}$ of which the last three are pairwise distinct. Then, for each fractional linear transformation T,*

$$\mathrm{DV}(\tau_0, \tau_1, \tau_2, \tau_3) = \mathrm{DV}(T(\tau_0), T(\tau_1), T(\tau_2), T(\tau_3)).$$

Proof. We consider the map

$$\begin{array}{rccc} S: & \hat{\mathbb{C}} & \longrightarrow & \hat{\mathbb{C}} \\ & z & \longmapsto & \mathrm{DV}(T(z), T(\tau_1), T(\tau_2), T(\tau_3)). \end{array}$$

The map S is a fractional linear transformation because $S = R \circ T$ with $R(w) = \mathrm{DV}(w, T(\tau_1), T(\tau_2), T(\tau_3))$. Then

$$S(\tau_1) = \mathrm{DV}(T(\tau_1), T(\tau_1), T(\tau_2), T(\tau_3)) = 0, \quad S(\tau_2) = 1, \quad S(\tau_3) = \infty.$$

Therefore S is the fractional linear transformation with

$$S(z) = \mathrm{DV}(z, \tau_1, \tau_2, \tau_3).$$

This proves Proposition 1.26. □

Chapter 2

Holomorphic functions of several variables

2.1. Holomorphic functions of several variables

We begin by transferring the concepts and results from the theory of complex functions of one complex variable to the multidimensional case.

In what follows we consider the n-dimensional complex number space:

$$\mathbb{C}^n := \{z = (z_1, \dots, z_n) \mid z_j \in \mathbb{C},\ j \in \{1, \dots, n\}\}.$$

We write

$$z_j = x_j + iy_j$$

for the decomposition of the coordinates z_j into real and imaginary parts. This gives us a bijection

$$\begin{array}{ccc} \mathbb{C}^n & \longrightarrow & \mathbb{R}^{2n} \\ (z_1, \dots, z_n) & \longmapsto & (x_1, y_1, \dots, x_n, y_n). \end{array}$$

This is an isomorphism of vector spaces. We use this isomorphism to introduce a norm and thus a topology on $\mathbb{C}^n$. Since all norms on $\mathbb{R}^{2n}$ are equivalent, all norms on $\mathbb{C}^n$ define the same topology. We shall often consider the following norms:

$$\begin{array}{lrl} \text{Euclidean norm:} & |z| := & \sqrt{\sum\limits_{j=1}^{n} z_j \bar{z}_j} = \sqrt{\sum\limits_{j=1}^{n} (x_j^2 + y_j^2)} \\ \text{maximum norm:} & \|z\|_{\max} := & \max_{j=1,\dots,n} |z_j|. \end{array}$$

Let $U \subset \mathbb{C}^n$ be an open subset. U is also called a *domain* of $\mathbb{C}^n$. An open and connected subset $G \subset \mathbb{C}^n$ is called a *region* of $\mathbb{C}^n$.

We first recall the case $n = 1$, $U \subset \mathbb{C}$. There are several equivalent characterizations of a holomorphic function $f : U \to \mathbb{C}$. Two of these characterizations are that a continuous function $f : U \to \mathbb{C}$ is called holomorphic when it satisfies one of the following equivalent conditions:

(i) f is complex differentiable for each $a \in U$,

(ii) f is analytic, i.e., f can be expanded as a convergent power series around every $a \in U$.

We now wish to generalize both these characterizations to the n-dimensional case.

Definition. A function $f : U \to \mathbb{C}$ is called *complex differentiable* at $a \in U$ if there are a complex linear map $T : \mathbb{C}^n \to \mathbb{C}$ and a function $r : U \to \mathbb{C}$, continuous at a, such that for all $z \in U$

$$f(z) = f(a) + T(z - a) + r(z)\|z - a\|$$

and $r(a) = 0$ (here $\| \ \|$ denotes any norm on $\mathbb{C}^n$).

If f is differentiable at $a \in U$, the linear map T is called the *derivative* of f at a. With respect to the canonical basis of $\mathbb{C}^n$ it can be described by a matrix with values in $\mathbb{C}$, in terms of a row vector $\operatorname{grad} f(a)$. The j-th component of $\operatorname{grad} f(a)$ is called the *partial derivative of f with respect to z_j at a* and is written as

$$\frac{\partial f}{\partial z_j}(a) \text{ or } f_{z_j}(a).$$

Then

$$f(z) = f(a) + \sum_{j=1}^{n} \frac{\partial f}{\partial z_j}(a)(z_j - a_j) + r(z)\|z - a\|$$

for all $z \in U$.

Proposition 2.1. *A function $f : U \to \mathbb{C}$ is complex differentiable at $a \in U$ if and only if there are functions $R_j : U \to \mathbb{C}$, $j = 1, \ldots, n$, continuous at a, such that for all $z \in U$*

$$f(z) = f(a) + \sum_{j=1}^{n} R_j(z)(z_j - a_j).$$

Remark 2.1. If $f : U \to \mathbb{C}$ is complex differentiable at $a \in U$, then

$$R_j(a) = \frac{\partial f}{\partial z_j}(a).$$

Proof of Proposition 2.1. In what follows let $\| \ \|$ be the Euclidean norm.

"$\Rightarrow$": Let $f : U \to \mathbb{C}$ be complex differentiable at $a \in U$, i.e., $(\partial f/\partial z_j)(a)$ exists for every $j = 1, \dots, n$ and for all $z \in U$

$$f(z) = f(a) + \sum_{j=1}^{n} \frac{\partial f}{\partial z_j}(a)(z_j - a_j) + r(z)\|z - a\|$$

for a function $r : U \to \mathbb{C}$ continuous at a with $r(a) = 0$. Put

$$R_j(z) = \begin{cases} \frac{\partial f}{\partial z_j}(a) - \frac{r(z)}{\|z-a\|}(\bar{z}_j - \bar{a}_j) & \text{for } z \neq a, \\ \frac{\partial f}{\partial z_j}(a) & \text{for } z = a. \end{cases}$$

Then R_j is continuous at a for $j = 1, \dots, n$ and for all $z \in U$

$$f(z) = f(a) + \sum_{j=1}^{n} R_j(z)(z_j - a_j).$$

"$\Leftarrow$": Conversely, if for all $z \in U$

$$f(z) = f(a) + \sum_{j=1}^{n} R_j(z)(z_j - a_j)$$

for functions $R_j : U \to \mathbb{C}$ continuous at a, then, for $z \neq a$, we put

$$r(z) := \sum_{j=1}^{n} \frac{(R_j(z) - R_j(a))(z_j - a_j)}{\|z - a\|}.$$

Then

$$\lim_{z \to a} r(z) = 0$$

and, with $r(a) := 0$, we have

$$f(z) = f(a) + \sum_{j=1}^{n} R_j(a)(z_j - a_j) + r(z)\|z - a\|$$

for all $z \in U$. Thus f is complex differentiable at a. □

Proposition 2.2. *Let $U \subset \mathbb{C}^n$ be an open subset and let $f : U \to \mathbb{C}$ be complex differentiable at $a \in U$. Then f is continuous at a.*

Proof. By Proposition 2.1

$$f(z) = f(a) + \sum_{j=1}^{n} R_j(z)(z_j - a_j)$$

for all $z \in U$ for some functions $R_j : U \to \mathbb{C}$ continuous at a. But the right side of this equation is a function of z continuous at a. □

Definition. Let $U \subset \mathbb{C}^n$ be open. A function $f : U \to \mathbb{C}$ is called *holomorphic* if and only if f is complex differentiable at every $a \in U$.

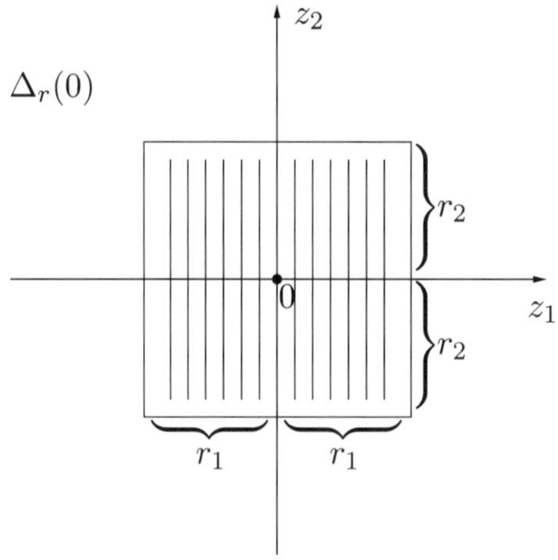

Figure 2.1. Polycylinder around $0 \in \mathbb{C}^2$

We next want to generalize Cauchy's integral formula for a disc in $\mathbb{C}$. Instead of a disc we consider a polycylinder.

Definition. Let

$$\mathbb{R}^n_{>0} := \{x = (x_1 \dots, x_n) \in \mathbb{R}^n \mid x_j > 0 \text{ for all } j\},$$

$r = (r_1, \dots, r_n) \in \mathbb{R}^n_{>0}$, $a \in \mathbb{C}^n$. Then

$$\Delta_r(a) := \{z \in \mathbb{C}^n \mid |z_j - a_j| < r_j, 1 \le j \le n\}$$

is called the *polycylinder around a with (poly-)radius r*.

A real picture of a polycylinder in $\mathbb{C}^2$ is represented in Figure 2.1.

The boundary of the closure of $\Delta_r(a)$ contains an n-dimensional torus

$$T_r(a) = \{z \in \mathbb{C}^n \mid |z_j - a_j| = r_j,\ 1 \le j \le n\}.$$

Now let $r \in \mathbb{R}^n_{>0}$ be given, $T := T_r(0)$. Let $h : T \to \mathbb{C}$ be a continuous function. We define

$$\int_T h(\zeta)d\zeta := \int_{|\zeta_n|=r_n} \cdots \int_{|\zeta_1|=r_1} h(\zeta_1, \dots, \zeta_n)d\zeta_1 \cdots d\zeta_n,$$

where the last integrals are complex path integrals.

Proposition 2.3 (Cauchy's integral formula). *Let $U \subset \mathbb{C}^n$ be open, $r \in \mathbb{R}^n_{>0}$, $\Delta := \Delta_r(0)$ a polycylinder around 0 with $\bar{\Delta} \subset U$ and $T = T_r(0)$. If $f : U \to \mathbb{C}$ is holomorphic then, for all $z \in \Delta$,*

$$f(z) = \left(\frac{1}{2\pi i}\right)^n \int_T \frac{f(\zeta)}{(\zeta_1 - z_1) \cdots (\zeta_n - z_n)} d\zeta\,.$$

Proof. We prove the proposition by induction on n.

For $n = 1$ this is Cauchy's integral formula for the disc for a holomorphic function of one complex variable.

Induction step: $n-1 \to n$: For fixed $(z_1, \ldots, z_{n-1}) \in \mathbb{C}^{n-1}$ with $|z_j| < r_j$ we consider the function $F : U_n \to \mathbb{C}$ with

$$F(\zeta_n) = f(z_1, \ldots, z_{n-1}, \zeta_n),$$

where $U_n := \{\zeta_n \in \mathbb{C} \,|\, (z_1, \ldots, z_{n-1}, \zeta_n) \in U\}$. Now F is holomorphic in U_n since f is holomorphic in U. The circle $\{\zeta_n \in \mathbb{C} \,|\, |\zeta_n| = r_n\}$ is contained in U_n. By Cauchy's integral formula for a holomorphic function of one complex variable we have

$$F(z_n) = \frac{1}{2\pi i} \int\limits_{|\zeta_n|=r_n} \frac{F(\zeta_n)}{\zeta_n - z_n} d\zeta_n. \tag{2.1}$$

By the induction hypothesis (now ζ_n is fixed and $z_1, \ldots, z_{n-1}$ are variable)

(2.2) $f(z_1, \ldots, z_{n-1}, \zeta_n)$

$$= \left(\frac{1}{2\pi i}\right)^{n-1} \int\limits_{|\zeta_{n-1}|=r_{n-1}} \cdots \int\limits_{|\zeta_1|=r_1} \frac{f(\zeta_1, \ldots, \zeta_{n-1}, \zeta_n)}{(\zeta_1 - z_1) \cdots (\zeta_{n-1} - z_{n-1})} d\zeta_1 \cdots d\zeta_{n-1}.$$

If we substitute the formula (2.2) into the formula (2.1), we obtain

$$f(z_1, \ldots, z_n) = \left(\frac{1}{2\pi i}\right)^{n} \int\limits_{|\zeta_n|=r_n} \cdots \int\limits_{|\zeta_1|=r_1} \frac{f(\zeta_1, \ldots, \zeta_n)}{(\zeta_1 - z_1) \cdots (\zeta_n - z_n)} d\zeta_1 \cdots d\zeta_n,$$

as was to be shown. □

Next we consider power series in several variables. A formal power series around 0 is an expression

$$\sum_{\nu_1, \ldots, \nu_n = 0}^{\infty} a_{\nu_1, \ldots, \nu_n} z_1^{\nu_1} \cdot \ldots \cdot z^{\nu_n}$$

with $a_{\nu_1, \ldots, \nu_n} \in \mathbb{C}$. To simplify the formulae, we introduce multiindices. Let $\nu_j, 1 \le j \le n$, be nonnegative integers, $z = (z_1, \ldots, z_n) \in \mathbb{C}^n$. Then we put

$$\begin{aligned} \nu &:= (\nu_1, \ldots, \nu_n), \\ |\nu| &:= \nu_1 + \ldots + \nu_n, \\ \nu! &:= \nu_1! \cdots \nu_n!, \\ z^\nu &:= z_1^{\nu_1} \cdots z_n^{\nu_n}. \end{aligned}$$

Definition. Let $w \in \mathbb{C}^n$, $a_\nu \in \mathbb{C}$, $|\nu| \ge 0$. A *formal power series around* w is an expression

$$\sum_{\nu=0}^{\infty} a_\nu (z - w)^\nu.$$

We now want to introduce a concept of convergence. Here we must pay attention to the fact that the multiindices can be ordered in many different ways.

Definition. Let $z' \in \mathbb{C}^n$. The power series

$$\sum_{\nu=0}^{\infty} a_\nu (z-w)^\nu$$

is said to be *convergent at the point* z' if the series

$$\sum_{\nu=0}^{\infty} a_\nu (z'-w)^\nu$$

converges with respect to every (arbitrary) order of summation.

From a theorem in analysis (see e.g. [**CJ89**, §7.1c]) this is equivalent to the last series being absolutely convergent.

Now let U be an open subset of $\mathbb{C}^n$ and let $f_m : U \to \mathbb{C}, m = 0, 1, 2, \ldots,$ be a sequence of continuous functions. The series

$$\sum_{m=0}^{\infty} f_m$$

is called *normally convergent* if the series

$$\sum_{m=0}^{\infty} \sup_{z \in K} |f_m(z)|$$

converges for every compact subset $K \subset U$. If $\sum_{m=0}^{\infty} f_m$ converges normally, then the series converges absolutely and uniformly on every compact subset $K \subset U$.

Definition. A power series

$$\sum_{\nu=0}^{\infty} a_\nu (z-w)^\nu$$

is called *normally convergent in* U if it is normally convergent with respect to any, and therefore with respect to every, order of summation. The individual terms $a_\nu (z-w)^\nu$ are to be understood as complex valued functions on U.

Proposition 2.4. *Let $w \in \mathbb{C}^n$ with $w_j \neq 0$ for $j = 1, \ldots, n$. If the power series $\sum_{\nu=0}^{\infty} a_\nu z^\nu$ converges at w, then it converges normally in the polycylinder*

$$\Delta := \{z \in \mathbb{C}^n \mid |z_j| < |w_j|, 1 \leq j \leq n\}.$$

This proposition follows from the following result:

Proposition 2.5 (Abel's lemma). *Let $w \in \mathbb{C}^n$ with $w_j \neq 0$ for $j = 1, \ldots, n$. If the set $\{a_\nu w^\nu \mid |\nu| \geq 0\}$ is bounded, then the power series $\sum_{\nu=0}^\infty a_\nu z^\nu$ converges normally in the polycylinder*

$$\Delta := \{z \in \mathbb{C}^n \mid |z_j| < |w_j|, 1 \leq j \leq n\}.$$

Proof. a) Since the set $\{a_\nu w^\nu \mid |\nu| \geq 0\}$ is bounded, there is an $M \in \mathbb{R}$ with $|a_\nu w^\nu| < M$ for all ν.

Now let $q \in \mathbb{R}$ with $0 < q < 1$. Let

$$\Delta_{(q)} := \{z \in \mathbb{C}^n \mid |z_j| < q|w_j|, 1 \leq j \leq n\}.$$

For $z \in \Delta_{(q)}$ we have

$$\begin{aligned} |a_\nu z^\nu| &= |a_\nu||z_1|^{\nu_1} \cdots |z_n|^{\nu_n} \\ &< |a_\nu||q \cdot w_1|^{\nu_1} \cdots |q \cdot w_n|^{\nu_n} \\ &= |a_\nu w^\nu| q^{|\nu|} \\ &< M \cdot q^{|\nu|} \end{aligned}$$

and

$$\begin{aligned} \sum_{\nu-0}^{\infty} M \cdot q^{|\nu|} &= M \cdot \sum_{\nu-0}^{\infty} q^{\nu_1 + \ldots + \nu_n} \\ &= M \cdot \left(\sum_{\nu_1=0}^{\infty} q^{\nu_1} \right) \cdots \left(\sum_{\nu_n=0}^{\infty} q^{\nu_n} \right) \\ &= M \cdot \left(\frac{1}{1-q} \right)^n . \end{aligned}$$

Thus the power series converges normally in the polycylinder $\Delta_{(q)}$.

b) Now let $K \subset \Delta$ be compact. The sets $\Delta_{(q)}$, $0 < q < 1$, form an open covering of Δ, so also of K. Since K is compact,

$$K \subset \Delta_{(q_1)} \cup \cdots \cup \Delta_{(q_l)}.$$

But $K \subset \Delta_{(q)}$ for $q = \max_{j=1,\ldots,l} q_j$. By a), $\sum_{\nu=0}^\infty a_\nu z^\nu$ converges uniformly absolutely on K. This power series is therefore normally convergent on Δ. □

Proposition 2.6. *Let $r \in \mathbb{R}^n_{>0}$, $\Delta := \Delta_r(0) \subset \mathbb{C}^n$ a polycylinder around 0, $T := T_r(0)$ and let $h : T \to \mathbb{C}$ be continuous. Then the function $f : \Delta \to \mathbb{C}$ with*

$$f(z) := \frac{1}{(2\pi i)^n} \int_T \frac{h(\zeta)}{(\zeta_1 - z_1) \cdots (\zeta_n - z_n)} d\zeta$$

can be expanded as a convergent power series around 0 in the whole of Δ.

Proof. Let $r = (r_1, \ldots, r_n)$. Let $z \in \Delta$ be chosen fixed. Then

$$q_j := \frac{|z_j|}{r_j} < 1 \text{ for } j = 1, \ldots, n.$$

Now let $\zeta \in T$. Then

$$\begin{aligned}
\frac{h(\zeta)}{(\zeta_1 - z_1)\cdots(\zeta_n - z_n)} &= \frac{h(\zeta)}{\zeta_1 \cdots \zeta_n} \frac{1}{\left(1 - \frac{z_1}{\zeta_1}\right) \cdots \left(1 - \frac{z_n}{\zeta_n}\right)} \\
&= \frac{h(\zeta)}{\zeta_1 \cdots \zeta_n} \left(\sum_{\nu_1=0}^{\infty} \left(\frac{z_1}{\zeta_1}\right)^{\nu_1} \right) \cdots \left(\sum_{\nu_n=0}^{\infty} \left(\frac{z_n}{\zeta_n}\right)^{\nu_n} \right) \\
&= \sum_{\nu=0}^{\infty} \frac{h(\zeta)}{\zeta_1 \cdots \zeta_n} \cdot \frac{1}{\zeta^\nu} \cdot z^\nu.
\end{aligned}$$

Since h is continuous and T is compact, the function

$$\zeta \mapsto \frac{h(\zeta)}{\zeta_1 \cdots \zeta_n}$$

is bounded on T; there is therefore an $M \in \mathbb{R}$ with

$$\left| \frac{h(\zeta)}{\zeta_1 \cdots \zeta_n} \right| < M$$

for all $\zeta \in T$. For fixed $z \in \Delta$ the sum $\sum_{\nu=0}^{\infty} M q^\nu$ is then a majorant on T for the series

$$\sum_{y=0}^{\infty} \frac{h(\zeta)}{\zeta_1 \cdots \zeta_n} \cdot \frac{1}{\zeta^\nu} \cdot z^\nu.$$

This series therefore converges absolutely and uniformly on T and we may interchange summation and integration to get

$$\begin{aligned}
f(z) &= \left(\frac{1}{2\pi i}\right)^n \int_T \frac{h(\zeta)}{(\zeta_1 - z_1)\cdots(\zeta_n - z_n)} \, d\zeta \\
&= \left(\frac{1}{2\pi i}\right)^n \int_T \left(\sum_{\nu=0}^{\infty} \frac{h(\zeta) z^\nu}{\zeta_1^{\nu_1+1} \cdots \zeta_n^{\nu_n+1}} \right) d\zeta \\
&= \sum_{\nu=0}^{\infty} z^\nu \cdot \left(\frac{1}{2\pi i}\right)^n \int_T \frac{h(\zeta)}{\zeta_1^{\nu_1+1} \cdots \zeta_n^{\nu_n+1}} \, d\zeta \\
&= \sum_{\nu=0}^{\infty} a_\nu z^\nu
\end{aligned}$$

with

$$a_\nu := \left(\frac{1}{2\pi i}\right)^n \int\limits_T \frac{h(\zeta)}{\zeta_1^{\nu_1+1}\cdots\zeta_n^{\nu_n+1}}d\zeta.$$

This last series converges for every $z \in \Delta$. □

Definition. Let $U \subset \mathbb{C}^n$ be open. A function $f : U \to \mathbb{C}$ is called *analytic in U* if for every $w \in U$ there are a neighborhood V of w in U and a power series $\sum_{\nu=0}^{\infty} a_\nu (z-w)^\nu$ that converges to f on V.

Proposition 2.7. *Let $U \subset \mathbb{C}^n$ be open. A function $f : U \to \mathbb{C}$ is holomorphic in U if and only if it is analytic in U.*

Proof. "$\Rightarrow$": Let f be holomorphic in U. Let $w \in U$. We assume that $w = 0$. Then there is a polycylinder $\Delta := \Delta_r(w)$ around w such that $\bar{\Delta} \subset U$. Let $T := T_r(w)$. By Proposition 2.3

$$f(z) = \left(\frac{1}{2\pi i}\right)^n \int\limits_T \frac{f(\zeta)}{(\zeta_1 - z_1)\cdots(\zeta_n - z_n)}d\zeta$$

for all $z \in \Delta$. By Proposition 2.2, f is continuous on T. From Proposition 2.6 it follows that f is analytic at w.

"$\Leftarrow$": Now let f be analytic in U. Let $w \in U$. Then there are a neighborhood V of w and a power series $\sum_{\nu=0}^{\infty} a_\nu (z-w)^\nu$ which converges to f in V. We assume furthermore that $w = 0$. Let

$$\begin{aligned} \Delta &:= \{z \in \mathbb{C}^n \mid |z_j| < \varepsilon, 1 \le j \le n\}, \\ T &:= \{z \in \mathbb{C}^n \mid |z_j| = \varepsilon, 1 \le j \le n\}, \end{aligned}$$

where $\varepsilon > 0$ is chosen so that $\bar{\Delta} \subset V$. Choose $v \in T$. Then $\sum_{\nu=0}^{\infty} a_\nu v^\nu$ converges. By Proposition 2.4, $\sum_{\nu=0}^{\infty} a_\nu z^\nu$ then converges normally in Δ. We can write

$$\begin{aligned} \sum_{\nu=0}^{\infty} a_\nu z^\nu &= a_{0,\dots,0} + z_1 \cdot \sum_{\substack{\nu = 0 \\ \nu_1 \ge 1}}^{\infty} a_\nu z_1^{\nu_1 - 1} z_2^{\nu_2} \cdots z_n^{\nu_n} \\ &\quad + z_2 \cdot \sum_{\nu_2 \ge 1} a_{0,\nu_2,\dots,\nu_n} z_2^{\nu_2 - 1} z_3^{\nu_3} \cdots z_n^{\nu_n} \\ &\quad + \dots \\ &\quad + z_n \cdot \sum_{\nu_n \ge 1} a_{0,\dots,0,\nu_n} z_n^{\nu_n - 1}. \end{aligned}$$

Since $|v_j| \neq 0$ for all j, all of the subseries in the above formula converge absolutely at v and thus normally in Δ. The limit functions are therefore continuous. If we denote them by $R_1, \dots, R_n$, then

$$f(z) = f(w) + z_1 R_1(z) + \dots + z_n R_n(z).$$

By Proposition 2.1, f is complex differentiable at w. □

Proposition 2.7 signifies that the concepts "holomorphic" and "analytic" are equivalent. We shall treat them as synonyms from now on.

Corollary 2.1. *Let $U \subset \mathbb{C}^n$ be open and $f : U \to \mathbb{C}$ holomorphic. Furthermore, let $w \in U$ and let $\Delta := \Delta_r(w)$ be a polycylinder around w with $\bar{\Delta} \subset U$, $T := T_r(w)$. Then f can be expanded as a power series*

$$f(z) = \sum_{\nu=0}^{\infty} a_\nu (z-w)^\nu$$

in a neighborhood of w, with coefficients

$$a_\nu = \left(\frac{1}{2\pi i}\right)^n \int_T \frac{f(\zeta)}{(\zeta_1 - w_1)^{\nu_1+1} \dots (\zeta_n - w_n)^{\nu_n+1}} d\zeta.$$

Proof. This follows from Proposition 2.7, Cauchy's integral formula (Proposition 2.3) and the proof of Proposition 2.6. □

Proposition 2.8 (Weierstrass theorem). *Let $U \subset \mathbb{C}^n$ be open and let $f_m : U \to \mathbb{C}$, $m = 0, 1, 2, \dots$, be holomorphic functions. If the sequence of functions (f_m) converges uniformly to a function $f : U \to \mathbb{C}$, then f is also holomorphic.*

Proof. Let $w \in U$. We again assume that $w = 0$. We shall show that f is analytic at w.

So let $\Delta := \Delta_r(w)$ be a polycylinder around w with $\bar{\Delta} \subset U$, $T := T_r(w)$ and $z \in \Delta$. Then

$$\begin{aligned} f(z) &= \lim_{m\to\infty} f_m(z) \\ &= \lim_{m\to\infty} \left(\frac{1}{2\pi i}\right)^n \int_T \frac{f_m(\zeta)}{(\zeta_1 - z_1)\cdots(\zeta_n - z_n)} d\zeta \qquad \text{(Proposition 2.3)} \\ &= (*)\,. \end{aligned}$$

We now need to interchange integrating and taking limits. To justify this, we must show that the sequence of functions $f_m(\zeta)/((\zeta_1 - z_1)\cdots(\zeta_n - z_n))$ converges uniformly to the function $f(\zeta)/((\zeta_1 - z_1)\cdots(\zeta_n - z_n))$. We put

$$g(\zeta) := (\zeta_1 - z_1)\cdots(\zeta_n - z_n).$$

Then g is continuous and does not vanish on T. Therefore $1/g$ is also continuous on T and there is an $M \in \mathbb{R}$ such that

$$\left|\frac{1}{g(\zeta)}\right| < M \text{ for all } \zeta \in T.$$

Then, on T,

$$\left\|\frac{f_m}{g} - \frac{f}{g}\right\| = \left\|\frac{1}{g}\right\| \|f_m - f\| < M \|f_m - f\| .$$

Therefore f_m/g converges uniformly to f/g on T. Hence

$$\begin{aligned}(*) &= \left(\frac{1}{2\pi i}\right)^n \int\limits_T \lim_{m\to\infty} \frac{f_m(\zeta)}{(\zeta_1 - z_1)\cdots(\zeta_n - z_n)} d\zeta \\ &= \left(\frac{1}{2\pi i}\right)^n \int\limits_T \frac{f(\zeta)}{(\zeta_1 - z_1)\cdots(\zeta_n - z_n)} d\zeta.\end{aligned}$$

Since all the f_m are continuous on T, so too is f continuous on T. It follows from Proposition 2.6 that f can be expanded around 0 in a power series that converges on all Δ. □

Proposition 2.9. *Let $f(z) := \sum_{\nu=0}^{\infty} a_\nu z^\nu$ be a formal power series that converges on a polycylinder Δ around 0. Then f is holomorphic in Δ.*

Proof. By Proposition 2.4 the series $\sum_{\nu=0}^{\infty} a_\nu z^\nu$ is normally convergent in Δ. We choose any ordering of the multiindices. Let f_m the m-th partial sum. Then f_m is holomorphic in Δ (even in all $\mathbb{C}^n$) and the sequence of functions (f_m) converges uniformly in Δ to the function f. By the Weierstrass theorem (Proposition 2.8) f is holomorphic in Δ. □

Proposition 2.10. *Let $U \subset \mathbb{C}^n$ be open and $f : U \to \mathbb{C}$ holomorphic. Then all the partial derivatives f_{z_j}, $1 \le j \le n$, are also holomorphic in U. If $\Delta \subset U$ is a polycylinder around 0 and*

$$f(z) = \sum_{\nu=0}^{\infty} a_\nu z^\nu$$

in Δ, then

$$f_{z_j}(z) = \sum_{\nu=0}^{\infty} a_\nu \nu_j z_1^{\nu_1} \cdots z_j^{\nu_j - 1} \cdots z_n^{\nu_n}$$

in Δ.

Proof. a) First we show that the series

$$\sum a_\nu \nu_j z_1^{\nu_1} \cdots z_j^{\nu_j - 1} \cdots z_n^{\nu_n}$$

converges.

Take $w \in \Delta$ with $w_j \neq 0$ for $j = 1, \ldots, n$, $0 < q < 1$ and $v := q \cdot w$. Then we can estimate as follows:

$$|a_\nu \nu_j v_1^{\nu_1} \cdots v_j^{\nu_j - 1} \cdots v_n^{\nu_n}| = \frac{\nu_j}{|v_j|} \cdot |a_\nu v^\nu| = \frac{\nu_j}{|v_j|} |a_\nu w^\nu| \cdot q^{|\nu|}.$$

Since $\sum a_\nu w^\nu$ converges, the set $\{|a_\nu w^\nu| \mid |\nu| \geq 0\}$ is bounded. Furthermore

$$\sum_{\nu=0}^{\infty} \nu_j \cdot q^{|\nu|} = \left(\sum_{\nu_1=0}^{\infty} q^{\nu_1}\right) \cdots \left(\sum_{\nu_j=0}^{\infty} \nu_j q^{\nu_j}\right) \cdots \left(\sum_{\nu_n=0}^{\infty} q^{\nu_n}\right).$$

Now $\sum_{\nu_k=0}^{\infty} q^{\nu_k}$ is a geometric series for $k \neq j$. The series $\sum_{\nu_j=0}^{\infty} \nu_j q^{\nu_j}$ converges too. One has

$$\lim_{\nu_j \to \infty} \frac{(\nu_j + 1) q^{\nu_j + 1}}{\nu_j q^{\nu_j}} = q \lim_{\nu_j \to \infty} \frac{\nu_j + 1}{\nu_j} = q < 1.$$

From the ratio test it follows that this series also converges. Thus $\sum_{\nu=0}^{\infty} \nu_j q^{|\nu|}$ converges, and, in particular, the set $\{\nu_j q^{|\nu|} \mid |\nu| \geq 0\}$ is also bounded. From Abel's lemma it follows that the series

$$\sum_{\nu=0}^{\infty} a_\nu \nu_j z_1^{\nu_1} \cdots z_j^{\nu_j - 1} \cdots z_n^{\nu_n}$$

converges normally in $\Delta_v = \{z \in \mathbb{C}^n \mid |z_k| < |v_k|, 1 \leq k \leq n\}$. Since Δ can be covered by polycylinders Δ_v, the above series converges in Δ to a holomorphic function g_j.

b) We now show that $g_j = f_{z_j}$. We put $h_\nu(z) := a_\nu z^\nu$. Then

$$f(z) = \sum_{\nu=0}^{\infty} h_\nu(z), \quad g_j(z) = \sum_{\nu=0}^{\infty} (h_\nu)_{z_j}(z).$$

Next we consider

$$\begin{aligned} & \int_0^{z_j} g_j(z_1, \ldots, z_{j-1}, \zeta, z_{j+1}, \ldots, z_n) d\zeta + f(z_1, \ldots, 0, \ldots, z_n) \\ = & \int_0^{z_j} \left(\sum_{\nu=0}^{\infty} (h_\nu)_{z_j}(z_1, \ldots, z_{j-1}, \zeta, z_{j+1}, \ldots, z_n) \right) d\zeta + f(z_1, \ldots, 0, \ldots, z_n) \\ = & \ (*). \end{aligned}$$

We are dealing here with a conventional path integral; the path is the interval from 0 to z_j in the z_j-plane and must lie completely in Δ. Since this path is a compact subset of Δ, the series under the integral sign converges uniformly there. We can therefore interchange summation and integration to obtain

$$\begin{aligned}(*) &= \sum_{\nu=0}^{\infty}\left(\int_0^{z_j}(h_\nu)_{z_j}(z_1,\ldots,z_{j-1},\zeta,z_{j+1},\ldots,z_n)d\zeta + h_\nu(z_1,\ldots,0,\ldots,z_n)\right)\\ &= \sum_{\nu=0}^{\infty} h_\nu(z)\\ &= f(z).\end{aligned}$$

Therefore

$$f_{z_j}(z) = g_j(z)$$

for all $z \in \Delta$. □

Let $U \subset \mathbb{C}^n$ be open, $f : U \to \mathbb{C}$ a holomorphic function. By Proposition 2.10, f is arbitrarily often (partially) complex differentiable. For $w \in U$ and a multiindex ν we put

$$\partial^\nu f(w) = \frac{\partial^{\nu_1+\ldots+\nu_n} f}{\partial z_1^{\nu_1}\cdots\partial z_n^{\nu_n}}(w).$$

Then

Proposition 2.11. *Let $U \subset \mathbb{C}^n$ be open, $f : U \to \mathbb{C}$ holomorphic. Then for every $w \in U$ there is a power series expansion*

$$f(z) = \sum_{\nu=0}^{\infty} a_\nu (z-w)^\nu$$

around w in a neighborhood of w, with coefficients

$$a_\nu = \frac{\partial^\nu f(w)}{\nu!}.$$

Proof. Let $\Delta = \{z \in \mathbb{C}^n \mid |z_j - w_j| < r_j, 1 \le j \le n\}$ be a polycylinder around w with $\bar{\Delta} \subset U$. Furthermore, let $T := T_r(w)$. By Corollary 2.1 there is a power series expansion

$$f(z) = \sum_{\nu=0}^{\infty} a_\nu (z-w)^\nu$$

around w with

$$\begin{aligned}a_\nu &= \left(\frac{1}{2\pi i}\right)^n \int_T \frac{f(\zeta)}{(\zeta_1-w_1)^{\nu_1+1}\cdots(\zeta_n-w_n)^{\nu_n+1}}d\zeta\\ &= \frac{1}{2\pi i}\int_{|\zeta_n|=r_n}\frac{1}{(\zeta_n-w_n)^{\nu_n+1}}\cdots\left[\frac{1}{2\pi i}\int_{|\zeta_1|=r_1}\frac{f(\zeta)}{(\zeta_1-w_r)^{\nu_1+1}}d\zeta_1\right]\cdots d\zeta_n.\end{aligned}$$

By Cauchy's integral formula for a single variable we now have

$$\begin{aligned} a_\nu &= \frac{1}{\nu_1!}\frac{1}{2\pi i}\int\limits_{|\zeta_n|=r_n}\frac{1}{(\zeta_n-w_n)^{\nu_n+1}}\cdots\left[\frac{1}{2\pi i}\int\limits_{|\zeta_2|=r_2}\frac{\frac{\partial^{\nu_1}f}{\partial z_1^{\nu_1}}(\zeta')}{(\zeta_2-w_2)^{\nu_2+1}}d\zeta_2\right]\cdots d\zeta_n \\ &= \ldots = \frac{\partial^\nu f(w)}{\nu!}, \end{aligned}$$

where $\zeta' = (w_1, \zeta_2, \ldots, \zeta_n)$. □

Many familiar results from the theory of functions of one variable can be generalized easily to functions of several complex variables. The identity theorem is such an example. It is, however, not true that two holomorphic functions $f, g : G \to \mathbb{C}$, $G \subset \mathbb{C}^n$ a region, must agree if the set $\{z \in G | f(z) = g(z)\}$ has an accumulation point in G. Even for $n = 2$ this is false in general: Consider $G = \mathbb{C}^2$, $f(z) := z_1$, $g(z) := z_1 z_2$. Then

$$\{z \in G \,|\, f(z) = g(z)\} = \{z \in \mathbb{C}^2 \,|\, z_1 = 0 \text{ or } z_2 = 1\},$$

but $f \neq g$ on $\mathbb{C}^2$. However we do have the following results.

Proposition 2.12 (Identity theorem for power series). *Let $G \subset \mathbb{C}^n$ be a region with $0 \in G$ and let $\sum_{\nu=0}^\infty a_\nu z^\nu$ and $\sum_{\nu=0}^\infty b_\nu z^\nu$ be two power series convergent in G. If the set*

$$\{z \in G \,|\, \sum_{\nu=0}^\infty a_\nu z^\nu = \sum_{\nu=0}^\infty b_\nu z^\nu\}$$

contains an open neighborhood V of 0, then $a_\nu = b_\nu$ for all ν.

Proof. Let $f(z) := \sum_{\nu=0}^\infty a_\nu z^\nu$ and $g(z) := \sum_{\nu=0}^\infty b_\nu z^\nu$ for $z \in G$. The functions f and g are by Proposition 2.9 holomorphic in G and by hypothesis agree on the open neighborhood V of 0. Then their higher derivatives, which, by Proposition 2.11, exist, also agree. Repeated application of Proposition 2.10 leads to

$$a_\nu = \frac{\partial^\nu f(0)}{\nu!} = \frac{\partial^\nu g(0)}{\nu!} = b_\nu$$

for all ν. □

Proposition 2.13 (Identity theorem for holomorphic functions). *Let $G \subset \mathbb{C}^n$ be a region, $f, g : G \to \mathbb{C}$ holomorphic. Let $U \subset G$ be open and not empty, and assume $f|_U = g|_U$. Then $f = g$ on G.*

Proof. Let M be the interior of the set $\{z \in G \,|\, f(z) = g(z)\}$. Then M is an open subset of G, and $M \neq \emptyset$ since $U \subset M$. It is enough to show that M is relatively closed in G. It will follow that $M = G$, since G is connected.

So let $w \in G \cap \bar{M}$. Take $r \in \mathbb{R}$, $r > 0$, so small that for the polycylinder $\Delta_\rho(w)$ around w with $\rho = (r, \ldots, r)$ we have $\bar{\Delta}_\rho(w) \subset G$. Since $w \in \bar{M}$, there exists a $w' \in M$ with $|w'_j - w_j| < r/2$ for $j = 1, \dots, n$. If we put $\rho/2 = (r/2, \dots, r/2)$, then $w \in \Delta_{\rho/2}(w')$ and $\Delta_{\rho/2}(w') \subset \Delta_\rho(w) \subset G$. Then f and g are holomorphic in $\Delta_{\rho/2}(w')$. Let

$$f(z) = \sum_{\nu=\infty}^{\infty} a_\nu (z - w')^\nu \text{ and } g(z) = \sum_{\nu=0}^{\infty} b_\nu (z - w')^\nu$$

be the power series expansions of f, resp. g, in $\Delta_{\rho/2}(w')$. These power series converge in $\Delta_{\rho/2}(w')$ and their values agree in a neighborhood of w'. By Proposition 2.12, $a_\nu = b_\nu$ for all ν. Thus $f(w) = g(w)$, and so $w \in M$. We have shown that M is relatively closed in G, as required. □

2.2. Holomorphic maps and the implicit function theorem

Next we consider holomorphic maps.

Definition. Let $U \subset \mathbb{C}^n$ be open. A map $f = (f_1, \dots, f_m) : U \to \mathbb{C}^m$ is called *holomorphic* if all the component functions f_μ are holomorphic in U.

Let $U \subset \mathbb{C}^n$ be open, $f = (f_1, \dots, f_m) : U \to \mathbb{C}^m$ a holomorphic map. The matrix

$$J_f(w) := \left(\frac{\partial f_\mu}{\partial z_j}(w) \right)_{j=1,\dots,n}^{\mu=1,\dots,m}$$

is called the *holomorphic functional matrix* or the *Jacobi matrix* of f at $w \in U$.

Next we put

$$\begin{aligned} u_\mu &:= \operatorname{Re} f_\mu, \\ v_\mu &:= \operatorname{Im} f_\mu, \\ x_j &:= \operatorname{Re} z_j, \\ y_j &:= \operatorname{Im} z_j. \end{aligned}$$

Then we have the Cauchy-Riemann differential equations

$$\frac{\partial u_\mu}{\partial x_j} = \frac{\partial v_\mu}{\partial y_j}, \quad \frac{\partial u_\mu}{\partial y_j} = -\frac{\partial v_\mu}{\partial x_j} \quad (\mu = 1, \dots, m;\ j = 1, \dots, n).$$

Let us introduce the following abbreviated notation:

$$\begin{aligned} \frac{\partial u}{\partial x} &:= \left(\frac{\partial u_\mu}{\partial x_j}\right)^{\mu=1,\dots,m}_{j=1,\dots,n}, \\ \frac{\partial u}{\partial y} &:= \left(\frac{\partial u_\mu}{\partial y_j}\right)^{\mu=1,\dots,m}_{j=1,\dots,n}, \\ \frac{\partial v}{\partial x} &:= \left(\frac{\partial v_\mu}{\partial x_j}\right)^{\mu=1,\dots,m}_{j=1,\dots,n}, \\ \frac{\partial v}{\partial y} &:= \left(\frac{\partial v_\mu}{\partial y_j}\right)^{\mu=1,\dots,m}_{j=1,\dots,n}, \\ \frac{\partial(u,\nu)}{\partial(x,y)} &:= \begin{pmatrix} \frac{\partial u}{\partial x} & \frac{\partial u}{\partial y} \\ \frac{\partial v}{\partial x} & \frac{\partial v}{\partial y} \end{pmatrix}. \end{aligned}$$

The matrix $\partial(u,v)/\partial(x,y)$ is the (real) functional matrix of the real differentiable map $f : U \to \mathbb{R}^{2m}$ determined by f through the identification

$$\begin{array}{ccc} \mathbb{C}^n & \to & \mathbb{R}^{2n} \\ (z_1,\dots,z_n) & \mapsto & (x_1,\dots,x_n,y_1,\dots,y_n), \quad z_j = x_j + iy_j, \end{array}$$

and the corresponding identification of $\mathbb{C}^m$ with $\mathbb{R}^{2m}$. From the Cauchy-Riemann differential equations

$$\frac{\partial(u,v)}{\partial(x,y)} = \begin{pmatrix} \frac{\partial u}{\partial x} & -\frac{\partial v}{\partial x} \\ \frac{\partial v}{\partial x} & \frac{\partial u}{\partial x} \end{pmatrix}.$$

Now let $m = n$. Then, for the determinant of this matrix,

$$\begin{aligned} \det \frac{\partial(u,v)}{\partial(x,y)} &= \det \begin{pmatrix} \frac{\partial u}{\partial x} + i\frac{\partial v}{\partial x} & i(\frac{\partial u}{\partial x} + i\frac{\partial v}{\partial x}) \\ \frac{\partial v}{\partial x} & \frac{\partial u}{\partial x} \end{pmatrix} \\ &= \det \begin{pmatrix} \frac{\partial u}{\partial x} + i\frac{\partial v}{\partial x} & 0 \\ \frac{\partial v}{\partial x} & \frac{\partial u}{\partial x} - i\frac{\partial v}{\partial x} \end{pmatrix} \\ &= |\det J_f|^2 . \end{aligned}$$

We next want to show that the implicit function theorem can be generalized to holomorphic maps.

To this end we consider

$$\mathbb{C}^n \times \mathbb{C}^m = \{(z,w) \mid z = (z_1,\dots,z_n) \in \mathbb{C}^n,\ w = (w_1,\dots,w_m) \in \mathbb{C}^m\},$$

$U \subset \mathbb{C}^n \times \mathbb{C}^m$ an open subset, and $f : U \to \mathbb{C}^m$ a holomorphic map. Then, given a point $c = (a,b) \in U$, the map

$$w \mapsto f(a,w)$$

is holomorphic at b. We put

$$\frac{\partial f}{\partial w}(a,b) = \begin{pmatrix} \frac{\partial f_1}{\partial w_1}(a,b) & \cdots & \frac{\partial f_1}{\partial w_m}(a,b) \\ \vdots & \ddots & \vdots \\ \frac{\partial f_m}{\partial w_1}(a,b) & \cdots & \frac{\partial f_m}{\partial w_m}(a,b) \end{pmatrix}.$$

Proposition 2.14 (Implicit function theorem). *Let $U \subset \mathbb{C}^n \times \mathbb{C}^m$ be open and $f : U \to \mathbb{C}^m$ holomorphic. Furthermore, let $f(a,b) = 0$ and* $\det(\partial f/\partial w)(a,b) \neq 0$. *Then there are an open neighborhood $V \subset \mathbb{C}^n$ of a and a uniquely determined holomorphic map $\varphi : V \to \mathbb{C}^m$ such that for all $z \in V$*

$$f(z, \varphi(z)) = 0.$$

Proof. We deduce this proposition from the implicit function theorem for real maps $f : U \to \mathbb{R}^{2m}$, $U \subset \mathbb{R}^{2n} \times \mathbb{R}^{2m}$.

We treat f as a map $f : U \to \mathbb{R}^{2m}$. For $(z,w) \in U$ we put

$$\begin{aligned} U_w &:= \{z \in \mathbb{C}^n \mid (z,w) \in U\}, \\ U_z &:= \{w \in \mathbb{C}^m \mid (z,w) \in U\}. \end{aligned}$$

Then the partial maps

$$\begin{aligned} f(z,\cdot) \quad &: \quad U_z \to \mathbb{R}^{2m} \qquad \text{for fixed } z \\ & \qquad w \mapsto f(z,w), \\ f(\cdot,w) \quad &: \quad U_w \mapsto \mathbb{R}^{2m} \qquad \text{for fixed } w \\ & \qquad z \mapsto f(z,w) \end{aligned}$$

are also complex differentiable. If we put

$$\begin{aligned} f &= u + iv, \\ w &= x + iy, \end{aligned}$$

then, as we have seen above,

$$\det \frac{\partial(u,v)}{\partial(x,y)}(a,b) = \left| \det \frac{\partial f}{\partial w}(a,b) \right|^2 \neq 0.$$

From the implicit function theorem of analysis (see e.g. [**Lan97**, XVIII, §4]) it follows that there are a neighborhood V of a in $\mathbb{C}^n$ and a continuous real differentiable map $\varphi : V \to \mathbb{R}^{2m}$ such that for all $z \in V$

$$f(z, \varphi(z)) = 0.$$

If we put

$$z = \xi + i\eta,$$

then for the derivative $\varphi'(z)$ for all $z \in V$ we have

$$\varphi'(z) = -\left(\frac{\partial(u,v)}{\partial(x,y)}(z,\varphi(z))\right)^{-1} \circ \left(\frac{\partial(u,v)}{\partial(\xi,\eta)}(z,\varphi(z))\right).$$

Since the partial maps $f(z,\cdot)$ and $f(\cdot,w)$ are complex differentiable, the Cauchy-Riemann differential equations are satisfied for these maps. It follows from the formula above that the Cauchy-Riemann differential equations are also satisfied for the map $\varphi : V \to \mathbb{R}^{2m}$. Therefore φ is holomorphic. $\square$

2.3. Local rings of holomorphic functions

We consider the solution set of a system of equations

$$f_1(z) = \ldots = f_m(z) = 0,$$

where $f_1, \ldots, f_m : U \to \mathbb{C}$, $U \subset \mathbb{C}^n$ open, are holomorphic functions. We shall study such solution sets locally, i.e., we shall investigate the solution sets in the neighborhood of a solution $z \in U$.

We are therefore interested in the behavior of holomorphic functions in arbitrarily small neighborhoods of a fixed point of the domain of definition. The concept of the germ of a holomorphic function $f : U \to \mathbb{C}$ at a point $a \in U$ has already been introduced, in §1.4, where U was an open subset of a Riemann surface. If $w \in \mathbb{C}^n$ is a fixed point and U, V are open neighborhoods of w, we define in complete analogy:

Definition. Two functions $f : U \to \mathbb{C}$, $g : V \to \mathbb{C}$ are called *equivalent at the point* w (symbolically, $f \sim_w g$) $:\Leftrightarrow$ there is an open neighborhood W of w such that $W \subset U \cap V$ and $f|_W = g|_W$. An equivalence class of such functions is called a *function germ at* w.

If U is an open neighborhood of w and $f : U \to \mathbb{C}$ is a function, then f belongs to an equivalence class. This class is called the *germ of the function* f and is denoted by $\bar{f}$. We shall, however, no longer continue to distinguish notationally between germs and their representatives, i.e., we shall drop the overline.

Two functions f, g equivalent at w certainly have the same value $f(w) = g(w)$ at w; this value is called the *value* of the function germ $\bar{f}$ at w.

Next we consider holomorphic functions in particular. Given an open subset $U \subset \mathbb{C}^n$, we put

$$\mathcal{O}(U) = \{f : U \to \mathbb{C} \text{ holomorphic}\}.$$

Sums and products of holomorphic functions are again holomorphic. With these operations $\mathcal{O}(U)$ forms a commutative ring with a unit element. If $w \in U$ we write

$$\mathcal{O}_w$$

for the set of all germs of holomorphic functions at w. If one defines the sum and product of germs representativewise, then $\mathcal{O}_w$ inherits the structure of a commutative ring with a unit element. The ring $\mathcal{O}_w$ is the *ring of holomorphic function germs at* w.

Notation. We often write $\mathcal{O}$ in place of $\mathcal{O}_0$. If the complex dimension of the underlying space $\mathbb{C}^n$ is to be indicated, we may write $\mathcal{O}_{n,w}$.

We now want to show that the ring $\mathcal{O}_w$ of holomorphic function germs is isomorphic to the ring $\mathbb{C}\{z-w\}$ of convergent power series around w.

We denote the ring of formal power series

$$\sum_{\nu=0}^{\infty} a_\nu z^\nu$$

in the unknowns $z_1, \ldots, z_n$ with complex coefficients by $\mathbb{C}[[z_1, \ldots, z_n]] = \mathbb{C}[[z]]$.

Definition. Let $w \in \mathbb{C}^n$. A formal power series

$$\sum_{\nu=0}^{\infty} a_\nu (z-w)^\nu \in \mathbb{C}[[z-w]]$$

around w is called *convergent* if it converges in some polycylinder around w (in the sense of the definition of §2.1).

We denote the ring of all convergent power series

$$\sum_{\nu=0}^{\infty} a_\nu (z-w)^\nu$$

around w with complex coefficients by $\mathbb{C}\{z-w\} = \mathbb{C}\{z_1 - w_1, \ldots, z_n - w_n\}$.

Proposition 2.15. *The ring $\mathcal{O}_w$ of holomorphic function germs at $w \in \mathbb{C}^n$ is isomorphic to the ring $\mathbb{C}\{z-w\}$ of convergent power series at w.*

Proof. We define a map

$$\Phi : \mathcal{O}_w \to \mathbb{C}\{z-w\}$$

by associating to a holomorphic function germ represented by a holomorphic function $f : U \to \mathbb{C}$ its power series expansion at w.

This map is well defined: Two representatives of the same germ agree in an open neighborhood of w and therefore by Proposition 2.2 have the same power series expansion at w (the coefficients a_ν are determined by the generalized Cauchy integral formula).

Φ is surjective: A power series $\sum_{\nu=0}^{\infty} a_\nu (z-w)^\nu$ that converges in a polycylinder around w defines a holomorphic function there, by Proposition 2.9, and so a holomorphic function germ.

Φ is also injective since this holomorphic function germ is uniquely determined.

It is clear that Φ respects the ring operations so it is a ring isomorphism. $\square$

For fixed $w \in \mathbb{C}^n$ the linear coordinate transformation

$$\zeta_j = z_j - w_j, \quad j = 1, \ldots, n,$$

induces an isomorphism between the rings $\mathcal{O}_w$ and $\mathcal{O}_0$ and so between the rings $\mathbb{C}\{z-w\}$ and $\mathbb{C}\{z\}$. It is therefore enough to study the ring

$$\mathcal{O}_0 \cong \mathbb{C}\{z\} = \mathbb{C}\{z_1, \ldots, z_n\}.$$

Proposition 2.16. *The ring $\mathcal{O}_0$ is an integral domain, i.e., for all $f, g \in \mathcal{O}_0$ it follows from $f \cdot g = 0$ that $f = 0$ or $g = 0$.*

Proof. Let $f, g \in \mathcal{O}_0$ with $f \cdot g = 0$. The germs f, g are represented by holomorphic functions $f, g : U \to \mathbb{C}$ where U is an open neighborhood of 0. From $f \cdot g = 0$ it follows that $f(z)g(z) = 0$ for all $z \in V \subset U$, where V is a connected open neighborhood of 0 with $V \subset U$.

If $f(w) \neq 0$ for some $w \in V$, then $f(z) \neq 0$ for all $z \in W$, with W an open neighborhood of w. Then $g(z) = 0$ for all $z \in W$. By the identity theorem (Proposition 2.13) for holomorphic functions it follows that $g(z) = 0$ for all $z \in V$, and therefore $g = 0$. $\square$

Let R be a ring. (For us a ring will always be a commutative ring with a unit element.) An element $f \in R$ is called a *unit* if and only if there is an element $g \in R$ with $f \cdot g = 1$.

Proposition 2.17. *The units of the ring $\mathcal{O}_0$ are precisely the function germs f with $f(0) \neq 0$.*

Proof. a) If $f \in \mathcal{O}_0$ is a unit, there is a $g \in \mathcal{O}_0$ with $f \cdot g = 1$, in particular $f(0) \cdot g(0) = 1$, so $f(0) \neq 0$.

b) Now, conversely, let f be the representative of a holomorphic function germ from $\mathcal{O}_0$ with $f(0) \neq 0$. Then, by continuity, $f(z) \neq 0$ for all z in an open neighborhood U of 0. Then $1/f$ is defined and continuous in U. Now $1/f$ is holomorphic in each variable. Thus $1/f$ is holomorphic in U, whence $1/f \in \mathcal{O}_0$. $\square$

Let R again be a ring. An *ideal* I in R is a subset $I \subset R$ for which

(i) I is a subgroup of the additive group underlying R,

(ii) for all $f \in R$ and $g \in I$ we have $f \cdot g \in I$.

An ideal $\mathfrak{m}$ is called *maximal* if $\mathfrak{m} \neq R$ and if there is no ideal I with $\mathfrak{m} \subsetneqq I \subsetneqq R$. A ring R is called *local* if it has a unique maximal ideal $\mathfrak{m}$.

Remark 2.2. A ring is local if and only if its nonunits form an ideal I. If R is local, then I is its maximal ideal. (Proof: Exercise.)

Proposition 2.18. *The ring $\mathcal{O}_0$ is a local ring.*

Proof. By the remark it is enough to show that those elements of $\mathcal{O}_0$ which are not units form an ideal in $\mathcal{O}_0$. By Proposition 2.17 these are just the holomorphic function germs f with $f(0) = 0$. If f is such a germ and $g \in \mathcal{O}_0$ is an arbitrary function germ, then $f \cdot g(0) = f(0) \cdot g(0) = 0$. Thus

$$\mathfrak{m} := \{f \in \mathcal{O}_0 \mid f(0) = 0\}$$

is an ideal in $\mathcal{O}_0$. Hence $\mathcal{O}_0$ is a local ring with maximal ideal $\mathfrak{m}$. □

2.4. The Weierstrass preparation theorem

We now want to derive a fundamental theorem in the further study of local rings of holomorphic function germs. This is the Weierstrass preparation theorem, which will play a central role in the theory of holomorphic functions of several complex variables.

In complex analysis of a single variable it is shown that a nonvanishing convergent power series $f(t) = \sum_{m=0}^{\infty} a_m t^m$ in one complex variable t has a unique representation in the form

$$f(t) = t^k u(t)$$

where $u(t)$ is a convergent power series with $u(0) \neq 0$, so it is a unit in $\mathcal{O}_{1,0}$. The number k here is called the order of the zero $0 \in \mathbb{C}$. The (general) Weierstrass preparation theorem asserts that a convergent power series

$$g(z, t) = g(z_1, \ldots, z_n, t) \in \mathbb{C}\{z_1, \ldots, z_n, t\}$$

can be represented uniquely in the form

$$g(z, t) = (t^k + c_1(z)t^{k-1} + \ldots + c_k(z)) \cdot u(z, t)$$

(possibly after a linear change of coordinates) with $c_j(z) \in \mathbb{C}\{z\}$, $c_j(0) = 0$, and a unit $u(z, t) \in \mathbb{C}\{z, t\}$. Before we are able to formulate this theorem, we need to introduce several concepts.

Let $\mathbb{C}\{z\} = \mathbb{C}\{z_1, \ldots, z_{n+1}\}$ be the ring of convergent power series in $n + 1$ variables. Let

$$g(z) = \sum_{\nu=0}^{\infty} a_\nu z^\nu \in \mathbb{C}\{z_1, \ldots, z_{n+1}\}.$$

Then $g(z)$ has a unique representation

$$g(z) = \sum_{m=0}^{\infty} p_m(z)$$

in terms of the homogeneous polynomials

$$p_m(z) := \sum_{|\nu|=m} a_\nu z^\nu$$

of degree m.

Definition. The number

$$\min\{|\nu| \mid a_\nu \neq 0\}$$

is called the *order* $o(g)$ of the power series.

It is thus the smallest number m for which $p_m \not\equiv 0$.

Definition. A power series $g(z) \in \mathbb{C}\{z\}$ is called *regular of order k in z_{n+1}* if and only if $g(0, \ldots, 0, z_{n+1})$ is a power series in z_{n+1} of order k.

Lemma 2.1. *If $g(z) \in \mathbb{C}\{z\}$ is a convergent power series of order $k < \infty$, then $g(z')$, where $z'_1, \ldots, z'_{n+1}$ are appropriate coordinates of $\mathbb{C}^{n+1}$ obtained by a linear change of coordinates from $z_1, \ldots, z_{n+1}$, is regular of order k in z'_{n+1}.*

Proof. Let

$$g(z) = \sum_{m=k}^{\infty} p_m(z), \; p_k \not\equiv 0,$$

be the expansion of $g(z)$ in homogeneous polynomials. Let

$$p_k(z) = \sum_{|\nu|=k} a_\nu z^\nu.$$

We next put

$$\begin{aligned} z'_j &= z_j - w_j \, z_{n+1}, \; j = 1, \ldots, n, \\ z'_{n+1} &= z_{n+1}, \end{aligned}$$

with $w_j \in \mathbb{C}$, $j = 1, \ldots, n$. Then the coefficient of z'^k_{n+1} in $p_k(z')$ is equal to

$$c = \sum_{\nu_1 + \ldots + \nu_{n+1} = k} a_{\nu_1, \ldots, \nu_{n+1}} w_1^{\nu_1} \cdots w_n^{\nu_n}.$$

Since $p_k \not\equiv 0$, there are $w_1, \ldots, w_n \in \mathbb{C}$ such that $c \neq 0$. For this choice of $w_1, \ldots, w_n$

$$g(0, \ldots, 0, z'_{n+1}) = c z'^k_{n+1} + \text{ higher terms.}$$

Thus $g(z')$ is regular of order k in z'_{n+1}. □

To distinguish, we now put $t = z_{n+1}$ and

$$\begin{aligned} \mathbb{C}\{z\} &= \mathbb{C}\{z_1, \ldots, z_n\}, \\ \mathbb{C}\{z, t\} &= \mathbb{C}\{z_1, \ldots, z_n, t\}. \end{aligned}$$

Definition. A *Weierstrass polynomial* of degree $k > 0$ is a polynomial

$$t^k + c_1(z)t^{k-1} + \ldots + c_k(z),$$

where $c_j \in \mathbb{C}\{z\},\ c_j(0) = 0,\ j = 1, \ldots, k.$

A Weierstrass polynomial is thus an element $h(z,t) \in \mathbb{C}\{z\}[t]$ with leading coefficient 1 and whose other coefficients are nonunits in $\mathbb{C}\{z\}$. In particular $h(z,t)$ is regular of order k in t. The Weierstrass preparation theorem now asserts that this is the general form of a convergent power series in $\mathbb{C}\{z,t\}$ that is regular of order k in t. More precisely:

Theorem 2.1 (Weierstrass preparation theorem). *Let $g(z,t)$ be a convergent power series in $\mathbb{C}\{z,t\}$ that is regular of order k in t. Then there are a uniquely determined Weierstrass polynomial $h(z,t) \in \mathbb{C}\{z\}[t]$ and a uniquely determined unit $u(z,t) \in \mathbb{C}\{z,t\}$ such that*

$$g(z,t) = h(z,t)u(z,t).$$

Remark 2.3. This theorem is called a "preparation theorem" because the power series $g(z,t)$ has been "prepared" for the investigation of its zeros. Since u does not vanish in a neighborhood of $0 \in \mathbb{C}^{n+1}$, it follows from the Weierstrass preparation theorem that the zero set of g in this neighborhood coincides with the zero set of the polynomial

$$t^k + c_1(z)t^{k-1} + \ldots + c_k(z).$$

We have already investigated such zero sets in Chapter 1 for the case $n = 1$, i.e., $z \in \mathbb{C}$: It concerns the Riemann surface of an algebraic function, which is a branched covering over the z-plane.

There are various proofs of the Weierstrass preparation theorem, some complex analytic, others more algebraic. A mainly algebraic proof was given by L. Stickelberger already in 1887, a purely algebraic variant of this proof by C.L. Siegel (1968). (Historical material on the Weierstrass preparation theorem and its proof can be found in [**GR71**, p. 35 f. and p. 29 ff.].) Proofs based on the ideas of Siegel and Stickelberger appear in many of the texts we cite. We present a proof following the book [**GG74**].

This proof, with modifications, can also be found in [**BK86**]. It employs elegant methods developed to prove the Malgrange preparation theorem, a version of Theorem 2.1 for differentiable functions. We shall in fact present a somewhat more general theorem.

Theorem 2.2 (Weierstrass division theorem). *Let $f, g \in \mathbb{C}\{z,t\}$ and let g be regular of order k in t. Then there are a uniquely determined power series $q \in \mathbb{C}\{z,t\}$ and a uniquely determined polynomial r in $\mathbb{C}\{z\}[t]$ of degree $\leq k-1$ with*

$$f = q \cdot g + r.$$

Ascribing this theorem to Weierstrass (it is also called the Weierstrass formula) is not completely justified historically, for it was first proved by Stickelberger in 1887 and independently by Späth in 1929.

Deduction of Theorem 2.1 from Theorem 2.2. We apply Theorem 2.2 to $f(z,t) := t^k$ and g as in Theorem 2.1. By Theorem 2.2 there exist $q \in \mathbb{C}\{z,t\}$ and

$$r(z,t) = \sum_{j=1}^{k} a_j(z)t^{k-j} \in \mathbb{C}\{z\}[t]$$

with

$$t^k = q \cdot g + r$$

or

$$q(z,t) \cdot g(z,t) = t^k - a_1(z)t^{k-1} - \ldots - a_k(z).$$

We substitute $z = 0$ into this equation and compare the coefficients of t^k: Since g is regular of order k in t, we have

$$g(0,t) = ct^k + \text{ higher terms}$$

with $c \neq 0$. Hence

$$q(0,0) \neq 0.$$

Hence q is a unit in $\mathbb{C}\{z,t\}$ and there is a $u \in \mathbb{C}\{z,t\}$ with $qu = 1$. The power series u is then also a unit. The polynomial

$$h(z,t) := t^k - a_1(z)t^{k-1} - \ldots - a_k(z)$$

is however a Weierstrass polynomial. From this it follows that

$$g = h \cdot u.$$

The uniqueness of h and u follows from the uniqueness of q and r. □

Theorem 2.1 and Theorem 2.2 are in fact equivalent: one can deduce Theorem 2.2 from Theorem 2.1 (Exercise).

Proof of uniqueness in Theorem 2.2. Assume

$$f = qg + r = \tilde{q}g + \tilde{r},$$

where $q, \tilde{q} \in \mathbb{C}\{z,t\}$ and $r, \tilde{r} \in \mathbb{C}\{z\}[t]$ are polynomials of degree $\leq k-1$. We may assume that $k > 0$, for otherwise $r = \tilde{r} \equiv 0$ and $q = \tilde{q}$ and we are done. Now

$$r - \tilde{r} = (\tilde{q} - q)g.$$

For fixed $z \in \mathbb{C}^n$ the function $(r - \tilde{r})(z,t)$ is a polynomial of degree $\leq k-1$ in t and has (counting multiplicity) at most $k-1$ zeros. We shall show that there is a neighborhood V of 0 in $\mathbb{C}^n$ such that for all $z \in V$ the power series $g(z,t)$ as a function of t has at least k zeros. Then $(r - \tilde{r})(z,t)$ also has at least k zeros, and so must be the zero polynomial. It follows that $r = \tilde{r}$ and thus also $q = \tilde{q}$ since g does not vanish identically in a neighborhood of 0.

By hypothesis the power series $g(z, t)$ is regular of order k in t, i.e., $g(0, t)$ has a k-th order zero at $t = 0 \in \mathbb{C}$. The power series $g(0, t)$ defines a holomorphic function in a neighborhood of $0 \in \mathbb{C}$ and its zeros are isolated. Therefore there is a $\delta > 0$ with $g(0, t) \neq 0$ for all t with $0 < |t| \leq \delta$. Put

$$\varepsilon := \inf_{|t|=\delta} |g(0, t)|.$$

Since g is continuous, there is a neighborhood V of $0 \in \mathbb{C}^n$ such that for all $z \in V$ and $t \in \mathbb{C}$ with $|t| = \delta$ we have

$$|g(z, t) - g(0, t)| < \varepsilon \leq |g(0, t)|.$$

Let $z \in V$ be fixed. From Rouché's theorem it follows that $g(z, \cdot)$ and $g(0, \cdot)$ then have the same number of zeros in the disc $|t| < \delta$ (counting multiplicity). Thus $g(z, \cdot)$ has at least k zeros for $z \in V$ (counting multiplicity). □

Using "Malgrange's trick" one can reduce the division theorem to the following one, a special case of Theorem 2.2.

Theorem 2.3 (Special division theorem). *Let $p_k(y, t) \in \mathbb{C}\{y_1, \cdots, y_k\}[t]$ be the "general polynomial of degree k", i.e.,*

$$p_k(y, t) = t^k + y_1 t^{k-1} + \ldots + y_k.$$

Then for each $f(z, y, t) \in \mathbb{C}\{z, y, t\}$ there exist $q(z, y, t) \in \mathbb{C}\{z, y, t\}$ and a polynomial $r(z, y, t) \in \mathbb{C}\{z, y\}[t]$ of degree $\leq k - 1$ such that

$$f = q \cdot p_k + r.$$

Remark 2.4. This is a special case of Theorem 2.2: Theorem 2.3 deals only with division by the "general polynomial" instead of by an arbitrary power series, regular in t, from $\mathbb{C}\{z, y, t\}$.

Deduction of Theorem 2.2 from Theorem 2.3. Let $g(z, t) \in \mathbb{C}\{z, t\}$ be regular of order k in t and let $f \in \mathbb{C}\{z, t\}$. By Theorem 2.3 there exist $\tilde{q}, \tilde{\tilde{q}}, \tilde{r}, \tilde{\tilde{r}}$ such that

$$g(z, t) = \tilde{q}(z, y, t) p_k(y, t) + \tilde{r}(z, y, t), \tag{2.3}$$

$$f(z, t) = \tilde{\tilde{q}}(z, y, t) p_k(y, t) + \tilde{\tilde{r}}(z, y, t), \tag{2.4}$$

where $\tilde{r}, \tilde{\tilde{r}} \in \mathbb{C}\{z, y\}[t]$ are polynomials of degree $\leq k - 1$. Let

$$\tilde{r}(z, y, t) = A_1(z, y) t^{k-1} + \ldots + A_k(z, y).$$

By hypothesis, $g(0, t)$ is a power series of the form

$$g(0, t) = c \cdot t^k + \text{ higher terms}$$

with $c \neq 0$. If we put $z = y = 0$ in (2.3) and compare the coefficients of $1, t, \ldots, t^k$, we find

$$A_j(0, 0) = 0 \text{ for } j = 1, \ldots, k, \quad \tilde{q}(0, 0, 0) = c.$$

The idea of the proof is now to substitute appropriate holomorphic functions $\varphi_j(z)$ for the "general coefficients" y_j of the polynomials p_k so that the remainder $\tilde{r}$ in (2.3) vanishes. To achieve this, we apply the implicit function theorem.

First we show

$$\frac{\partial A_j}{\partial y_l}(0,0) = \begin{cases} 0 & \text{for } j > l, \\ -c & \text{for } j = l. \end{cases}$$

Indeed, on differentiating both sides of the equation (2.3) with respect to the variable y_l and putting $y = z = 0$, we obtain

$$0 = \frac{\partial \tilde{q}}{\partial y_l}(0,0,t)t^k + \tilde{q}(0,0,t) \cdot t^{k-l} + \frac{\partial A_1}{\partial y_l}(0,0) \cdot t^{k-1} + \ldots + \frac{\partial A_k}{\partial y_l}(0,0).$$

On comparing the coefficients of t^{k-j} for $k \geq j \geq l$, we find

$$\frac{\partial A_k}{\partial y_l}(0,0) = 0, \frac{\partial A_{k-1}}{\partial y_l}(0,0) = 0, \ldots, \frac{\partial A_{l+1}}{\partial y_l}(0,0) = 0$$

and

$$\frac{\partial A_l}{\partial y_l}(0,0) = -\tilde{q}(0,0,0) = -c.$$

The matrix

$$\frac{\partial A}{\partial y}(0,0) = \left(\frac{\partial A_j}{\partial y_l}(0,0)\right)_{l=1,\ldots,k}^{j=1,\ldots,k}$$

is therefore an upper triangular matrix with the entries $-c$ on the main diagonal. Thus

$$\det \frac{\partial A}{\partial y}(0,0) = (-c)^k \neq 0.$$

The convergent power series $A_1, \ldots, A_k \in \mathbb{C}\{z, y\}$ define a holomorphic map $A : U \to \mathbb{C}^k$ on an open neighborhood $U \subset \mathbb{C}^n \times \mathbb{C}^k$ of the point $(0,0)$, with $A(0,0) = 0$ and $\det(\partial A/\partial y)(0,0) \neq 0$. By the implicit function theorem (Proposition 2.14) there must be an open neighborhood $V \subset \mathbb{C}^n$ of $0 \in \mathbb{C}^n$ and a holomorphic map $\varphi : V \to \mathbb{C}^k$ such that

$$A(z, \varphi(z)) = 0$$

for all $z \in V$. For all $z \in V$ and $t \in \mathbb{C}$ we now have

$$\tilde{r}(z, \varphi(z), t) = 0$$

and so

$$g(z,t) = \tilde{q}(z, \varphi(z), t) p_k(\varphi(z), t).$$

Now $\tilde{q}$ is a unit, since $\tilde{q}(0,0,0) = c \neq 0$, so it is invertible, and from equation (2.4) we get

$$f(z,t) = \tilde{\tilde{q}}(z, \varphi(z), t) p_k(\varphi(z), t) + \tilde{\tilde{r}}(z, \varphi(z), t),$$

and so

$$f = \tilde{\tilde{q}} \cdot \tilde{q}^{-1} \cdot g + \tilde{\tilde{r}}.$$

If we put

$$\begin{aligned} q(z,t) &:= \tilde{\tilde{q}}(z,\varphi(z),t)\cdot \tilde{q}^{-1}(z,\varphi(z),t), \\ r(z,t) &:= \tilde{\tilde{r}}(z,\varphi(z),t), \end{aligned}$$

we obtain the desired representation

$$f = q\cdot g + r.$$

The uniqueness of this representation has already been shown. □

Proof of Theorem 2.3. Let $f(z,y,t)\in\mathbb{C}\{z,y,t\}$ be given. We must show that there are $q(z,y,t)\in\mathbb{C}\{z,y,t\}$ and a polynomial $r(z,y,t)\in\mathbb{C}\{z,y\}[t]$ of degree $\le k-1$ such that

$$f(z,y,t) = q(z,y,t)p_k(y,t) + r(z,y,t).$$

We can consider $p_k(y,t)$ as an element of $\mathbb{C}\{z,y,t\}$. There exists a polycylinder $\Delta_1\subset\mathbb{C}^n\times\mathbb{C}^k\times\mathbb{C}$ around $(0,0,0)$, on which $f(z,y,t)$ and $p_k(y,t)$ define holomorphic functions. The polynomial $p_k(0,t)$ has a zero of order k in $0\in\mathbb{C}$. Since the zeros of a polynomial are isolated, there is a $\delta>0$ such that $p_k(0,t)\ne 0$ for $|t|=\delta$. Since p_k is continuous, there are $r_1,\dots,r_n;\rho_1,\dots,\rho_k>0$ such that for the polycylinder

$$\Delta = \left\{ (z,y,t)\in\mathbb{C}^n\times\mathbb{C}^k\times\mathbb{C} \;\middle|\; \begin{array}{l} |z_j|<r_j,\ j=1,\dots,n, \\ |y_l|<\rho_l,\ l=1,\dots,k, \\ |t|<\delta \end{array} \right\}$$

we have $\bar\Delta\subset\Delta_1$ and for $(z,y,t)\in\bar\Delta$ with $|t|-\delta$ we have

$$p_k(y,t)\ne 0.$$

By Cauchy's integral formula in a single variable the function

$$q(z,y,t) := \frac{1}{2\pi i}\int\limits_{|\tau|=\delta}\frac{f(z,y,\tau)}{p_k(y,\tau)(\tau-t)}d\tau$$

is holomorphic in Δ. The function

$$r(z,y,t) := f(z,y,t) - q(z,y,t)p_k(y,t)$$

is likewise holomorphic in Δ and has the following integral representation in Δ:

$$\begin{aligned} r(z,y,t) &= \frac{1}{2\pi i}\int\limits_{|\tau|=\delta}\frac{f(z,y,\tau)}{\tau-t}d\tau - \frac{1}{2\pi i}\int\limits_{|\tau|=\delta}\frac{f(z,y,\tau)p_k(y,t)}{p_k(y,\tau)(\tau-t)}d\tau \\ &= \frac{1}{2\pi i}\int\limits_{|\tau|=\delta}\frac{f(z,y,\tau)}{p_k(y,\tau)}\left(\frac{p_k(y,\tau)-p_k(y,t)}{\tau-t}\right)d\tau. \end{aligned}$$

Now

$$\begin{aligned}\frac{p_k(y,\tau)-p_k(y,t)}{\tau-t} &= \frac{(\tau^k-t^k)+y_1(\tau^{k-1}-t^{k-1})+\ldots+y_{k-1}(\tau-t)}{\tau-t}\\ &= s_1(y,\tau)t^{k-1}+\ldots+s_k(y,\tau),\end{aligned}$$

where $s_1,\ldots,s_k\in\mathbb{C}\{y\}[\tau]$. Hence

$$r(z,y,t)=\sum_{j=1}^{k}\left(\frac{1}{2\pi i}\int_{|\tau|=\delta}\frac{f(z,y,\tau)}{p_k(y,\tau)}s_j(y,\tau)d\tau\right)t^{k-j}.$$

If we put

$$A_j(z,y):=\frac{1}{2\pi i}\int_{|\tau|=\delta}\frac{f(z,y,\tau)}{p_k(y,\tau)}s_j(y,\tau)d\tau,$$

then $A_j(z,y)\in\mathbb{C}\{z,y\}$ for $j=1,\ldots,k$. The polynomial $r(z,y,t)$ is thus a polynomial in $\mathbb{C}\{z,y\}[t]$ of degree $\leq k-1$. With this we have established the desired division formula. □

Remark 2.5. We have given a complex analytic proof of Theorem 2.3. An algebraic proof can be found in [**BK86**]. This proof, however, employs the theorem on symmetric functions for convergent power series which must also be proved. The proof there makes clear that preparation and division theorems hold for formal power series too. An analogue of the preparation theorem for germs of differentiable real functions is a deep result proved by Malgrange in 1964: the *Malgrange preparation theorem.*

As an application of Theorems 2.1 and 2.2 we now demonstrate a few algebraic properties of the local rings $\mathcal{O}_{n,0}$.

We recall that a ring R is called *Noetherian* when every ideal in R is finitely generated.

Proposition 2.19 (Rückert basis theorem). *The ring $\mathcal{O}_{n,0}\cong\mathbb{C}\{z_1,\ldots,z_n\}$ is Noetherian.*

Proof. We prove the theorem by induction on n.

$\mathcal{O}_{0,0}=\mathbb{C}$ is a field when $n=0$, so it is trivially a Noetherian ring.

Let $I\subset\mathcal{O}_{n,0}$ be an ideal and $g\in I$, $g\neq 0$. By Lemma 2.1 we can assume that, after a change of coordinates if necessary, g is regular of order k in z_n. By the induction hypothesis $\mathcal{O}_{n-1,0}$ is Noetherian. The *Hilbert basis theorem* (see e.g. [**Lan02**, IV, §4]) asserts that the polynomial ring over a Noetherian ring is again Noetherian. Thus the ring $\mathcal{O}_{n-1,0}[z_n]$ is Noetherian. It follows that the ideal $I\cap\mathcal{O}_{n-1,0}[z_n]$ is generated by finitely many elements $g_1,\ldots,g_l\in I\cap\mathcal{O}_{n-1,0}[z_n]$.

Now let $f \in I$. By the Weierstrass division theorem f can be written as

$$f = q \cdot g + r \quad \text{with } r \in \mathcal{O}_{n-1,0}[z_n]\,.$$

Since $f \in I$ and $g \in I$, we have

$$r = f - q \cdot g \in I,$$

so $r \in I \cap \mathcal{O}_{n-1,0}[z_n]$. Thus there are $a_1, \ldots, a_l \in \mathcal{O}_{n-1,0}[z_n]$ with

$$r = a_1 g_1 + \ldots + a_l g_l.$$

It follows that

$$f = q \cdot g + a_1 \cdot g_1 + \ldots + a_l g_l.$$

We have proved that $g, g_1, \ldots, g_l$ generate the ideal I. □

Remark 2.6. The theorem means geometrically that the common zero sets of arbitrarily many holomorphic functions can always be described locally by finitely many equations.

Let R be an integral domain. An element $f \in R$, $f \neq 0$, is called *irreducible over* R if from $f = g_1 g_2$ with $g_1, g_2 \in R$ it follows that g_1 is unit or g_2 is unit; otherwise f is called *reducible over* R.

Lemma 2.2. *A Weierstrass polynomial* $h \in \mathbb{C}\{z_1, \ldots, z_{n-1}\}[z_n]$ *is reducible over* $\mathbb{C}\{z_1, \ldots, z_{n-1}\}[z_n]$, *if and only if it is reducible over* $\mathbb{C}\{z_1, \ldots, z_n\}$. *If* h *is reducible, then up to units in* $\mathbb{C}\{z_1, \ldots, z_{n-1}\}[z_n]$ *all its factors are Weierstrass polynomials.*

Proof. a) First let h be reducible over $\mathbb{C}\{z_1, \ldots, z_n\}$, i.e., $h = g_1 g_2$, where $g_1, g_2 \in \mathbb{C}\{z_1, \ldots, z_n\}$ are nonunits. Since h is a Weierstrass polynomial, it is regular in z_n. As is easy to see, g_1 and g_2 are then also regular in z_n. By the Weierstrass preparation theorem we have $g_i = u_i h_i$, where u_i is a unit in $\mathbb{C}\{z_1, \ldots, z_n\}$ and h_i is a Weierstrass polynomial in $\mathbb{C}\{z_1, \ldots, z_{n-1}\}[z_n]$ $(i = 1, 2)$. Thus

$$h = (u_1 u_2) \cdot (h_1 h_2).$$

One can now show (Exercise) that $h_1 h_2$ is also a Weierstrass polynomial. We thus have two representations of h as a product of a unit and of a Weierstrass polynomial, namely

$$h = 1 \cdot h \text{ and } h = (u_1 u_2) \cdot (h_1 h_2).$$

The uniqueness assertion of the Weierstrass preparation theorem yields

$$h = h_1 \cdot h_2.$$

The Weierstrass polynomials h_1, h_2 cannot be units in $\mathbb{C}\{z_1, \ldots, z_{n-1}\}[z_n]$ for otherwise g_1 and g_2 would be units.

b) Now let h be reducible over $\mathbb{C}\{z_1, \ldots, z_{n-1}\}[z_n]$, i.e., $h = g_1 g_2$ where $g_1, g_2 \in \mathbb{C}\{z_1, \ldots, z_{n-1}\}[z_n]$ are nonunits.

We need to show that g_1, g_2 are nonunits in $\mathbb{C}\{z_1, \ldots, z_n\}$.

Let $z = (z_1, \ldots, z_{n-1})$ and

$$\begin{aligned} g_1(z, z_n) &= a_0(z)z_n^r + a_1(z)z_n^{r-1} + \ldots + a_r(z), \\ g_2(z, z_n) &= b_0(z)z_n^s + b_1(z)z_n^{s-1} + \ldots + b_s(z), \end{aligned}$$

with $a_0(z), \ldots, a_r(z), b_0(z), \ldots, b_s(z) \in \mathbb{C}\{z\}$. Comparison of the coefficients in $h = g_1 g_2$ yields

$$a_0 b_0 = 1.$$

We can therefore assume without loss of generality that $a_0 = b_0 = 1$. Then g_1 and g_2 are Weierstrass polynomials.

If g_1 were a unit in $\mathbb{C}\{z_1, \ldots, z_n\}$, then

$$g_2 = g_1^{-1} h$$

would be a decomposition of the Weierstrass polynomial g_2 as the product of a unit and of a Weierstrass polynomial. The uniqueness assertion of Theorem 2.1 again gives

$$g_1 = 1, h = g_2.$$

The assertion $g_1 = 1$, however, is a contradiction to the fact that g_1 is a nonunit in $\mathbb{C}\{z\}[z_n]$. □

Now let R again be an arbitrary integral domain, $f \in R \setminus \{0\}$, $g \in R$. One says that f *divides* g (symbolically, $f|g$) if there is an $h \in R$ such that $g = f \cdot h$. Let $f \in R \setminus \{0\}$ be a nonunit. The element f is called *prime* if from $f|g_1 \cdot g_2$, $g_1, g_2 \in R$, it follows that $f|g_1$ or $f|g_2$.

Remark 2.7. If f is prime, then f is also irreducible. The converse does not hold in general.

An integral domain R is called *factorial* (or a *unique factorization domain*) if each nonunit $f \in R\backslash\{0\}$ can be written as the product of finitely many prime elements. This decomposition is then unique up to the order of the factors and multiplication by units.

Remark 2.8. It can be shown (see e.g. [**Lan02**, II, §5]) that an integral domain R is factorial if and only if each nonunit $f \in R\backslash\{0\}$ can be written as the product of finitely many irreducible elements, and this decomposition is unique up to the order of the factors and multiplication by units.

Proposition 2.20. *The ring* $\mathcal{O}_{n,0} \cong \mathbb{C}\{z_1, \ldots, z_n\}$ *is factorial.*

Proof (by induction over n).

Start of induction $n = 0$: $\mathcal{O}_{0,0} \cong \mathbb{C}$ is a field and so trivially factorial.

Induction step $n - 1 \to n$: Let $f \in \mathbb{C}\{z_1, \ldots, z_n\}$, $f \neq 0$, be a nonunit. By Lemma 2.1 we can again assume without loss of generality that f is regular in z_n. By the Weierstrass preparation theorem there exist a unit $u \in \mathbb{C}\{z_1, \ldots, z_n\}$ and a Weierstrass polynomial $h \in \mathbb{C}\{z_1, \ldots, z_{n-1}\}[z_n]$ such that

$$f = u \cdot h.$$

By the induction hypothesis $\mathbb{C}\{z_1, \ldots, z_{n-1}\}$ is factorial. By a lemma due to Gauss (see e.g. [**Lan02**, IV, §2]) $\mathbb{C}\{z_1, \ldots, z_{n-1}\}[z_n]$ is also factorial. Therefore there exists a unique decomposition of h,

$$h = h_1 \cdots h_k,$$

into irreducible elements $h_j \in \mathbb{C}\{z_1, \ldots, z_{n-1}\}[z_n]$. By Lemma 2.2 the $h_1, \ldots, h_k$ are then Weierstrass polynomials and

$$f = u \cdot h_1 \cdots h_k$$

is a decomposition of f into factors $h_1, \ldots, h_k$ that are irreducible in $\mathbb{C}\{z_1, \ldots, z_n\}$. This decomposition is unique up to multiplication by units and up to the order of the factors. □

Another useful consequence of the Weierstrass preparation theorem is the following theorem, known as *Hensel's lemma.* Let again $z = (z_1, \ldots, z_n)$ and $\mathbb{C}\{z\} := \mathbb{C}\{z_1, \ldots, z_n\}$. We call a polynomial $p(z,t) \in \mathbb{C}\{z\}[t]$ with leading coefficient 1 a *monic* polynomial.

Proposition 2.21 (Hensel's lemma). *Let*

$$p(z,t) = t^k + a_1(z)t^{k-1} + \ldots + a_k(z) \in \mathbb{C}\{z\}[t]$$

be a monic polynomial with $k \geq 1$ and suppose that

$$p(0,t) = (t - c_1)^{k_1} \cdots (t - c_r)^{k_r}$$

with pairwise distinct $c_j \in \mathbb{C}$, $j = 1, \ldots, r$. Then there are monic polynomials $p_1(z,t), \ldots, p_r(z,t) \in \mathbb{C}\{z\}[t]$ with

$$p(z,t) = p_1(z,t) \cdots p_r(z,t),$$

where $p_j(z,t)$ has degree k_j and $p_j(0,t) = (t - c_j)^{k_j}$ for $j = 1, \ldots, r$.

Proof. We deduce this theorem from the Weierstrass preparation theorem by induction on r. For $r = 1$ we put $p_1(z,t) = p(z,t)$.

For $r > 1$ we consider

$$g(z,t) := p(z, t + c_r) \in \mathbb{C}\{z\}[t].$$

Then $g(z,t)$ is regular of order k_r in t. By the Weierstrass preparation theorem (Theorem 2.1) there are a Weierstrass polynomial $h(z,t) \in \mathbb{C}\{z\}[t]$ of degree k_r and a unit $u(z,t) \in \mathbb{C}\{z,t\}$ such that

$$g(z,t) = h(z,t)u(z,t). \tag{2.5}$$

In fact $u(z,t)$ is an element of $\mathbb{C}\{z\}[t]$: Division with remainder in $\mathbb{C}\{z\}[t]$ gives

$$g(z,t) = h(z,t)q(z,t) + r(z,t), \tag{2.6}$$

where $q(z,t), r(z,t) \in \mathbb{C}\{z\}[t]$ and $r(z,t)$ is a polynomial of degree $\leq k_r - 1$ in t. We can consider the decompositions (2.5) and (2.6) as decompositions according to the Weierstrass division theorem (Theorem 2.2). From the uniqueness assertion of this theorem it follows that $r(z,t) = 0$ and $u(z,t) = q(z,t) \in \mathbb{C}\{z\}[t]$.

We now put

$$\tilde{p}(z,t) := u(z, t - c_r), \quad p_r(z,t) := h(z, t - c_r).$$

Then $p_r(z,t)$ is monic, $p(z,t) = \tilde{p}(z,t)p_r(z,t)$, and

$$\tilde{p}(0,t) = (t - c_1)^{k_1} \cdots (t - c_{r-1})^{k_{r-1}}.$$

The existence of $p_1(z,t), \ldots, p_{r-1}(z,t)$ follows by the application of the induction hypothesis to $\tilde{p}(z,t)$. □

2.5. Analytic sets

We now come to our core theme, investigating the zero sets of holomorphic functions. For holomorphic functions in a single variable the situation is particularly simple: The zeros of such a function form a discrete set. In several variables the situation is by far more complicated. We now present the fundamental definition of an analytic set.

Definition. Let $U \subset \mathbb{C}^n$ be open, $X \subset U$ a set.

(i) If $x \in U$, then X is called *analytic at x* if there are an open neighborhood $V \subset U$ of x and finitely many holomorphic functions $f_1, \ldots, f_r$ on V such that

$$X \cap V = \{z \in V \mid f_1(z) = \ldots = f_r(z) = 0\}.$$

(ii) The subset X is called a *locally analytic subset* of U if X is analytic at all $x \in X$.

(iii) The subset X is called an *analytic subset* of U if X is analytic at all $x \in U$.

Examples of analytic subsets of $U = \mathbb{C}^n$ are U, $\emptyset$ and algebraic subsets, i.e., the zero sets of finitely many polynomials.

Proposition 2.22. *An analytic subset X of U is closed in U.*

Proof. Let $x \in U \backslash X$, with V as in the definition. Then $X \cap V$ is closed in V. Therefore there exists an open neighborhood $W \subset V$ of x with $W \cap X = \emptyset$. So $U \backslash X$ is open. □

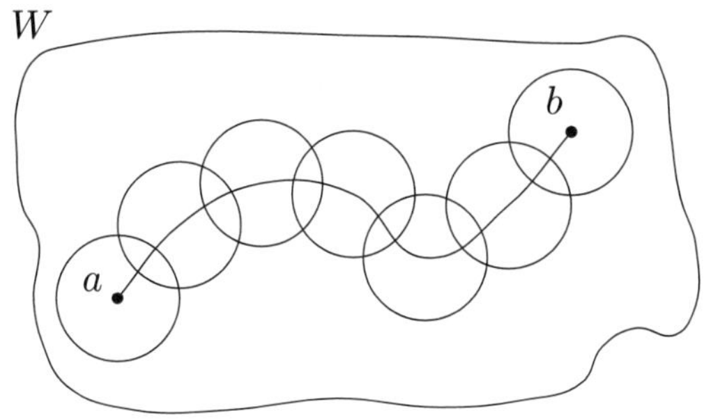

Figure 2.2. Choice of the balls $B_1, \ldots, B_t$

Exercise 2.1. Show that a subset $X \subset U$ is analytic if and only if it is locally analytic and closed in U. Give an example of an open subset $U \subset \mathbb{C}^n$ and a locally analytic but not analytic subset $X \subset U$.

A subset M of a topological space T is called *nowhere dense* in T if $\overline{M}$ contains no interior points. A subset M of a connected topological space T is called *nowhere separating* in T if $V \backslash M$ is also connected for all open and connected subsets V of T.

Proposition 2.23. *Let G be a region of $\mathbb{C}^n$. An analytic subset X of G, $X \neq G$, is nowhere dense in G and nowhere separating in G.*

Proof. Let $x \in G$, and let $V \subset G$ be an open connected neighborhood of x with

$$X \cap V = \{z \in V \mid f_1(z) = \ldots = f_r(z) = 0\}$$

for holomorphic functions $f_1, \ldots, f_r$ on V. It is enough to show that $Y = X \cap V$ is nowhere dense and nowhere separating in V.

From the identity theorem for holomorphic functions it follows that Y contains no interior points.

Let

$$B := B_r(x) := \{z \in \mathbb{C}^n \mid \|x - z\| < r\}$$

be a ball around x with radius r such that $B \subset V$. Then $B \setminus (Y \cap B)$ is connected: Let $a, b \in B \setminus (Y \cap B)$, and let Γ be a complex line joining a and b. Then $\Gamma \cap B$ is 1-dimensional, and $\Gamma \cap B \cap Y$ is a proper analytic subset of $\Gamma \cap B$, so discrete. Thus a and b can be connected by a path in $\Gamma \cap (B \setminus (Y \cap B))$.

Now let $W \subset V$ be open and connected, $a, b, \in W \setminus (W \cap Y)$. Since W is connected, a and b can be connected by a path in W. Let $B_1, \ldots, B_t$ be balls with centers on the path, $B_1 = B_{r_1}(a)$, $B_t = B_{r_t}(b)$, $B_j \cap B_{j+1} \neq \emptyset$ for $j = 1, \ldots, t-1$ (see Figure 2.2). Since Y contains no interior points, we have

$$(B_j \setminus (B_j \cap Y)) \cap (B_{j+1} \setminus (B_{j+1} \cap Y)) \neq \emptyset.$$

Thus

$$\bigcup_{j=1}^{t} (B_j \setminus (B_j \cap Y))$$

is a connected subset of W containing a and b. Hence $W \setminus (W \cap Y)$ is connected. □

Definition. We say that an analytic subset X in U has *codimension s at $x \in X$* (symbolically, $s = \operatorname{codim}_x X$) if an s-dimensional, but no $(s+1)$-dimensional, affine subspace Γ of $\mathbb{C}^n$ exists such that x is an isolated point of $\Gamma \cap X$. For $X \neq \emptyset$ we define

$$\operatorname{codim} X := \min_{x \in X} \operatorname{codim}_x X.$$

Proposition 2.24. *If $G \subset \mathbb{C}^n$ is a region and X is a proper analytic subset of G, then* $\operatorname{codim} X \geq 1$.

Proof. Let $x \in X$ and let $V \subset G$ be an open neighborhood of x with

$$X \cap V = \{z \in V \mid f_1(z) = \ldots = f_r(z) = 0\}$$

for holomorphic functions $f_1, \ldots, f_r$ in V. Without loss of generality let $V = B$ be a ball around x. Let $y \in B \setminus (X \cap B)$ and let Γ be a complex affine line joining x and y. Then $X \cap \Gamma \cap B$ is a proper analytic subset of $\Gamma \cap B \subset \Gamma$, so it is discrete. Hence $\operatorname{codim}_x X \geq 1$. □

2.6. Analytic set germs

Just as we introduced function germs in §2.3 to study the local properties of holomorphic functions, so we now define analytic set germs.

Definition. Let U and U' be open in $\mathbb{C}^n$, and let $X \subset U$, $X' \subset U'$ be analytic subsets. The analytic subsets X and X' define the same *analytic set germs* at $x \in U \cap U'$ if there is an open neighborhood $V \subset U \cap U'$ of x such that

$$X \cap V = X' \cap V.$$

We write (X, x) for the set germ of X at x. (More exactly, (X, x) is the equivalence class of all pairs (X', U) ($U' \subset \mathbb{C}^n$ open, $X' \subset U'$ an analytic subset, $x \in U'$) under the above equivalence relation given by the pair (X, U).) The analytic subset X is also called a *representative* of the set germ (X, x).

Analytic set germs stand in correspondence to ideals of $\mathcal{O}_{n,w}$, $w \in \mathbb{C}^n$. Let $I \subset \mathcal{O}_w$ be an ideal. Since $\mathcal{O}_w$ is Noetherian (Proposition 2.19), I is generated by finitely many holomorphic function germs $f_1, \dots, f_r$. Let $\tilde{f}_1, \dots, \tilde{f}_r : U \to \mathbb{C}$ be holomorphic functions, where U is an open neighborhood of w, that represent these germs. We define

$$V(\tilde{f}_1, \dots, \tilde{f}_r) := \{z \in U \mid \tilde{f}_1(z) = \dots = \tilde{f}_r(z) = 0\}.$$

Proposition 2.25. *The analytic set germ*

$$(V(\tilde{f}_1, \dots, \tilde{f}_r), w)$$

does not depend on the choice of $\tilde{f}_1, \dots, \tilde{f}_r$.

Proof. Let $(g_1, \dots, g_s)$ be another generating system of I, and $\tilde{g}_1 \dots, \tilde{g}_s : V \to \mathbb{C}$ holomorphic functions that represent $g_1, \dots, g_s$, with V an open neighborhood of w. Then there are holomorphic function germs $a_{jk} \in \mathcal{O}_w$ such that

$$f_j = \sum_{k=1}^{s} a_{jk}\, g_k, \quad j = 1, \dots, r.$$

Let $\tilde{a}_{jk} : W \to \mathbb{C}$ be suitable holomorphic representatives of a_{jk} on an open neighborhood $W \subset U \cap V$ of w. Then, on W,

$$\tilde{f}_j = \sum_{k=1}^{s} \tilde{a}_{jk} \tilde{g}_k, \quad j = 1, \dots, r.$$

From this it follows that

$$V(\tilde{g}_1, \dots, \tilde{g}_s) \cap W \subset V(\tilde{f}_1, \dots, \tilde{f}_r) \cap W.$$

On reversing the rôles of $f_1, \dots, f_r$ and $g_1, \dots, g_s$ we can show that there is an open neighborhood W' of w such that

$$V(\tilde{f}_1, \dots, f_r) \cap W' \subset V(\tilde{g}_1, \dots, \tilde{g}_s) \cap W'.$$

Thus

$$V(\tilde{g}_1, \dots, \tilde{g}_s) \cap W \cap W' = V(\tilde{f}_1, \dots, \tilde{f}_r) \cap W \cap W'.$$

Hence $g_1, \dots, g_s$ define the same analytic set germs as $f_1, \dots, f_r$. □

Following Proposition 2.25 we can define:

Definition. The analytic set germ

$$V(I) := (V(\tilde{f}_1, \dots, \tilde{f}_r), w)$$

is called the analytic set germ defined by the ideal I.

Definition. Let (X, x) be an analytic set germ. The set $I(X)$ of all germs $f \in \mathcal{O}_x$ represented by holomorphic functions $\tilde{f} : U \to \mathbb{C}$, U an open neighborhood of x, that vanish on a representative $\tilde{X} \subset U$ of X, is called the *ideal* of the set germ X.

Proposition 2.26. *The following connections hold for a set germ and its ideal and for an ideal and its corresponding set germ:*

(i) $I_1 \subset I_2 \Rightarrow V(I_1) \supset V(I_2)$.

(ii) $X_1 \subset X_2 \Rightarrow I(X_1) \supset I(X_2)$.

(iii) $V(I(X)) = X$.

(iv) $I(V(I)) \supset I$.

Proof. (i) Each representative of an element $f \in I_1$ vanishes on a representative $\widetilde{V(I_1)}$ of $V(I_1)$. Since $I_1 \subset I_2$, it also vanishes on a representative $\widetilde{V(I_2)}$ of $V(I_2)$. Hence $V(I_2) \subset V(I_1)$.

(ii) Let $f \in I(X_2)$, $\tilde{f}$ a representative of f. Then $\tilde{f}$ vanishes on a representative $\tilde{X}_2$ of X_2. Since $X_1 \subset X_2$, then $\tilde{f}$ vanishes also on a representative $\tilde{X}_1$ of X_1. Hence $f \in I(X_1)$.

(iii) $X \subset V(I(X))$ since each representative of an $f \in I(X)$ vanishes on a representative of X.

We show that $V(I(X)) \subset X$. Since X is an analytic set germ, there are holomorphic functions $\tilde{f}_1, \ldots, \tilde{f}_r : U \to \mathbb{C}$, U open neighborhood of x, such that

$$X = (V(\tilde{f}_1, \ldots, \tilde{f}_r), x).$$

Since $\tilde{f}_j$ vanishes on $V(\tilde{f}_1, \ldots, \tilde{f}_r)$, we have $f_j \in I(X)$ for the germ f_j of $\tilde{f}_j$. Thus $\tilde{f}_j$ vanishes on a representative $\tilde{Y}_j$ of $V(I(X))$. Hence

$$\tilde{Y}_j \subset V(\tilde{f}_j), \quad j = 1, \ldots, r.$$

It follows that

$$\tilde{Y}_1 \cap \ldots \cap \tilde{Y}_r \subset V(\tilde{f}_1, \ldots, \tilde{f}_r),$$

for the corresponding germ

$$V(I(X)) \subset X.$$

(iv) This inclusion is clear and is left to the reader. □

Remark 2.9. In general we do not have $I = I(V(I))$. Example: $I = (z^2) \subset \mathbb{C}\{z\} = \mathcal{O}_0$, $V(I) = \{0\}$, $I(V(I)) = (z) \neq (z^2)$.

Definition. Let R be a ring and $I \subset R$ an ideal. Then

$$\operatorname{rad}(I) := \{f \in R \mid f^k \in I \text{ for some } k \in \mathbb{N}\}$$

is called the *radical* of I.

In analogy to Hilbert's Nullstellensatz (see e.g. [**Lan02**, IX, §1]) we have

Proposition 2.27 (Rückert's Nullstellensatz). *For the ideal of an analytic set germ $V(I)$ we have*

$$I(V(I)) = \operatorname{rad} I.$$

Proof (for the case that I is a principal ideal $I = (f)$ in $\mathcal{O}_{n,0}$).

a) $\operatorname{rad}(f) \subset I(V(f))$ follows from Proposition 2.26(iv).

b) $I(V(f)) \subset \operatorname{rad}(f)$: We must show that if a representative $\tilde{g}$ of $g \in \mathcal{O}_0$ vanishes on a representative $\widetilde{V(f)}$ (in brief: $g|_{V(f)} = 0$), there exists a k with $f|g^k$.

Without loss of generality we may assume that f is irreducible. For, assume that

$$f = f_1 \cdots f_r$$

is the decomposition of f into irreducible factors. Then it follows from $g|_{V(f)} = 0$ that $g|_{V(f_j)} = 0$ for $j = 1, \ldots, r$. But this means that $f_j|g^{k_j}$ for some k_j and all $j = 1, \ldots, r$, once we have demonstrated the result for irreducible function germs. It follows finally that $f|g^{k_1+\ldots+k_r}$.

So we assume that f is irreducible.

The proof is by contradiction. Suppose that our assertion is false. Then there is a $g \in \mathcal{O}_0$ with $g|_{V(f)} = 0$ but f does not divide g^k for any $k \in \mathbb{N}$. Then f and g are relatively prime. By Lemma 2.1 we can assume that the coordinates $z_1, \ldots, z_n$ in $\mathbb{C}^n$ are chosen so that representatives $\tilde{f}$ of f and $\tilde{g}$ of g are regular in z_n. (The proof of Lemma 2.1 shows that one can find these coordinates for two different functions $\tilde{f}$ and $\tilde{g}$ too.) By the Weierstrass preparation theorem

$$f = u \cdot p, \quad g = v \cdot q,$$

where u, v are units and p, q are Weierstrass polynomials. But since we are interested only in the zero sets of f and g, and in divisibility properties, we can without loss of generality assume that f and g are already Weierstrass polynomials in $\mathcal{O}_{n-1,0}[z_n]$.

Since f and g are relatively prime in $\mathcal{O}_{n,0}$, they are also, by Lemma 2.2, relatively prime in $\mathcal{O}_{n-1,0}[z_n]$. Let K be the quotient field of $\mathcal{O}_{n-1,0}$. By the Gauss lemma it follows that f and g are also relatively prime in $K[z_n]$. Hence there are $\alpha, \beta \in K[z_n]$ such that

$$\alpha f + \beta g = 1.$$

The polynomials α, β are of the form

$$\alpha = \frac{a}{c}, \quad \beta = \frac{b}{c'}$$

with $a, b \in \mathcal{O}_{n-1,0}[z_n]$, $c, c' \in \mathcal{O}_{n-1,0}$, $c \neq 0$, $c' \neq 0$. Since we can scale the denominators as necessary, we can assume that $c' = c$. Then

$$a \cdot f + b \cdot g = c \in \mathcal{O}_{n-1,0}.$$

According to our assumption f is a Weierstrass polynomial, so of the form

$$f(z_1, \ldots, z_n) = z_n^k + c_1(z_1, \ldots, z_{n-1}) z_n^{k-1} + \ldots + c_k(z_1, \ldots, z_{n-1})$$

for a $k > 0$ and $c_j(0, \ldots, 0) = 0$, $j = 1, \ldots, k$. As is easy to see, for each $\varepsilon > 0$ there is a $\delta > 0$ such that for all

$$(z_1, \ldots, z_{n-1}) \in U := \{(z_1, \ldots, z_{n-1}) \,|\, |z_j| < \delta,\ j = 1, \ldots, n-1\}$$

the polynomial $f(z_1, \ldots, z_{n-1}, z_n)$ in the variables z_n has exactly k roots (counting multiplicity) in the disc $\{z_n \,|\, |z_n| < \varepsilon\}$. In particular, for each $(z_1, \ldots, z_{n-1}) \in U$ there is a z_n with $|z_n| < \varepsilon$ such that

$$f(z_1, \ldots, z_n) = 0.$$

Since by hypothesis g has a representative $\tilde{g}$ that vanishes on a representative $\widetilde{V(f)}$, for each $(z_1, \ldots, z_{n-1}) \in U$ there is a $|z_n| < \varepsilon$ (where δ and ε may have to be chosen smaller) such that

$$\begin{aligned} c(z_1, \ldots, z_{n-1}) &= a(z_1, \ldots, z_n) \cdot f(z_1, \ldots, z_n) + b(z_1, \ldots, z_n) \cdot g(z_1, \cdots, z_n) \\ &= 0\,. \end{aligned}$$

Thus c vanishes on U, whence $c = 0$. This contradicts the assumption $c \neq 0$. Hence our original assumption is false. □

An ideal I with rad $I = I$ is called a *radical ideal.* From Proposition 2.26 and Proposition 2.27 it follows that a map

$$\begin{aligned} \left\{ \begin{array}{c} \text{analytic set germs} \\ \text{at } x \end{array} \right\} &\to \left\{ \begin{array}{c} \text{radical ideals} \\ \text{in } \mathcal{O}_x \end{array} \right\} \\ (X, x) &\mapsto I(X) \end{aligned}$$

is defined and is surjective. From the following proposition it follows that this map is in fact a bijection.

Proposition 2.28. *Given analytic set germs (X_1, x), (X_2, x): if $(X_1, x) \neq (X_2, x)$, then $I(X_1) \neq I(X_2)$.*

Proof. If $(X_1, x) \neq (X_2, x)$, then for each representative X_1 of (X_1, x) and X_2 of (X_2, x) we have $X_1 \neq X_2$. Let

$$X_1 = V(f_1, \ldots, f_r), \quad \text{resp. } X_2 = V(g_1, \ldots, g_s),$$

be representatives of (X_1, x), resp. (X_2, x), in a neighborhood U of x. Then for each neighborhood $W \subset U$ of x

$$V(f_1, \ldots, f_r) \cap W \neq V(g_1, \ldots, g_s) \cap W.$$

Thus there is a

$$z_W \in ((X_1 \cap W) - (X_2 \cap W)) \cup ((X_2 \cap W) - (X_1 \cap W))\,.$$

If $z_W \in X_2 \cap W$, then there is an $f_{j(W)}$ such that $f_{j(W)}(z_W) \neq 0$. If $z_W \in X_1 \cap W$, there is a $g_{k(W)}$ with $g_{k(W)}(z_W) \neq 0$. This holds for all neighborhoods W of x. Since there are only finitely many f_j, g_k, there must be an f_j with $f_j|_{X_2 \cap W} \not\equiv 0$ for all W or a g_k with $g_k|_{X_1 \cap W} \not\equiv 0$ for all W. Hence $f_j \notin I(X_2)$ for some $j \in \{1, \dots, r\}$ or $g_k \notin I(X_1)$ for some $k \in \{1, \dots, s\}$. It follows that $I(X_1) \neq I(X_2)$. □

Next we consider holomorphic function germs on an analytic set germ.

Let $X \subset \mathbb{C}^n$ be an analytic subset, $x \in \mathbb{C}^n$ and (X, x) the corresponding analytic set germ. A holomorphic function germ $f \in \mathcal{O}_{n,x}$ is defined by restricting a function germ to (X, x). Two holomorphic function germs $f, g \in \mathcal{O}_{n,x}$ define the same function germ on (X, x) if $f - g$ vanishes on X, i.e., if $f - g \in I(X)$. Hence, the ring of holomorphic function germs on (X, x) is the ring

$$\mathcal{O}_{n,x}/I(X).$$

We denote this ring by $\mathcal{O}_{X,x}$.

Definition. An *analytic algebra* is an algebra over $\mathbb{C}$ of the form

$$\mathbb{C}\{z_1, \dots, z_n\}/I,$$

where I is an ideal in $\mathbb{C}\{z_1, \dots, z_n\}$.

A $\mathbb{C}$-algebra R is called a *local algebra* if R is local as a ring and the composition of the canonical maps

$$\mathbb{C} \cdot 1 \hookrightarrow R \twoheadrightarrow R/\mathfrak{m}$$

is a field isomorphism. If in a local algebra R one identifies the field $\mathbb{C}$ with its image $\mathbb{C} \cdot 1 \subset R$, then R is, as a $\mathbb{C}$-vector space, isomorphic to $\mathbb{C} \oplus \mathfrak{m}$.

Proposition 2.29. *An analytic algebra A is a Noetherian local algebra.*

Proof. a) A is a local algebra: Let

$$\pi : \mathbb{C}\{z_1, \dots, z_n\} \to A \cong \mathbb{C}\{z_1, \dots, z_n\}/I$$

be the canonical projection. If $\mathfrak{m}$ is a maximal ideal of $\mathbb{C}\{z_1, \dots, z_n\}$, then $\pi(\mathfrak{m})$ is a maximal ideal of A. From the vector space decomposition

$$\mathbb{C}\{z_1, \dots, z_n\} = \mathbb{C} \cdot 1 \oplus \mathfrak{m}$$

it follows that

$$\mathbb{C}\{z_1, \dots, z_n\}/I \cong (\mathbb{C} \cdot 1 \oplus \mathfrak{m})/(0 \oplus I) \cong \mathbb{C} \oplus (\mathfrak{m}/I).$$

b) A is Noetherian: This follows from the fact that $\mathbb{C}\{z_1, \dots, z_n\}$ is Noetherian (Exercise). □

Let A be an analytic algebra. An element $f \in A$ is called *nilpotent* if $f^k = 0$ for a sufficiently large k. The set $\mathfrak{n}_A$ of all nilpotent elements of A is the radical rad$\,(0)$ of the null ideal so $\mathfrak{n}_A$ is called the *nilradical* of A. The algebra A is called *reduced* if and only if A contains no nilpotent elements $\neq 0$. The algebra A is thus reduced if and only if $\mathfrak{n}_A = 0$.

Remark 2.10. (i) $\mathbb{C}\{z_1, \dots, z_n\}/I$ is reduced $\Leftrightarrow$ I is a radical ideal.

(ii) If (X, x) is an analytic set germ, then the algebra $\mathcal{O}_{X,x}$ is a reduced analytic algebra.

From the bijection above it follows that there is a bijection

$$\left\{ \begin{array}{c} \text{analytic set germs} \\ \text{at } x \end{array} \right\} \to \left\{ \begin{array}{c} \text{reduced analytic} \\ \text{algebras} \end{array} \right\}$$
$$(X, x) \mapsto \mathcal{O}_{X,x}\,.$$

Next we consider the local decomposition of analytic set germs into irreducible components.

Proposition 2.30. *If (X, w) and (Y, w) are analytic set germs at $w \in \mathbb{C}^n$, then $(X \cap Y, w)$ and $(X \cup Y, w)$ are also analytic set germs at w.*

Proof. Let

$$X = V(f_1, \dots, f_r), \quad Y = V(g_1, \dots, g_s).$$

Then

$$\begin{aligned} X \cap Y &= V(f_1, \dots, f_r, g_1, \dots, g_s), \\ X \cup Y &= V(f_j g_k; 1 \le j \le r, 1 \le k \le s). \end{aligned}$$

Hence $(X \cap Y, w)$ and $(X \cup Y, w)$ are analytic set germs at w. □

Definition. An analytic set germ (X, x) is called *irreducible* if from $X = X_1 \cup X_2$ for analytic set germs (X_1, x), (X_2, x) it follows that $(X, x) = (X_1, x)$ or $(X, x) = (X_2, x)$. Otherwise X is called *reducible.*

Proposition 2.31. *An analytic set germ (X, x) is irreducible if and only if its ideal $I(X)$ is a prime ideal.*

Proof. "$\Leftarrow$": Suppose that $X = X_1 \cup X_2$ with $X \neq X_1$, $X \neq X_2$. Then from Proposition 2.28 it follows that $I(X) \neq I(X_1)$, $I(X) \neq I(X_2)$. From $X \supset X_j$ it follows that $I(X) \subset I(X_j)$. Hence there are elements $f_j \in I(X_j)$ with $f_j \notin I(X)$, $j = 1, 2$. Then

$$f_1 f_2 \in I(X_1) \cap I(X_2) = I(X), f_1 \notin I(X), f_2 \notin I(X).$$

Hence $I(X)$ is not a prime ideal.

"$\Rightarrow$": Suppose that $I(X)$ is not a prime ideal. Then there are $f, g \in \mathcal{O}_{n,x}$, $f, g \notin I(X)$, such that $fg \in I(X)$. Since $f, g \notin I(X)$, we have

$$f|_X \not\equiv 0, \quad g|_X \not\equiv 0,$$

whence $X_1 := X \cap V(f) \subsetneqq X, X_2 := X \cap V(g) \subsetneqq X$. Now

$$\begin{aligned} X_1 \cup X_2 &= (X \cap V(f)) \cup (X \cap V(g)) \\ &= X \cap (V(f) \cup V(g)) \\ &= X \cap (V(fg)) \\ &= X, \end{aligned}$$

since $fg \in I(X)$. Thus X is reducible. □

Proposition 2.32. *Every analytic set germ (X, x) in $x \in \mathbb{C}^n$ has a uniquely determined decomposition*

$$X = X_1 \cup \cdots \cup X_r$$

into irreducible germs (X_i, x) with $X_i \not\subset X_j$ for $i \neq j$. The analytic set germs $(X_1, x), \ldots, (X_r, x)$ are called the irreducible components *of (X, x).*

Proof. a) The existence of the decomposition follows from the fact that $\mathcal{O}_{n,x}$ is Noetherian: If X is not irreducible, then

$$X = X_1 \cup X_2, \quad X_1 \subsetneqq X, \quad X_2 \subsetneqq X.$$

If X_1 and X_2 are irreducible, we have finished. Otherwise

$$X_1 = X_{11} \cup X_{12}, \quad X_{11} \subsetneqq X_1, \quad X_{12} \subsetneqq X_1,$$

etc. If this process were not to end after finitely many steps, there would be an infinite decreasing sequence of analytic set germs

$$Y_1 \supsetneqq Y_2 \supsetneqq \ldots.$$

To this sequence there would correspond an infinite sequence of strictly increasing ideals (by Proposition 2.26)

$$I(Y_1) \subsetneqq I(Y_2) \subsetneqq \ldots$$

in $\mathcal{O}_{n,x}$. But this cannot happen because $\mathcal{O}_{n,x}$ is Noetherian.

b) Uniqueness: Let

$$X = X_1' \cup \cdots \cup X_s'$$

be another decomposition of X into irreducible components with $X_j' \not\subset X_k'$ for $j \neq k$. Then

$$X_j' \subset X_1 \cup \cdots \cup X_r,$$

and so

$$X_j' = (X_1 \cap X_j') \cup \cdots \cup (X_r \cap X_j').$$

Now by hypothesis X_j' is irreducible. It follows that $X_j' = X_{k(j)} \cap X_j'$ for some $k(j)$ and so $X_j' \subset X_{k(j)}$. Since $X_{k(j)}$ is irreducible, it follows that $X_j' = X_{k(j)}$. The map $j \mapsto k(j)$ must be injective since $X_j' \not\subset X_l'$ for $j \neq l$. Hence the map $j \mapsto k(j)$ is bijective and the uniqueness of the decomposition is proved. □

We are particularly interested in the zero set of a single equation.

Definition. An analytic set germ $V(I)$ at $x \in \mathbb{C}^n$ is called a *hypersurface* if $I = (f)$ is a principal ideal in $\mathcal{O}_{n,x}$.

Remark 2.11. If $f = f_1^{k_1} \cdots f_r^{k_r}$ is the decomposition of f into distinct irreducible factors, then

$$\operatorname{rad}(f) = (f_1 \cdots f_r).$$

The equation $f_1(z) \cdots f_r(z) = 0$ is called the *equation* of the hypersurface and is uniquely determined up to a unit.

Proposition 2.33. *If $X = V(f)$ is a hypersurface and $f = f_1^{k_1} \cdots f_r^{k_r}$ is the decomposition of f into pairwise distinct irreducible factors, then*

$$V(f) = V(f_1) \cup \cdots \cup V(f_r)$$

is the decomposition of $V(f)$ into irreducible components.

Proof. Since the f_j are irreducible and the ring $\mathcal{O}_{n,x}$ is factorial, the f_j are prime elements. Hence the principal ideals (f_j) are prime ideals. By Proposition 2.31 the analytic set germs $V(f_j)$ are irreducible. Hence

$$V(f) = V(f_1) \cup \cdots \cup V(f_r)$$

is a decomposition of $V(f)$ into irreducible components. Suppose that $V(f_j) \subset V(f_k)$. Then $I(V(f_k)) \subset I(V(f_j))$. By Proposition 2.27 it follows that $\operatorname{rad}(f_k) \subset \operatorname{rad}(f_j)$ and so $(f_k) \subset (f_j)$. Hence $f_j | f_k$ and so $j = k$. Thus $V(f_j) \not\subset V(f_k)$ if $j \neq k$. □

Remark 2.12. One can define the concepts "irreducible" and "reducible" analogously also for analytic subsets $X \subset G$ in a region $G \subset \mathbb{C}^n$. In general one then has a decomposition into infinitely many irreducible components. Nevertheless the system of irreducible components is *locally finite*.

2.7. Regular and singular points of analytic sets

We now want to define regular and singular points of analytic sets. First we consider the concept of a complex submanifold.

Definition. Let $U \subset \mathbb{C}^n$ be open. A subset M of U is called a *complex submanifold* of U if for each point $a \in M$ there are an open neighborhood V of a in U and a biholomorphic map $\Phi : V \to \Delta$ onto a polycylinder Δ around $\Phi(a) = 0$ in $\mathbb{C}^n$ such that

$$M \cap V = \Phi^{-1}(\Delta^s \times \{0\})$$

with respect to the decomposition $\Delta = \Delta^s \times \Delta^{n-s}$ (see Figure 2.3). The number s is called the dimension of M at a; symbolically, $\dim_a M = s$.

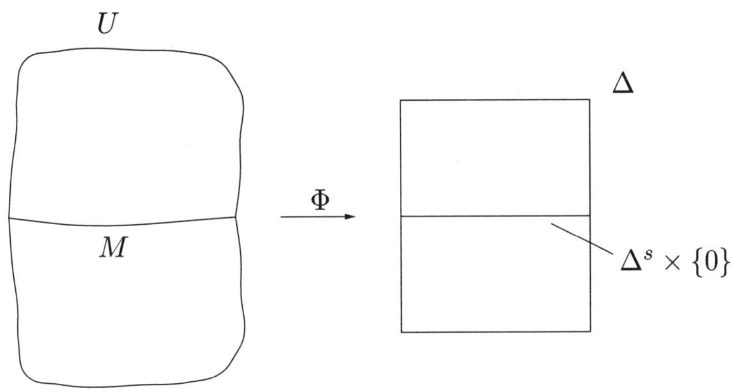

Figure 2.3. The chart Φ

Definition. Let $G \subset \mathbb{C}^n$ be a region and $X \subset G$ an analytic subset. A point $x \in X$ is called a *regular point* of X if there is a neighborhood U of x in $\mathbb{C}^n$ such that $X \cap U$ is a complex submanifold of U. Otherwise x is called a *singular point* of X. Let $S(X)$ be the set of singular points of X.

We shall use the following proposition in deriving a criterion for a regular point.

Proposition 2.34 (Rank theorem). *Let $U \subset \mathbb{C}^n$ be open and $f : U \to \mathbb{C}^m$ a holomorphic map for which J_f has constant rank r in a neighborhood of $a \in U$.*

Then there are open neighborhoods V of a and W of $b = f(a)$, a polycylinder $\Delta^n \subset \mathbb{C}^n$, $\Delta^m \subset \mathbb{C}^m$ around 0 and biholomorphic maps $\varphi : \Delta^n \to V$, $\psi : W \to \Delta^m$ with $\varphi(0) = a$ and $\psi(b) = 0$ such that the map $\tilde{f} = \psi \circ f \circ \varphi : \Delta^n \to \Delta^m$ satisfies

$$\tilde{f}(z_1, \dots, z_n) = (z_1, \dots, z_r, 0, \dots, 0).$$

Proof. Without loss of generality we can assume $a = b = 0$. By hypothesis there is an invertible $(r \times r)$-submatrix of $J_f(0)$. By permuting the coordinates of $\mathbb{C}^n$ and $\mathbb{C}^m$, if necessary, we can assume that the matrix

$$\left(\frac{\partial f_j}{\partial z_k}(0)\right)_{k=1,\dots,r}^{j=1,\dots,r}$$

is invertible.

Next we consider the map

$$\begin{aligned} g : U &\to \mathbb{C}^n \\ z &\mapsto (f_1(z), \dots, f_r(z), z_{r+1}, \dots, z_n). \end{aligned}$$

Then

$$J_g = \left(\begin{array}{c|c} \dfrac{\partial f_j}{\partial z_k} & \dfrac{\partial f_j}{\partial z_l} \\ \hline \underbrace{\qquad 0 \qquad}_{r} & \begin{matrix}1 & & \\ & \ddots & \\ & & 1\end{matrix}\end{array}\right)\begin{matrix}\left.\vphantom{\dfrac{\partial f_j}{\partial z_k}}\right\} r \\ \vphantom{\begin{matrix}1\\ \ddots\\ 1\end{matrix}}\end{matrix}$$

and so $\det J_g(0) \neq 0$. By the local invertibility theorem (cf. §2.2) there is an open neighborhood V of 0 in $\mathbb{C}^n$ which is mapped biholomorphically by g onto a polycylinder $\Delta^n \subset \mathbb{C}^n$ around 0. Put

$$\varphi := (g|_V)^{-1}.$$

Then the diagram

$$\begin{array}{ccc} V & \xrightarrow{\qquad\qquad f \qquad\qquad} & \mathbb{C}^m \\ & {\scriptstyle g}\searrow \qquad\qquad \nearrow{\scriptstyle f\circ\varphi} & \\ & \Delta^n & \end{array}$$

$$\begin{array}{ccc} (z_1, \ldots, z_n) & \longmapsto & (f_1(z), \ldots, f_m(z)) \\ & \searrow \qquad\qquad \nearrow & \\ & (f_1(z), \ldots, f_r(z), z_{r+1}, \ldots, z_n) & \\ & \| & \\ & (w_1, \ldots, w_n) & \end{array}$$

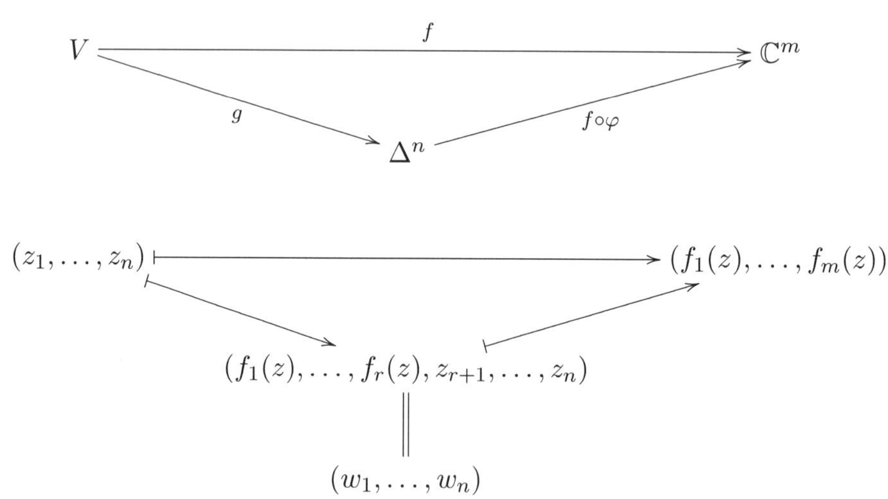

shows that for $w \in \Delta^n$

$$f \circ \varphi(w_1, \ldots, w_n) = (w_1, \ldots, w_r, h_{r+1}(w), \ldots, h_m(w)),$$

where each h_j is holomorphic. The Jacobi matrix of $f \circ \varphi$ therefore has the form

$$J_{(f\circ\varphi)} = \left(\begin{array}{ccc|c} 1 & & & \\ & \ddots & & \overbrace{\quad 0 \quad}^{n-r} \\ & & 1 & \\ \hline & * & & \dfrac{\partial h_j}{\partial w_k} \end{array}\right) \begin{array}{l} \left.\vphantom{\begin{array}{c}1\\ \ddots\\ 1\end{array}}\right\} r \\ \left.\vphantom{\dfrac{\partial h_j}{\partial w_k}}\right\} m-r \end{array}$$

But now, since rank $J_f(z) = r$ for z in a neighborhood of 0, without loss of generality for $z \in \Delta^n$, we must have

$$\frac{\partial h_j}{\partial w_k}(z) = 0 \quad \text{for } r+1 \leq j \leq m, \quad r+1 \leq k \leq n,$$

in this neighborhood. The h_j must therefore be independent of the variables $w_{r+1}, \ldots, w_n$. Next we consider the map

$$\begin{aligned} \gamma : \Delta^r \times \mathbb{C}^{m-r} &\rightarrow \Delta^r \times \mathbb{C}^{m-r} \\ (u_1, \ldots, u_m) &\mapsto (u_1, \ldots, u_r, u_{r+1} - h_{r+1}(u_1, \ldots, u_r, 0, \ldots, 0), \ldots, \\ &\qquad u_m - h_m(u_1, \ldots, u_r, 0, \ldots, 0)) \end{aligned}$$

for some appropriate polycylinder Δ^r. The Jacobi matrix of γ has the form

$$J_\gamma = \left(\begin{array}{ccc|ccc} 1 & & & & & \\ & \ddots & & & 0 & \\ & & 1 & & & \\ \hline & & & 1 & & \\ & * & & & \ddots & \\ & & & & & 1 \end{array}\right) \begin{array}{l} \left.\vphantom{\begin{array}{c}1\\ \ddots\\ 1\end{array}}\right\} r \\ \left.\vphantom{\begin{array}{c}1\\ \ddots\\ 1\end{array}}\right\} m-r \end{array},$$

so γ is biholomorphic. Now let $\Delta^{m-r} \subset \mathbb{C}^{m-r}$ be a sufficiently large polycylinder around 0 such that $\gamma \circ f \circ \varphi(\Delta^n) \subset \Delta^r \times \Delta^{m-r} =: \Delta^m$, $W := \gamma^{-1}(\Delta^m)$, $\psi := \gamma|_W$. Then

$$\begin{aligned} \psi \circ f \circ \varphi(w) &= \gamma(w_1, \ldots, w_r, h_{r+1}(w), \ldots, h_m(w)) \\ &= (w_1, \ldots, w_r, 0, \ldots, 0). \end{aligned}$$

This proves Proposition 2.34. □

Proposition 2.35. *Let $U \subset \mathbb{C}^n$ be open, $M \subset U$. Then M is a complex submanifold of U if and only if for each $a \in M$ there are an open neighborhood V of a in U and a holomorphic map $f : V \rightarrow \mathbb{C}^m$ such that*

(i) $V \cap M = \{z \in V \mid f_1(z) = \ldots = f_m(z) = 0\}$,

(ii) *J_f has constant rank on V.*

If M is a complex submanifold of U, then

$$\dim_a M = n - \operatorname{rank} J_f(a).$$

Proof. "$\Rightarrow$": By definition there exist for $a \in M$ an open neighborhood V of a in U and a biholomorphic map $\Phi : V \to \Delta^n$ such that

$$M \cap V = \Phi^{-1}(\Delta^s \times \{0\}).$$

Put $f := \pi \circ \Phi$, with $\pi : \Delta^n = \Delta^s \times \Delta^{n-s} \to \Delta^{n-s}$ the canonical projection. Then

(i) $V \cap M = \{z \in V \mid f_1(z) = \ldots = f_{n-s}(z) = 0\}$,

(ii) $\operatorname{rank} J_f = \operatorname{rank} J_{(\pi\circ\Phi)} = \operatorname{rank} J_\pi = n - s$ on V.

"$\Leftarrow$": Let r be the rank of J_f on the neighborhood V of a. By the rank theorem there is a commutative diagram

$$\begin{array}{ccc} V & \xrightarrow{\ f\ } & W \\ {\scriptstyle\varphi}\uparrow{\scriptstyle\cong} & & {\scriptstyle\cong}\downarrow{\scriptstyle\psi} \\ \Delta^n & \longrightarrow & \Delta^m \end{array}$$

$$(z_1, \ldots, z_n) \mapsto (z_1, \ldots, z_r, 0, \ldots, 0),$$

where φ and ψ are biholomorphic. With $\Phi := \varphi^{-1}$ we have

$$\begin{aligned} M \cap V &= \{z \in V \mid f(z) = 0\} \\ &= \{z \in V \mid \psi \circ f(z) = 0\} \\ &= \{z \in V \mid \Phi_1(z) = \ldots = \Phi_r(z) = 0\} \\ &= \Phi^{-1}(\{0\} \times \Delta^{n-r}). \end{aligned}$$

This completes the proof of Proposition 2.35. □

Let X be an analytic subset of a region $G \subset \mathbb{C}^n$ and let $x \in X$. Let $I_x \subset \mathcal{O}_{n,x}$ be the ideal of the analytic set germ (X, x), and let $f_1, \ldots, f_r$ be generators of I_x. Then we put

$$\rho_{X,x} := \operatorname{rank} \left(\frac{\partial f_j}{\partial z_k}(x)\right).$$

Remark 2.13. The number $\rho_{X,x}$ is independent of the choice of the generators $f_1, \ldots, f_r$ of I_x (Exercise).

It follows from Proposition 2.35 that x is a regular point of X if and only if $\rho_{X,y}$ is constant for all y in a neighborhood of x.

Corollary 2.2. *Let X be an analytic subset of a region $G \subset \mathbb{C}^n$ and let $x \in X$. Then x is a regular point of X if and only if $\rho_{X,y} = \rho$ for all y in a neighborhood of x.*

Proof. This follows from Proposition 2.35. □

Proposition 2.36. *Let $G \subset \mathbb{C}^n$ be a region, $f : G \to \mathbb{C}$ holomorphic, $f \not\equiv 0$, $X := \{z \in G \mid f(z) = 0\}$. Suppose that the function germs of f do not have multiple factors at the points of X. Then*

$$S(X) = \{z \in X \mid \frac{\partial f}{\partial z_1}(z) = \ldots = \frac{\partial f}{\partial z_n}(z) = 0\}$$

for the set of singular points of the hypersurface X.

Proof. Let $z \in X$. If some partial derivative $(\partial f/\partial z_j)(z) \neq 0$, it follows from the implicit function theorem that z is a regular point of X.

Now suppose all partial derivatives $(\partial f/\partial z_j)(z) = 0$ for $j = 1, \ldots, n$. By Corollary 2.2 we need to show that there is no neighborhood V of z for which $(\partial f/\partial z_j)(y) = 0$ for all $y \in V \cap X$, $j = 1, \ldots, n$. Suppose there were such a V. Let $I(X)$ be the ideal of the set germ (X, z). Then for the function germs of the partial derivatives in z we have

$$\frac{\partial f}{\partial z_j} \in I(X), \quad j = 1, \ldots, n.$$

Since by hypothesis the function germ $\bar{f}$ of f in z has no multiple factors, we have $(\bar{f}) = \text{rad}\,(\bar{f})$, whence $I(X) = I(V(\bar{f})) = (\bar{f})$. Thus there are elements $a_j \in \mathcal{O}_{n,z}$ with

$$\frac{\partial f}{\partial z_j} = a_j \bar{f} \quad \text{for } j = 1, \ldots, n.$$

On differentiating, it follows that

$$\frac{\partial^2 f}{\partial z_j \partial z_k} = \frac{\partial a_j}{\partial z_k} \cdot \bar{f} + a_j \cdot \frac{\partial f}{\partial z_k} \in I(X), \quad j, k = 1, \ldots, n.$$

It follows by induction that all higher partial derivatives of f also lie in $I(X)$, so they vanish in a neighborhood of z in X. Hence $f \equiv 0$, in contradiction to our assumption. □

2.8. Map germs and homomorphisms of analytic algebras

We now want to define maps between analytic set germs.

Definition. Let $G \subset \mathbb{C}^n$, $G' \subset \mathbb{C}^m$ be regions, $X \subset G$, $Y \subset G'$ analytic subsets, $x \in X$. A map $f : X \to Y$ is called *holomorphic at $x \in X$* if there are an open neighborhood U of x in G and a holomorphic map $F : U \to \mathbb{C}^m$ with $F|_{U \cap X} = f|_{U \cap X}$.

Definition. Let $X \subset \mathbb{C}^n$, $Y \subset \mathbb{C}^m$ be analytic subsets, $x \in X$, $y \in Y$, and let $f : X \to Y$, $g : X \to Y$ be holomorphic maps with $f(x) = g(x) = y$. Then f and g define the same *map germ* $\varphi : (X, x) \to (Y, y)$ if there is an open neighborhood $U \subset \mathbb{C}^n$ of x such that

$$f|_{X \cap U} = g|_{X \cap U}.$$

We saw in §2.6 that reduced analytic algebras $\mathcal{O}_{X,x}$ and $\mathcal{O}_{Y,y}$ belong to the analytic set germs (X, x) and (Y, y). A map germ $\varphi : (X, x) \to (Y, y)$ induces a homomorphism

$$\varphi^* : \mathcal{O}_{Y,y} \to \mathcal{O}_{X,x}$$

of the corresponding analytic algebras: φ^* is defined by $\varphi^*(f) = f \circ \varphi$ for all $f \in \mathcal{O}_{Y,y}$.

We now want to show that each homomorphism $\Phi : \mathcal{O}_{Y,y} \to \mathcal{O}_{X,x}$ of reduced analytic algebras arises from a map germ $\varphi : (X, x) \to (Y, y)$. So we next consider arbitrary analytic algebras and algebra homomorphisms between analytic algebras.

Let A, B be analytic algebras and $\varphi : A \to B$ an algebra homomorphism. We note that $\varphi(1) = 1$, from the definition. It is easy to see that φ is *local*, i.e., if $\mathfrak{m}_A$, resp. $\mathfrak{m}_B$, are the maximal ideals of A, resp. B, then

$$\varphi(\mathfrak{m}_A) \subset \mathfrak{m}_B$$

(Proof: Exercise).

Now let $A = \mathcal{O}_{m,0}/I$ and $B = \mathcal{O}_{n,0}/J$. Our aim is to show that $\varphi : A \to B$ can be lifted to an algebra homomorphism $\Phi : \mathcal{O}_{m,0} \to \mathcal{O}_{n,0}$. We need some preparation.

Lemma 2.3. *Let* $\mathfrak{m}$ *be the maximal ideal of* $\mathcal{O}_{n,0} = \mathbb{C}\{z_1, \ldots, z_n\}$. *Then*

(i) $\mathfrak{m}^k = \{f \mid o(f) \geq k\}$,

(ii) $\bigcap_{k=1}^{\infty} \mathfrak{m}^k = \{0\}$.

Proof. (i) $\mathfrak{m}^k \subset \{f \mid o(f) \geq k\}$: By Proposition 2.17 we have $\mathfrak{m} = \{f | o(f) \geq 1\}$. The result follows from $o(fg) = o(f) + o(g)$.

$\mathfrak{m}^k \supset \{f \mid o(f) \geq k\}$: Suppose that $f \in \mathcal{O}_{n,0}$ with $o(f) \geq k$. Then f can be written as

$$f(z) = \sum_{|\nu|=k} P_\nu \cdot z^\nu$$

with $P_\nu \in \mathcal{O}_{n,0}$. Since $z^\nu \in \mathfrak{m}^{|\nu|} = \mathfrak{m}^k$, it follows that $f \in \mathfrak{m}^k$.

(ii) This follows immediately from (i). □

Let $f(z) = \sum_{\nu \geq 0} a_\nu z^\nu$ be a formal power series in $z = (z_1, \ldots, z_n)$. For $r \in \mathbb{R}^n_{>0}$ we define

$$\|f\|_r := \sum_{\nu \geq 0} |a_\nu| r^\nu \in \mathbb{R} \cup \{\infty\}.$$

A sequence $(f_j)_{j \in \mathbb{N}}$ of formal power series is called *summable* if for all $q \in \mathbb{N}$ there are at most finitely many indices $j \in \mathbb{N}$ with $o(f_j) \leq q$.

Lemma 2.4. (i) *$f(z) = \sum_{\nu \geq 0} a_\nu z^\nu$ is convergent if and only if there is an $r \in \mathbb{R}^n_{>0}$ such that $\|f\|_r < \infty$.*

(ii) *For a summable sequence $(f_j)_{j \in \mathbb{N}}$ of formal power series we have*

$$\|\sum_{j=0}^{\infty} f_j\|_r \leq \sum_{j=0}^{\infty} \|f_j\|_r.$$

(iii) *For formal power series f and g we have*

$$\|f \cdot g\|_r \leq \|f\|_r \|g\|_r.$$

(iv) *If $f(z) = \sum_{\nu \geq 0} a_\nu z^\nu$ is convergent, then*

$$\lim_{r \to 0} \|f\|_r = |a_0| = |f(0)|.$$

Proof. (i) follows from Proposition 2.4 and the definition of convergence.

(ii) Let $f_j(z) = \sum_{\nu \geq 0} a_{j\nu} z^\nu$. Since the value of a series with only positive members is independent under a permutation of its terms,

$$\begin{aligned} \|\sum_{j=0}^{\infty} f_j\|_r &= \sum_{\nu \geq 0} |\sum_{j=0}^{\infty} a_{j\nu}| r^\nu \\ &\leq \sum_{\nu \geq 0} \sum_{j=0}^{\infty} |a_{j\nu}| r^\nu = \sum_{j=0}^{\infty} \sum_{\nu \geq 0} |a_{j\nu}| r^\nu = \sum_{j=0}^{\infty} \|f_j\|_r. \end{aligned}$$

(iii) follows from a similar calculation.

(iv) Without loss of generality let $a_0 = f(0) = 0$. Put

$$f_j(z) := \sum_{\nu_j > 0} a_{0 \ldots 0 \nu_j \nu_{j+1} \ldots \nu_n} z_j^{\nu_j - 1} z_{j+1}^{\nu_{j+1}} \cdots z_n^{\nu_n}.$$

Then

$$f(z) = z_1 f_1(z) + \ldots + z_n f_n(z).$$

From (ii) and (iii) we have

$$\|f\|_r \leq \sum_{j=1}^{n} \|f_j z_j\|_r \leq \sum_{j=1}^{n} \|f_j\|_r r_j.$$

If f is convergent, then, by (i), there is an $s \in \mathbb{R}^n_{>0}$ with $\|f\|_s < \infty$, and for $r \le s$ we have

$$\|f\|_r \le \sum_{j=1}^{n} \|f_j\|_r r_j \le \sum_{j=1}^{n} \|f_j\|_s r_j \le \max(r_1, \ldots, r_n) \cdot \sum_{j=1}^{n} \|f_j\|_s.$$

It follows that $\lim_{r\to 0} \|f\|_r = 0$. □

Proposition 2.37. *Let $\mathfrak{m}_n$ be the maximal ideal of $\mathbb{C}\{w_1, \ldots, w_n\}$ and let $g_1, \ldots, g_m \in \mathfrak{m}_n$. Then there is precisely one algebra homomorphism*

$$\Phi : \mathbb{C}\{z_1, \ldots, z_m\} \to \mathbb{C}\{w_1, \ldots, w_n\}$$

with $\Phi(z_j) = g_j$ for $j = 1, \ldots, m$. Moreover

$$\Phi\left(\sum_{\nu\ge 0} a_\nu z^\nu\right) = \sum_{\nu \ge 0} a_\nu g^\nu.$$

Proof. a) Existence: Suppose $f(z) = \sum_{\nu\ge 0} a_\nu z^\nu \in \mathbb{C}\{z_1, \ldots, z_m\}$. We define Φ by $\Phi(f) = \sum_{\nu\ge 0} a_\nu g^\nu$. We have to show that $\Phi(f) \in \mathbb{C}\{w_1, \ldots, w_n\}$.

To do this, let $r \in \mathbb{R}^m_{>0}$ with $\|f\|_r < \infty$. Since $g_j(0) = 0$ for $j = 1, \ldots, m$, there is, by Lemma 2.4(iv), an $s \in \mathbb{R}^n_{>0}$ with $\|g_j\|_s < r_j$ for $j = 1, \ldots, m$. Let

$$f = \sum_{k=0}^{\infty} p_k$$

be the decomposition of f into homogeneous polynomials. Then, by Lemma 2.4(ii),(iii),

$$\begin{aligned}
\|\Phi(f)\|_s &= \|\sum_{k=0}^{\infty} \Phi(p_k)\|_s \\
&\le \sum_{k=0}^{\infty} \|\Phi(p_k)\|_s \quad \text{(Lemma 2.4(ii))} \\
&= \sum_{k=0}^{\infty} \| \sum_{|\nu|=k} a_{\nu_1\ldots\nu_m} \Phi(z_1)^{\nu_1} \cdots \Phi(z_m)^{\nu_m} \|_s \\
&\le \sum_{\nu\ge 0} |a_{\nu_1\ldots\nu_m}| \; \|\Phi(z_1)\|_s^{\nu_1} \cdots \|\Phi(z_m)\|_s^{\nu_m} \\
&\le \sum_{\nu\ge 0} |a_\nu| r^\nu = \|f\|_r < \infty.
\end{aligned}$$

By Lemma 2.4(i), $\Phi(f)$ is convergent, so it is in $\mathbb{C}\{w_1, \ldots, w_n\}$. It is clear that Φ is an algebra homomorphism.

b) Uniqueness: Suppose that Ψ is another algebra homomorphism with $\Psi(z_j) = g_j$, $j = 1, \ldots, m$. Then Φ and Ψ agree on the polynomial ring

$\mathbb{C}[z_1, \ldots, z_m]$. For each $f \in \mathbb{C}\{z_1, \ldots, z_m\}$ then $(\Phi - \Psi)f \in \mathfrak{m}_n^k$ for every $k \geq 1$, so $(\Phi - \Psi)(f) \in \bigcap_{k=1}^{\infty} \mathfrak{m}_n^k = \{0\}$, by Lemma 2.3. $\square$

We now need a few results from commutative algebra.

In what follows R is a local ring with maximal ideal $\mathfrak{m}$.

Proposition 2.38 (Nakayama's lemma). *If M is a finitely generated R-module with $M \subset \mathfrak{m}M$, then $M = 0$.*

Proof. Assume that $M \neq 0$. Since M is finitely generated, there is a natural number r such that M is generated by r elements. Let r the smallest such number and let $e_1, \ldots, e_r$ be a generating system for M. By assumption $r \geq 1$. From $M \subset \mathfrak{m}M$ it follows that e_r can be represented in the form

$$e_r = \sum_{j=1}^{r} a_j e_j$$

with $a_j \in \mathfrak{m}$. From this we obtain

$$(1 - a_r)e_r = \sum_{j=1}^{r-1} a_j e_j.$$

However $1 - a_r \notin \mathfrak{m}$, since $a_r \in \mathfrak{m}$ but $1 \notin \mathfrak{m}$. Thus $1 - a_r$ is a unit in R. It follows that M is already generated by $e_1, \ldots, e_{r-1}$, a contradiction. $\square$

The following corollary of Proposition 2.38 is often referred to as Nakayama's lemma too.

Corollary 2.3. *Let M be a finitely generated R-module and N a submodule of M. If $M = N + \mathfrak{m}M$, then $M = N$.*

Proof. Apply Proposition 2.38 to the module M/N. $\square$

Proposition 2.39 (Krull's intersection theorem). *Let R be a Noetherian local ring with maximal ideal $\mathfrak{m}$ and M a finitely generated module over R. Then for each submodule N of M*

$$\bigcap_{k=1}^{\infty} \left(N + \mathfrak{m}^k M \right) = N.$$

Proof (for the case $N = 0$, all that we need; see e.g. [**GR71**, p. 212] for the case of an arbitrary submodule N).

Let $N = 0$. Put

$$D := \bigcap_{k=1}^{\infty} \mathfrak{m}^k M.$$

To show: $D = 0$.

We consider the set of all R-submodules L of M with $L \cap D = \mathfrak{m}D$. This set is not empty (it contains $\mathfrak{m}D$) and therefore has a maximal element T, since M is Noetherian.

It is enough to show that for each $g \in \mathfrak{m}$ there is an l with $g^l M \subset T$: Since $\mathfrak{m}$ is finitely generated, there is also a k with $\mathfrak{m}^k M \subset T$. Since $D \subset \mathfrak{m}^k M$, it follows that $D \subset T$, and so

$$D = D \cap T = \mathfrak{m}D.$$

Since D is finitely generated, it follows from Nakayama's lemma (Proposition 2.38) that $D = 0$. By the maximality of T it is enough to show that for each $g \in \mathfrak{m}$ there is an l with

$$(g^l M + T) \cap D \subset \mathfrak{m}D.$$

To do this, we consider the increasing sequence (for fixed $g \in \mathfrak{m}$)

$$M_j := \{x \in M \mid g^j x \in T\}, \quad j = 1, 2, \ldots,$$

of submodules of M. Since M is Noetherian, this sequence becomes stationary, i.e., there exists an index l with $M_l = M_{l+1}$. We show $(g^l M + T) \cap D \subset \mathfrak{m}D$. Let $x \in (g^l M + T) \cap D$, $x = g^l y + t$, $y \in M$, $t \in T$. Then

$$g^{l+1} y = gx - gt \in gD + T \subset \mathfrak{m}D + T,$$

so $g^{l+1} y \in T$ because $\mathfrak{m}D \subset T$, thus $y \in M_{l+1} = M_l$ and hence $g^l y \in T$ and $x \in T + T = T$. It follows that $(g^l M + T) \cap D \subset T$, hence $(g^l M + T) \cap D \subset T \cap D = \mathfrak{m}D$, as was to be shown. □

Corollary 2.4. *Let $\mathfrak{m}_A$ be the maximal ideal of an analytic algebra A. Then*

$$\bigcap_{k=1}^{\infty} \mathfrak{m}_A^k = \{0\}.$$

Proposition 2.40 (Lifting theorem). *Let $A = \mathcal{O}_{m,0}/I$, $B = \mathcal{O}_{n,0}/J$ be analytic algebras and $\varphi : A \to B$ an algebra homomorphism. Then there is an algebra homomorphism $\Phi : \mathcal{O}_{m,0} \to \mathcal{O}_{n,0}$ such that the following diagram commutes:*

$$\begin{array}{ccc} \mathcal{O}_{m,0} & \overset{\Phi}{\dashrightarrow} & \mathcal{O}_{n,0} \\ {\scriptstyle \pi_A}\downarrow & & \downarrow{\scriptstyle \pi_B} \\ A & \overset{\varphi}{\longrightarrow} & B \end{array}$$

where $\pi_A : \mathcal{O}_{m,0} \to A$ and $\pi_B : \mathcal{O}_{n,0} \to B$ are the canonical projections.

Proof. Let $\mathcal{O}_{m,0} = \mathbb{C}\{z_1, \ldots, z_m\}$. We denote the maximal ideals of A, resp. B, by $\mathfrak{m}_A$, resp. $\mathfrak{m}_B$, and the maximal ideals of $\mathcal{O}_{m,0}$, resp. $\mathcal{O}_{n,0}$, by $\mathfrak{m}_m$, resp. $\mathfrak{m}_n$. Since π_A and φ are local,

$$(\varphi \circ \pi_A)(z_j) \in \mathfrak{m}_B \text{ for } j = 1, \ldots, m.$$

Since $\pi_B(\mathfrak{m}_n) = \mathfrak{m}_B$, there is a $g_j \in \mathfrak{m}_n$ with $\pi_B(g_j) = (\varphi \circ \pi_A)(z_j)$, $j = 1, \ldots, m$. By Proposition 2.37 there is exactly one algebra homomorphism $\Phi : \mathcal{O}_{m,0} \to \mathcal{O}_{n,0}$ with $\Phi(z_j) = g_j$, $j = 1, \ldots, m$. From $\varphi \circ \pi_A(z_j) = \pi_B \circ \Phi(z_j)$ we can conclude, with the aid of Corollary 2.4, as in part b) of the proof of Proposition 2.37, that $\varphi \circ \pi_A = \pi_B \circ \Phi$. □

Proposition 2.41. (i) *To each reduced analytic algebra A there corresponds an analytic set germ $(X, 0)$ with $\mathcal{O}_{X,0} = A$.*

(ii) *If A, B are reduced analytic algebras, $(Y, 0)$, $(X, 0)$ the corresponding analytic set germs, and $\gamma : A \to B$ is an algebra homomorphism, there is exactly one map germ $\varphi : (X, 0) \to (Y, 0)$ with $\varphi^* = \gamma$.*

Proof. (i) If $A = \mathcal{O}_{n,0}/I$, we put $(X, 0) := V(I)$.

(ii) Existence: Let $\gamma : A \to B$ be given, $A = \mathcal{O}_{m,0}/J$, $B = \mathcal{O}_{n,0}/I$, $(Y, 0) = V(J)$, $(X, 0) = V(I)$. By Proposition 2.40, γ can be lifted to an algebra homomorphism $\Gamma : \mathcal{O}_{m,0} \to \mathcal{O}_{n,0}$. If $\mathcal{O}_{m,0} = \mathbb{C}\{z_1, \ldots, z_m\}$, then, by Proposition 2.37, $g_j := \Gamma(z_j)$ lies in the maximal ideal $\mathfrak{m}_n$ of $\mathcal{O}_{n,0}$ for $j = 1, \ldots, m$. There exists an open neighborhood U of 0 in $\mathbb{C}^n$ such that for each $j = 1, \ldots, m$ the function germ g_j can be represented by a holomorphic function $\tilde{g}_j : U \to \mathbb{C}$. Then

$$\tilde{g} = (\tilde{g}_1, \ldots, \tilde{g}_m) : U \to \mathbb{C}^m$$

is a holomorphic map with $\tilde{g}(0) = 0$. If $g : (\mathbb{C}^n, 0) \to (\mathbb{C}^m, 0)$ is the map germ represented by $\tilde{g} : U \to \mathbb{C}^m$, then $g^* = \Gamma$. From the commutative diagram

$$\begin{array}{ccc} \mathcal{O}_{m,0} & \xrightarrow{\Gamma} & \mathcal{O}_{n,0} \\ {\scriptstyle \pi_A}\downarrow & & \downarrow{\scriptstyle \pi_B} \\ A = \mathcal{O}_{m,0}/J & \xrightarrow{\gamma} & B = \mathcal{O}_{n,0}/I \end{array}$$

it follows that $\Gamma(J) \subset I$. Thus $\tilde{g}$ induces a map germ

$$\varphi : (X, 0) \to (Y, 0).$$

Hence $\gamma = \varphi^*$.

Uniqueness: Let $\varphi, \psi : (X, 0) \to (Y, 0)$ be two map germs with $\varphi^* = \psi^* = \gamma$, let U be an open neighborhood of 0 in $\mathbb{C}^n$, and $\tilde{g} : U \to \mathbb{C}^m$, resp. $\tilde{h} : U \to \mathbb{C}^m$, holomorphic representatives of φ, resp. ψ. The holomorphic maps $\tilde{g}$ and $\tilde{h}$ likewise represent map germs $g : (\mathbb{C}^n, 0) \to (\mathbb{C}^m, 0)$, resp. $h : (\mathbb{C}^n, 0) \to (\mathbb{C}^m, 0)$. It follows from $\gamma \circ \pi_A = \pi_B \circ g^*$, $\gamma \circ \pi_A = \pi_B \circ h^*$ that

$$\pi_B \circ g^* = \pi_B \circ h^*.$$

If $(z_1, \ldots, z_m)$ are the coordinates of $\mathbb{C}^m$, then

$$\pi_B \circ g^*(z_j) = \pi_B \circ h^*(z_j),$$

i.e.,

$$g^*(z_j) - h^*(z_j) \in I.$$

From this it follows that

$$z_j \circ \tilde{g} - z_j \circ \tilde{h} = \tilde{g}_j - \tilde{h}_j$$

vanishes on a representative $\tilde{X}$ of $V(I)$. Therefore $\tilde{g}$ and $\tilde{h}$ induce the same map germ $\varphi = \psi : (X, 0) \to (Y, 0)$, as was to be shown. □

Proposition 2.41 can also be phrased as: The correspondence $(X, x) \mapsto \mathcal{O}_{X,x}$ defines an antiequivalence between the category of analytic set germs and the category of reduced analytic algebras.

Finally we introduce the concept of an isomorphism of analytic set germs.

Definition. A map germ $\varphi : (X, x) \to (Y, y)$ is called an *isomorphism* if there is a map germ $\psi : (Y, y) \to (X, x)$ with $\psi \circ \varphi = \mathrm{id}$ and $\varphi \circ \psi = \mathrm{id}$.

By Proposition 2.41, $\varphi : (X, x) \to (Y, y)$ is an isomorphism if and only if $\varphi^* : \mathcal{O}_{Y,y} \to \mathcal{O}_{X,x}$ is an isomorphism of analytic algebras.

2.9. The generalized Weierstrass preparation theorem

Our aim in this section is to generalize the Weierstrass preparation theorem to a result on finitely generated modules. To be able to formulate this assertion, we introduce yet another concept.

Let A be an analytic algebra and M an A-module.

Definition. The A-module M is called *finite over* A if M is a finitely generated A-module.

Now let B be a another analytic algebra and $\varphi : A \to B$ an algebra homomorphism. If M is a B-module, then M becomes an A-module by the action

$$\begin{array}{rcl} A \times M & \to & M \\ (a, x) & \mapsto & \varphi(a)x. \end{array}$$

In particular, B is an A-module in canonical manner.

Definition. The algebra homomorphism φ is called *finite* if B is finite over A.

Remark 2.14. (i) If $\varphi : A \to B$ and $\psi : B \to C$ are finite, then $\psi \circ \varphi : A \to C$ is also finite.

(ii) If $\varphi : A \to B$ is surjective, then φ is finite. For 1 is a generator of B as an A-module: Let $b \in B$. Since φ is surjective, there is an $a \in A$ with $b = \varphi(a) = \varphi(a) \cdot 1$.

The Weierstrass preparation theorem in its general form runs as follows:

Theorem 2.4 (Weierstrass preparation theorem for modules). *Let A, B be analytic algebras, $\mathfrak{m}_A$, $\mathfrak{m}_B$ their maximal ideals, $\varphi : A \to B$ an algebra homomorphism and M a finitely generated B-module. Then M is finite over A if and only if $M/M\varphi(\mathfrak{m}_A)$ is finite over $A/\mathfrak{m}_A \cong \mathbb{C}$.*

This proposition can be made more precise as follows:

Corollary 2.5. *Let A, B, φ, M be as in Theorem 2.4 and let $e_1, \dots, e_p$ be elements in M. If the images of $e_1, \dots, e_p$ in $M/M\varphi(\mathfrak{m}_A)$ generate this module over $\mathbb{C}$, then $e_1, \dots, e_p$ generate M over A.*

Proof. To prove Corollary 2.5, we consider the A-submodule N of M generated by the elements $e_1, \dots, e_p$. Since by hypothesis $M/M\varphi(\mathfrak{m}_A)$ is generated by the images of $e_1, \dots, e_p$ as a $\mathbb{C}$-vector space, we have

$$M - N + M\varphi(\mathfrak{m}_A)$$

as an A-module. By Theorem 2.4, M is finite over A. From Nakayama's lemma (Corollary 2.3) it follows that $M = N$, whence the conclusion. □

Proof of Theorem 2.4 (after an idea of J. Mather; cf. [**Mal68**]).

"$\Rightarrow$": Let M be finite over A. Then $M/M\varphi(\mathfrak{m}_A)$ is also finite over $A/\mathfrak{m}_A$.

"$\Leftarrow$": Now let $M/M\varphi(\mathfrak{m}_A)$ be finite over $\mathbb{C}$. We show that M is finite over A.

We divide the proof into three stages:

1) We first prove the result for the special case

$$A = \mathbb{C}\{z_1, \dots, z_{m-1}\}, \quad B = \mathbb{C}\{z_1, \dots, z_m\}, \quad \varphi = \pi^*,$$

where

$$\begin{aligned} \pi \ : \ & \mathbb{C}^m \to \mathbb{C}^{m-1} \\ & (z_1, \dots, z_m) \mapsto (z_1, \dots, z_{m-1}) \end{aligned}$$

is the natural projection (φ is then the natural inclusion $A \hookrightarrow B$). We first establish the

Claim. There are finitely many elements $e_1, \dots, e_p$ in M such that each $x \in M$ can be written as follows:

$$x = \sum_{j=1}^{p} b_j e_j \text{ with } b_j \in \varphi(A) + B\varphi(\mathfrak{m}_A).$$

Proof. Since M is finite over B, there are elements $\varepsilon_1, \dots, \varepsilon_q \in M$ that generate M over B. Since $M/M\varphi(\mathfrak{m}_A)$ is, by hypothesis, finite over $\mathbb{C}$, there are $\eta_1, \dots, \eta_r \in M$ such that the residue classes $\bar{\eta}_1, \dots, \bar{\eta}_r$ modulo

$M\varphi(\mathfrak{m}_A)$ generate the $\mathbb{C}$-vector space $M/M\varphi(\mathfrak{m}_A)$. Thus for every $x \in M$ there are complex numbers $\gamma_j \in \mathbb{C}$ with

$$x - \sum_{j=1}^{r} \gamma_j \eta_j \in M\varphi(\mathfrak{m}_A)$$

and elements $b_k \in B\varphi(\mathfrak{m}_A)$ with

$$x - \sum_{j=1}^{r} \gamma_j \eta_j = \sum_{k=1}^{q} b_k \varepsilon_k.$$

Now put $p = q + r$, $(e_1, \dots, e_p) = (\eta_1, \dots, \eta_r, \varepsilon_1, \dots, \varepsilon_q)$. □

We next show that M is finite over A. Let $e_1, \dots, e_p \in M$ be as in the Claim. Then for $1 \leq j \leq p$

$$z_m e_j = \sum_{k=1}^{p} \nu_{jk} e_k, \quad \nu_{jk} \in \varphi(A) + B\varphi(\mathfrak{m}_A).$$

We put

$$\begin{aligned} N: &= (z_m \delta_{jk} - \nu_{jk})_{k=1,\dots,p}^{j=1,\dots,p}, \\ \Delta: &= \det N. \end{aligned}$$

The matrix N_j arises from N by replacing all elements in the j-th column by zeros ($j = 1, \dots, p$). Then, by Cramer's rule,

$$\Delta e_j = \det N_j = 0, \quad j = 1, \dots, p,$$

and $\Delta M = 0$. It follows that M is a $B/(\Delta)$-module and is finitely generated (e.g. by $e_1, \dots, e_p$). Now $\nu_{jk} \in \varphi(A) + B\varphi(\mathfrak{m}_A)$ signifies that

$$\nu_{jk} = f_{jk} + b_{jk} g_{jk}, \ f_{jk}, g_{jk} \in \mathbb{C}\{z_1, \dots, z_{m-1}\}, \ b_{jk} \in B,$$

with $g_{jk}(0) = 0$, whence

$$\nu_{jk}(0, \dots, 0, z_m) = f_{jk}(0) = \text{ const.}$$

It follows that $\Delta(0, \dots, 0, z_m)$ is a monic polynomial in z_m of degree p. Hence Δ is regular in z_m of degree $l \leq p$. Now let $f \in B$. By the Weierstrass division theorem (Theorem 2.2) there are then a $q \in B$ and elements $a_1, \dots, a_l \in A$ with

$$f = q \cdot \Delta + \sum_{j=1}^{l} a_j z_m^{l-j}.$$

This means that $B/(\Delta)$ is generated by the residue classes of $1, z_m, \dots, z_m^{l-1}$ modulo (Δ). Thus $B/(\Delta)$ is finite over A. By Remark 2.14(i), M is then finite over A.

2) Now let $A = \mathbb{C}\{z_1, \ldots, z_m\}$, $B = \mathbb{C}\{w_1, \ldots, w_n\}$, and let $\varphi : A \to B$ be any algebra homomorphism. We can then factorize φ in the following way:

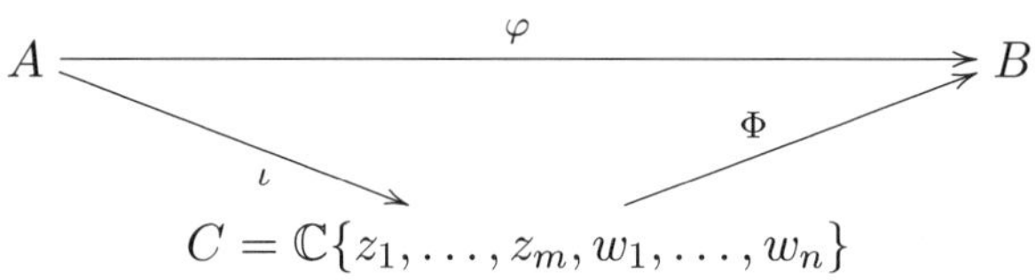

where ι is the natural injection and Φ is defined by

$$\Phi(z_i) = \varphi(z_i), \; \Phi(w_j) = w_j.$$

By hypothesis M is finite over B. Since Φ is surjective, Φ is finite, by Remark 2.14(ii). From Remark 2.14(i) it follows that M is then also finite over C. It is therefore enough to demonstrate the theorem for an inclusion $\iota : A \hookrightarrow C$. But this follows by induction from stage 1).

3) Now we consider the general case: $A = \mathcal{O}_{m,0}/I$, $B = \mathcal{O}_{n,0}/J$, $\varphi : A \to B$ an algebra homomorphism. By Proposition 2.40 we can lift $\varphi : A \to B$ to an algebra homomorphism $\Phi : \mathcal{O}_{m,0} \to \mathcal{O}_{n,0}$ so that the following diagram commutes:

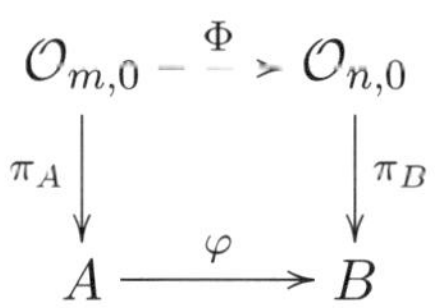

By hypothesis M is finite over B and $M/M\varphi(\mathfrak{m}_A)$ is finite over $\mathbb{C}$. Let $\mathfrak{m}_m$ be the maximal ideal of $\mathcal{O}_{m,0}$. Now $\mathfrak{m}_A = \pi_A(\mathfrak{m}_m)$ so $M/M\varphi(\mathfrak{m}_A)$ is isomorphic to

$$M/M(\varphi \circ \pi_A)(\mathfrak{m}_m) = M/M(\pi_B \circ \Phi)(\mathfrak{m}_m).$$

Since π_B surjective, M is also finite over $\mathcal{O}_{n,0}$ and $M/M\Phi(\mathfrak{m}_m)$ is finite over $\mathbb{C}$. Since π_A is surjective, it suffices, by Remark 2.14(ii), to show that M is finite over $\mathcal{O}_{m,0}$, and this follows from 2). □

By exploiting the correspondence between homomorphisms of reduced analytic algebras and map germs introduced in §2.8, we now give a geometric interpretation of the concept of finiteness.

Definition. A map germ $\varphi : (X, x) \to (Y, y)$ is called *finite* if $\varphi^* : \mathcal{O}_{Y,y} \to \mathcal{O}_{X,x}$ is finite.

For the following proposition we recall yet another concept from algebra. Let S be a ring and $R \subset S$ a subring. An element $s \in S$ is called *integral over* R if it satisfies an equation

$$s^p + a_1 s^{p-1} + \ldots + a_p = 0 \text{ with } a_j \in R.$$

We require the following result from algebra (see e.g. [**Lan02**, VII, §1]): If S is a finitely generated R-module, then every $s \in S$ is integral over R.

Proposition 2.42. *Let $G \subset \mathbb{C}^n$, $G' \subset \mathbb{C}^m$ be regions, let $X \subset G$, $Y \subset G'$ be analytic subsets, $f : X \to Y$ a holomorphic map, and $x \in X$. Then $f^* : \mathcal{O}_{Y,f(x)} \to \mathcal{O}_{X,x}$ is finite if and only if x is an isolated point of $f^{-1}f(x)$.*

Proof. Without loss of generality take $x = 0$, $f(x) = 0$.

"$\Rightarrow$": Let $f^* : \mathcal{O}_{Y,0} \to \mathcal{O}_{X,0}$ be finite. This means that $\mathcal{O}_{X,0}$ is a finitely generated module over $f^*(\mathcal{O}_{Y,0})$. From the remark before Proposition 2.42 it follows that each element of $\mathcal{O}_{X,0}$ is integral over $f^*(\mathcal{O}_{Y,0})$. If $g \in \mathcal{O}_{X,0}$, there exist $a_1, \dots, a_r \in \mathcal{O}_{Y,0}$ with

$$g^r + f^*(a_1)g^{r-1} + \ldots + f^*(a_r) = 0.$$

If $(z_1, \dots, z_n)$ are the coordinates of $\mathbb{C}^n$, then, in particular, for each k with $1 \le k \le n$ there are elements $a_{k1}, \dots, a_{kr} \in \mathcal{O}_{Y,0}$ with

$$z_k^r + f^*(a_{k1})z_k^{r-1} + \ldots + f^*(a_{kr}) = 0$$

for $k = 1, \dots, n$ and for a suitable r. Now let U be an open neighborhood of 0 in G and $F : U \to \mathbb{C}^m$ a holomorphic map with $F|_{U \cap X} = f|_{U \cap X}$. Furthermore, let V be a suitable open neighborhood of 0 in G' with $F(U) \subset V$, and let $\tilde{a}_{kj} : V \to \mathbb{C}$ be representatives of a_{kj}, $k = 1, \dots, n$, $j = 1, \dots, r$. By shrinking U, if necessary, for all $z \in U$ we have

$$z_k^r + \tilde{a}_{k1}(F(z))z_k^{r-1} + \ldots + \tilde{a}_{kr}(F(z)) = 0, \; k = 1, \dots, n.$$

Thus if $z \in X \cap U$ and $F(z) = f(z) = 0$, then z_k is a root of a polynomial for $k = 1, \dots, n$. Therefore there can be at most finitely many $z \in U$ with $f(z) = 0$. Thus 0 is an isolated point of $f^{-1}f(0)$.

"$\Leftarrow$": Let 0 be an isolated point of $f^{-1}f(0)$. Let I be the ideal of $\mathcal{O}_{n,0}$ generated by $I(X, 0)$ and $\bar{f}_1, \dots, \bar{f}_m$, where $\bar{f}_j$ is the germ of the component function f_j, $j = 1, \dots, m$. Then $V(I) = \{0\}$. Since $z_k \in I(\{0\})$, there is, by the Rückert Nullstellensatz, an r_k with $z_k^{r_k} \in I$, for $k = 1, \dots, n$. Thus for each k there are an r_k and elements $a_{k1}, \dots, a_{km} \in \mathcal{O}_{n,0}$ such that

$$z_k^{r_k} \equiv \sum_{j=1}^{m} a_{kj}\bar{f}_j \pmod{I(X,0)}.$$

It follows that there is a natural number $q > 0$ such that

$$\mathfrak{m}_{X,0}^q \subset \mathcal{O}_{X,0} f^*(\mathfrak{m}_{Y,0}),$$

where $\mathfrak{m}_{X,0}$, resp. $\mathfrak{m}_{Y,0}$, is the maximal ideal of $\mathcal{O}_{X,0}$, resp. $\mathcal{O}_{Y,0}$. This inclusion induces a surjective homomorphism

$$\mathcal{O}_{X,0}/\mathfrak{m}_{X,0}^q \to \mathcal{O}_{X,0}/\mathcal{O}_{X,0} f^*(\mathfrak{m}_{Y,0}).$$

But $\mathcal{O}_{X,0}/\mathfrak{m}_{X,0}^q$ is a finite-dimensional $\mathbb{C}$-vector space: For if the elements $x_1, \ldots, x_n$ are generators of $\mathfrak{m}_{X,0}$, then for each $g \in \mathcal{O}_{X,0}$ there is a polynomial $p \in \mathbb{C}[x_1, \ldots, x_n]$ of degree $\leq q-1$ with

$$g - p \in \mathfrak{m}_{X,0}^q.$$

Thus $\mathcal{O}_{X,0}/\mathfrak{m}_{X,0}^q$ is generated by the residue classes of the monomials in $x_1, \ldots, x_n$ of degree $\leq q-1$. Hence $\mathcal{O}_{X,0}/\mathcal{O}_{X,0}f^*(\mathfrak{m}_{Y,0})$ is also a finite-dimensional $\mathbb{C}$-vector space. It follows from Theorem 2.4 that f^* is finite. □

We next record a proposition that we shall need in Chapter 3. Let $z = (z_1, \ldots, z_n)$ and $\mathbb{C}\{z\} := \mathbb{C}\{z_1, \ldots, z_n\}$.

Proposition 2.43. *Let A be an analytic algebra with maximal ideal $\mathfrak{m}_A$ and suppose that $\mathbb{C}\{z\} \subset A$. Let $g \in \mathfrak{m}_A$ be integral over $\mathbb{C}\{z\}$ and let*

$$p(z,t) = t^k + a_1(z)t^{k-1} + \ldots + a_k(z) \in \mathbb{C}\{z\}[t]$$

be a polynomial of smallest degree with $p(z,g) = 0$. Then $a_j(0) = 0$ for $j = 1, \ldots, k$.

Proof. Since $g \in \mathfrak{m}_A$, we have

$$a_k(z) = -g(g^{k-1} + a_1(z)g^{k-2} + \ldots + a_{k-1}(z)) \in \mathfrak{m}_A \cap \mathbb{C}\{z\} \subset \mathfrak{m}_n,$$

where $\mathfrak{m}_n$ is the maximal ideal of $\mathbb{C}\{z\}$. Suppose there were a j, $1 \leq j < k$, with $a_j(0) \neq 0$. Let l the largest index with $a_l(0) \neq 0$. Then

$$p(0,t) = t^{k-l}(t^l + a_1(0)t^{l-1} + \ldots + a_l(0)).$$

By Hensel's lemma (Proposition 2.21) there are two monic polynomials $p_1(z,t), p_2(z,t) \in \mathbb{C}\{z\}[t]$ with

$$p_1(0,t) = t^{k-l}, \quad p_2(0,t) = t^l + a_1(0)t^{l-1} + \ldots + a_l(0).$$

If we had $p_2(z,g) \in \mathfrak{m}_A$, then, since $g \in \mathfrak{m}_A$, the constant term $p_2(z,0)$ of $p_2(z,t)$ would also have to lie in $\mathfrak{m}_A$; and then $p_2(0,0) = 0$. But this would contradict $a_l(0) \neq 0$. Hence $p_2(z,g)$ is a unit in A and from $p(z,g) = 0$ it follows that $p_1(z,g) = 0$. Since $p_1(z,t)$ is of degree $k - l < k$ in t, this contradicts the assumption that $p(z,t)$ is of smallest degree. Hence $a_j(0) = 0$ for $j = 1, \ldots, k$. □

2.10. The dimension of an analytic set germ

We now want to introduce a concept of dimension for analytic set germs. There are various definitions of dimension. We restrict ourselves here essentially to the definition of dimension after Weierstrass. First we show:

Proposition 2.44 (Noether normalization theorem). *For every analytic algebra A there exist a natural number d and a finite injective algebra homomorphism $\varphi : \mathcal{O}_{d,0} \hookrightarrow A$.*

Proof. By definition, A is of the form $\mathcal{O}_{n,0}/I$. The quotient map $\pi : \mathcal{O}_{n,0} \to A = \mathcal{O}_{n,0}/I$ is surjective, hence finite. Thus there is a finite algebra homomorphism $\sigma_m : \mathcal{O}_{m,0} \to A$. Let d be the smallest m that occurs, and let $\varphi : \mathcal{O}_{d,0} \to A$ be a corresponding finite algebra homomorphism.

We show that φ is injective, i.e., $\ker \varphi = \{0\}$. Suppose that $f \in \ker \varphi$, $f \neq 0$. By Lemma 2.1 there are then coordinates $z_1, \dots, z_d$ such that f is regular of order k in z_d. Now let $g \in \mathcal{O}_{d,0}$. By the Weierstrass division theorem (Theorem 2.2) there are then a $q \in \mathcal{O}_{d,0}$ and elements $a_1, \dots, a_k \in \mathcal{O}_{d-1,0}$ with

$$g = q \cdot f + \sum_{j=1}^{k} a_j z_d^{k-j}.$$

This means that $\mathcal{O}_{d,0}/(f)$ is generated by the residue classes of $1, z_d, \dots, z_d^{k-1}$ modulo (f). Hence $\mathcal{O}_{d,0}/(f)$ is finite over $\mathcal{O}_{d-1,0}$. Now $f \in \ker \varphi$ so A is finite over $\mathcal{O}_{d,0}/(f)$, so also finite over $\mathcal{O}_{d-1,0}$. But this contradicts the minimality of d. □

Remark 2.15. Let $\varphi : (X, x) \to (\mathbb{C}^d, 0)$ be a map germ. One can show that $\varphi^* : \mathcal{O}_{d,0} \to \mathcal{O}_{X,x}$ is finite and injective if and only if $\varphi : (X, x) \to (\mathbb{C}^d, 0)$ is the germ of a holomorphic branched covering $f : X \to \mathbb{C}^d$, hence it is a holomorphic, open and discrete map $f : X \to \mathbb{C}^d$. The existence of such a branched covering of $\mathbb{C}^d$ for an analytic set germ (X, x) is the geometric interpretation of the Noether normalization theorem.

Definition. Let A be an analytic algebra. The *(Weierstrass) dimension* of A, symbolically, $\dim A$, is the smallest number $d \geq 0$ for which there is a finite algebra homomorphism $\varphi : \mathcal{O}_{d,0} \to A$.

Let $G \subset \mathbb{C}^n$ be a region, $X \subset G$ an analytic subset, $x \in X$. The *dimension of* X *at* x, symbolically, $\dim_x X$, is the dimension of $\mathcal{O}_{X,x}$, where $\mathcal{O}_{X,x} = \mathcal{O}_{n,x}/I(X, x)$ is the structure algebra of the analytic set germ (X, x).

Proposition 2.45 (Weierstrass dimension = Chevalley dimension)**.** *The dimension of A is also the smallest number $d \geq 0$ for which there are elements $f_1, \dots, f_d \in \mathfrak{m}_A$ such that*

$$\dim_{\mathbb{C}} A/A(f_1, \dots, f_d) < \infty.$$

(This number d is called the Chevalley dimension *of A and such a system $f_1, \dots, f_d$ is called a* parameter system *of A.)*

Remark 2.16. By Proposition 2.42, $\dim_{\mathbb{C}} A/(f_1, \dots, f_d)A < \infty$ means that 0 lies isolated in the intersection of X with the hypersurfaces $\{f_1 = 0\}, \dots, \{f_d = 0\}$. The Chevalley concept of dimension captures the intuitive preconception that in a d-dimensional space the solution set of a single equation is a $(d-1)$-dimensional subspace and that one should need at least d equations to describe finite sets, i.e., 0-dimensional subspaces.

Proof of Proposition 2.45. If $\varphi : \mathcal{O}_{m,0} \to A$ is an algebra homomorphism and $\mathcal{O}_{m,0} = \mathbb{C}\{z_1, \dots, z_m\}$, then

$$A\varphi(\mathfrak{m}_m) = A(\varphi(z_1), \dots, \varphi(z_m)).$$

By Corollary 2.5, φ is finite if and only if

$$\dim_{\mathbb{C}} A/A\varphi(\mathfrak{m}_m) = \dim_{\mathbb{C}} A/A(\varphi(z_1), \dots, \varphi(z_m)) < \infty.$$

Now let d be the number from Proposition 2.45. If $\varphi : \mathcal{O}_{m,0} = \mathbb{C}\{z_1, \dots, z_m\} \to A$ is finite, we put $f_j = \varphi(z_j)$, $j = 1, \dots, m$. This shows $d \leq \dim A$.

Conversely, if $f_1, \dots, f_d \in \mathfrak{m}_A$ with $\dim_{\mathbb{C}} A/A(f_1, \dots, f_d) < \infty$, there is, by Proposition 2.37, exactly one $\varphi : \mathbb{C}\{z_1, \dots, z_d\} \to A$ with $\varphi(z_j) = f_j$ for $j = 1, \dots, d$. It follows that $\dim A \leq d$. □

Remark 2.17. Let A, B be analytic algebras. From the transitivity of finiteness it follows that $\dim A \geq \dim B$ if there is a finite homomorphism $\varphi : A \to B$.

Lemma 2.5. *Let A be an analytic algebra with maximal ideal $\mathfrak{m}_A$. Then for $f \in \mathfrak{m}_A$*

$$\dim(A/(f)) \geq \dim A - 1.$$

Proof. Let $\bar{A} := A/(f)$. Then for $f_1, \dots, f_d \in \mathfrak{m}_A$

$$A/A(f, f_1, \dots, f_d) \cong \bar{A}/\bar{A}(\bar{f}_1, \dots, \bar{f}_d).$$

Thus the claim follows from Proposition 2.45. □

Lemma 2.6. *Let $\mathfrak{m}_n$ be the maximal ideal of $\mathcal{O}_{n,0}$. Then, for $f \in \mathfrak{m}_n$, $f \neq 0$,*

$$\dim(\mathcal{O}_{n,0}/(f)) \leq n - 1.$$

Proof. Let $(z_1, \dots, z_n)$ be coordinates of $\mathbb{C}^n$ such that f is regular in z_n. By the Weierstrass division theorem the homomorphism ψ making the following diagram commutative is finite:

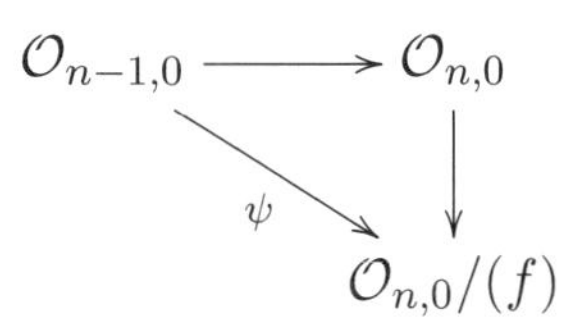

(cf. the proof of Proposition 2.44). It follows from the remark above that

$$\dim(\mathcal{O}_{n,0}/(f)) \leq \dim \mathcal{O}_{n-1,0} \leq n - 1.$$

This proves Lemma 2.6. □

We now want to study when equality holds in Lemma 2.5. We need the following definition.

Definition. Let A be an analytic algebra with maximal ideal $\mathfrak{m}_A$ and nilradical $\mathfrak{n}_A$. An element $f \in \mathfrak{m}_A$ is called *active* if for all $g \in A$ we have $f \cdot g \in \mathfrak{n}_A \Rightarrow g \in \mathfrak{n}_A$.

Remark 2.18. If $f \in \mathfrak{m}_A$ is a nonzerodivisor, then f is active: From $f \cdot g \in \mathfrak{n}_A$ it follows that there is a k with $(f \cdot g)^k = 0$. Hence $g^k = 0$ and so $g \in \mathfrak{n}_A$.

Proposition 2.46 (Active lemma). *If $f \in \mathfrak{m}_A$ is active, then*

$$\dim A/(f) = \dim A - 1.$$

Proof. By Lemma 2.5 it is enough to show that $\dim A/(f) \leq \dim A - 1$.

Let $n := \dim A$ and let $\varphi : \mathcal{O}_{n,0} \to A$ be a finite homomorphism. By the proof of Proposition 2.44, φ is injective. Thus we can assume that $\mathcal{O}_{n,0} \subset A$. It is now enough to show that there is a $0 \neq h \in A \cdot f \cap \mathcal{O}_{n,0}$. For consider the induced homomorphism ψ in the following diagram:

$$\begin{array}{ccc} \mathcal{O}_{n,0} & \xrightarrow{\varphi} & A \\ \downarrow & & \downarrow \\ \mathcal{O}_{n,0}/(h) & \xrightarrow{\psi} & A/(f) \end{array}$$

Since φ is finite and both the vertical homomorphisms are surjective, ψ is also finite. It follows from Remark 2.17 and Lemma 2.6 that

$$\dim A/(f) \leq \dim \mathcal{O}_{n,0}/(h) \leq n - 1.$$

It remains to show that there exists $0 \neq h \in A \cdot f \cap \mathcal{O}_{n,0}$. Now A is finite over $\mathcal{O}_{n,0}$ so f is integral over $\mathcal{O}_{n,0}$, i.e., it satisfies an equation

$$f^p + a_1 f^{p-1} + \ldots + a_p = 0 \text{ with } a_j \in \mathcal{O}_{n,0}.$$

If $a_p \neq 0$, we may put $h = a_p$. Otherwise let s be maximal with $a_s \neq 0$. Then

$$f^{p-s}(f^s + a_1 f^{s-1} + \ldots + a_s) = 0.$$

Since f is active, $f^s + a_1 f^{s-1} + \ldots + a_s$ is nilpotent. Hence there is a $q \geq 1$ such that

$$(f^s + a_1 f^{s-1} + \ldots + a_s)^q = f^{qs} + \ldots + a_s^q = 0.$$

It follows that $0 \neq a_s^q \in A \cdot f \cap \mathcal{O}_{n,0}$. So in this case we can put $h = a_s^q$. □

Corollary 2.6. $\dim \mathcal{O}_n = n$.

Proof (by induction on n).

$n = 0$: The identity $\mathrm{id} : \mathcal{O}_{0,0} \to \mathcal{O}_{0,0}$ is a finite homomorphism, so $\dim \mathcal{O}_{0,0} = 0$.

$n-1 \to n$: Let $n \geq 1$, $\mathcal{O}_{n,0} = \mathbb{C}\{z_1, \ldots, z_n\}$. Now $z_n \in \mathfrak{m}_n$ is a nonzerodivisor in $\mathcal{O}_{n,0}$, so z_n is active. Since $\mathcal{O}_{n,0}/(z_n) \cong \mathcal{O}_{n-1,0}$, it follows from Proposition 2.46 and the induction hypothesis that

$$n-1 = \dim \mathcal{O}_{n-1,0} = \dim\ \mathcal{O}_{n,0} - 1,$$

whence $\dim\ \mathcal{O}_{n,0} = n$. □

In §2.7 we considered regular and singular points of analytic sets. Let $G \subset \mathbb{C}^n$ be a region, $X \subset G$ an analytic subset, and $x \in X$ a regular point. Then the analytic set germ (X, x) is also called *regular*. It follows from the definition that (X, x) is isomorphic to $(\mathbb{C}^s, 0)$ for some $s \leq n$. From §2.8 this is equivalent to $\mathcal{O}_{X,x}$ being isomorphic to $\mathcal{O}_{s,0}$ for some $s \leq n$.

Corollary 2.7. *Let $(X, x) \subset (\mathbb{C}^n, x)$ be a regular analytic set germ isomorphic to $(\mathbb{C}^s, 0)$. Then*

$$\dim_x X = s.$$

We next show that one can characterize hypersurfaces by their dimension.

Proposition 2.47. *Let $(X, x) \subset (\mathbb{C}^n, x)$ be an irreducible analytic set germ. Then (X, x) is a hypersurface if and only if $\dim_x X = n - 1$.*

Proof. "⇒": If (X, x) is a hypersurface, the ideal $I(X)$ of (X, x) is a principal ideal. Let $I(X) = (f)$, $f \in \mathcal{O}_{n,x}$. Since $\mathcal{O}_{n,x}$ is an integral domain, f is not a zero divisor in $\mathcal{O}_{n,x}$. Then f is active and by the active lemma (Proposition 2.46)

$$\dim_x X = \dim \mathcal{O}_{n,x}/(f) = \dim \mathcal{O}_{n,x} - 1 = n - 1.$$

"⇐": Suppose conversely that $\dim_x X = n - 1$. Let $f \in I(X)$, $f \neq 0$, and let $f = f_1^{k_1} \cdots f_r^{k_r}$ be the decomposition of f into irreducible factors. Since (X, x) is irreducible, $I(X)$ is a prime ideal. Hence $f_j \in I(X)$ for some j.

We now show that $I(X) = (f_j)$, for otherwise there would be a $g \in I(X) \setminus (f_j)$. Since f_j is irreducible, the analytic algebra

$$A = \mathcal{O}_{n,x}/(f_j)$$

is an integral domain. So g would be a nonzerodivisor in A and hence active. By the active lemma we would then have

$$\dim A/Ag = \dim A - 1 = n - 2.$$

Since $f_j, g \in I(X)$, it would follow that $\dim_x X \leq n-2$, a contradiction. □

Complete intersections are a generalization of hypersurfaces.

Definition. An analytic set germ (X, x) is called a *complete intersection* if $\mathcal{O}_{X,x}$ is isomorphic to $\mathcal{O}_{n,0}/\mathcal{O}_{n,0}(f_1, \ldots, f_m)$ with $n - m = \dim \mathcal{O}_{X,x}$ and $f_1, \ldots, f_m \in \mathfrak{m}_n$.

By Proposition 2.45 an analytic set germ (X, x) is a complete intersection if and only if $I(X)$ is generated by m elements of a parameter system in $\mathcal{O}_n$

Together with the concept of a parameter system there is the concept of a prime sequence.

Definition. Let A be an analytic algebra. An m-tuple $(f_1, \ldots, f_m)$ of elements $f_j \in \mathfrak{m}_A$ is called a *prime sequence* (or *A-sequence*) if f_1 is a nonzerodivisor in A and if for all $j = 2, \ldots, m$ the residue class of f_j in $A/A(f_1, \ldots, f_{j-1})$ is a nonzerodivisor. The number m is called the *length* of the prime sequence.

Proposition 2.48. *If A is an analytic algebra and $(f_1, \ldots, f_m)$ is a prime sequence in A, then*

$$\dim A/A(f_1, \ldots, f_\mu) = \dim A - \mu$$

for $\mu = 1, 2, \ldots, m$.

Proof. It is enough to deal with the case $\mu = 1$. By the definition of a prime sequence, f_1 is a nonzerodivisor in A. Hence f_1 is active and by Proposition 2.46

$$\dim A/(f) = \dim A - 1,$$

as was to be shown. □

In particular, it follows from Proposition 2.48 that a prime sequence $(f_1, \ldots, f_m)$ in $\mathcal{O}_{n,0}$ determines a complete intersection $V(\mathcal{O}_{n,0}(f_1, \ldots, f_m))$ in $(\mathbb{C}^n, 0)$. We next want to demonstrate the converse. We need some preparation for this.

Proposition 2.49. *If $(f_1, \ldots, f_m)$ is a prime sequence in A and if π is a permutation of the set $\{1, \ldots, m\}$, then $(f_{\pi(1)}, \ldots, f_{\pi(m)})$ is also a prime sequence in A.*

This proposition follows from the following lemma.

Lemma 2.7. *If (f_1, f_2) is a prime sequence in A, so also is (f_2, f_1).*

Proof. By hypothesis, f_2 is a nonzerodivisor in $A/(f_1)$. We show that f_2 is then also a nonzerodivisor in A. Suppose $f_2 g = 0$ for some $g \in A$. Since f_2 is a nonzerodivisor in $A/(f_1)$, there is a $g_1 \in A$ with $g = f_1 g_1$. From $f_1(f_2 g_1) = f_2 g = 0$ it follows that $f_2 g_1 = 0$ since f_1 is a nonzerodivisor of A. Now g_1 can again be written as $g_1 = f_1 g_2$ with $g_2 \in A$. In this way we obtain successive elements $g_1, g_2, \dots$ from A with $g_j = f_1 g_{j+1}$ and $f_2 g_j = 0$ for $j = 1, 2, \dots$. It follows that

$$g = f_1^j g_j \in \bigcap_{k=1}^{\infty} \mathfrak{m}_A^k = 0.$$

We now show that f_1 is a nonzerodivisor in $A/(f_2)$. Let $g_1 \in A$ with $f_1 g_1 \in (f_2)$. Then there is a $g_2 \in A$ with $f_1 g_1 = f_2 g_2$. Then $f_2 g_2 \in (f_1)$ too. Since by hypothesis f_2 is a nonzerodivisor in $A/(f_1)$, it follows that $g_2 \in (f_1)$. Hence there is an $h \in A$ with $g_2 = f_1 h$. Then $f_1(g_1 - f_2 h) = 0$. Since f_1 is nonzerodivisor of A, it follows that $g_1 = f_2 h$, whence $g_1 \in (f_2)$. □

We now introduce several more algebraic concepts. Let R be a Noetherian local ring and $M \neq 0$ an R-module. For each subset $N \subset M$ the set

$$\operatorname{Ann}(N) := \{r \in R \,|\, rx = 0 \text{ for all } x \in N\}$$

is an ideal in R. One calls $\operatorname{Ann}(N)$ the *annihilator ideal* of N. A prime ideal $\mathfrak{p} \subset R$ is said to be *associated to* M if there is an $x \in M$ with $\mathfrak{p} = \operatorname{Ann}(x)$. Since R is Noetherian, every maximal element of the set $\{\operatorname{Ann}(x) \,|\, x \in M\}$ is a prime ideal and therefore associated to M. All elements of a prime ideal associated to M are zero divisors of M. If, conversely, $r \in R$ is a zero divisor of M, then r belongs to some prime ideal associated to M.

Proposition 2.50. *If $(f_1, \dots, f_m)$ is a prime sequence in A, then*

$$m \leq \dim A/\mathfrak{p} \leq \dim A$$

for all prime ideals $\mathfrak{p}$ associated to A.

Proof. For brevity we write

$$M_\mu := A/A(f_1, \dots, f_\mu), \quad \mu = 1, \dots, m.$$

Let $\mathfrak{p}$ be a prime ideal associated to A. We construct a chain of prime ideals

$$\mathfrak{p} = \mathfrak{p}_0 \subsetneqq \mathfrak{p}_1 \subsetneqq \dots \subsetneqq \mathfrak{p}_m$$

recursively in A so that $\mathfrak{p}_\mu$ is a prime ideal of A associated to M_μ. Suppose that $\mathfrak{p}_0, \dots, \mathfrak{p}_\mu$, $0 \leq \mu < m$, have already been constructed. We show that then $\mathfrak{p}_\mu$ is contained in a prime ideal $\mathfrak{p}_{\mu+1}$ associated to $M_{\mu+1}$. Suppose this is not the case. Then, by the preliminary remarks, $\mathfrak{p}_\mu$ must contain a nonzerodivisor f of $M_{\mu+1}$. Then $(f_1, \dots, f_{\mu+1}, f)$ is a prime sequence in A. By Proposition 2.48, $(f_1, \dots, f_\mu, f, f_{\mu+1})$ is then a prime sequence in A, and so too is $(f_1, \dots, f_\mu, f)$. This means that f is a nonzerodivisor of M_μ. But then f cannot lie in $\mathfrak{p}_\mu$ and we have obtained a contradiction. Therefore there is a prime ideal $\mathfrak{p}_{\mu+1}$ associated to $M_{\mu+1}$ with $\mathfrak{p}_\mu \subset \mathfrak{p}_{\mu+1}$. Since $f_{\mu+1}$ is a nonzerodivisor of M_μ, we have $f_{\mu+1} \notin \mathfrak{p}_\mu$. But since on the other hand

$f_{\mu+1}$ lies in the annihilator ideal of each element of $M_{\mu+1}$, we must have $f_{\mu+1}$ in $\mathfrak{p}_{\mu+1}$. Hence $\mathfrak{p}_\mu \neq \mathfrak{p}_{\mu+1}$. We have constructed the prime ideal chain.

To complete the proof, it is now enough to show that for every two prime ideals $\mathfrak{p}', \mathfrak{p}''$ of A with $\mathfrak{p}' \subsetneqq \mathfrak{p}''$ we have

$$\dim A/\mathfrak{p}' > \dim A/\mathfrak{p}''.$$

But this follows from the fact that every $\bar{f}$ with $f \notin \mathfrak{p}'$ is active in the integral domain $\bar{A} = A/\mathfrak{p}'$, for it follows from the active lemma that

$$\dim A/\mathfrak{p}' = \dim \bar{A} > \dim \bar{A}/(\bar{f}) \geq \dim A/\mathfrak{p}''.$$

This completes the proof of Proposition 2.50. □

It follows from Proposition 2.50 that the length of a prime sequence in A is always less than or equal to the dimension of A.

Definition. An analytic algebra A is called a *Cohen-Macaulay ring* (briefly, a *CM-ring*) if A has a prime sequence of length $\dim A$.

Example 2.1. The algebra $\mathcal{O}_{n,0} = \mathbb{C}\{z_1, \ldots, z_n\}$ of holomorphic function germs is a CM-ring, for $(z_1, \ldots, z_n)$ forms a prime sequence in $\mathcal{O}_{n,0}$. Indeed

$$\mathcal{O}_{n,0}/\mathcal{O}_{n,0}(z_1, \ldots, z_\mu) \cong \mathbb{C}\{z_{\mu+1}, \ldots, z_n\}$$

for $\mu = 1, \ldots, n$.

It follows immediately from Proposition 2.50 that

Corollary 2.8. *If A is a CM-ring, then $\dim A = \dim A/\mathfrak{p}$ for all prime ideals associated to A.*

Proposition 2.51. *Let A be a CM-ring and let $f_1, \ldots, f_m$ be elements of $\mathfrak{m}_A$ such that $\dim A/A(f_1, \ldots, f_m) = \dim A - m$. Then $(f_1, \ldots, f_m)$ is a prime sequence and $A/A(f_1, \ldots, f_m)$ is also a CM-ring.*

Proof. It is enough to treat the case $m = 1$. We must show that f_1 is a nonzerodivisor of A. Suppose that f_1 is a zero divisor of A. Then there is a prime ideal $\mathfrak{p}$ associated to A with $f_1 \in \mathfrak{p}$. From Corollary 2.8 it follows that

$$\dim A > \dim A/Af_1 \geq \dim A/\mathfrak{p} = \dim A,$$

a contradiction. □

Corollary 2.9. *Each parameter system $(f_1, \ldots, f_n)$ of $\mathcal{O}_{n,0}$ is also a prime sequence.*

Proof. By the above example $\mathcal{O}_{n,0}$ is a CM-ring. The result now follows from Proposition 2.51. □

We need the following proposition for later applications.

Proposition 2.52. *Let A be a CM-ring of dimension n and $(f_1, \ldots, f_n)$ a prime sequence. Let*

$$R = \mathbb{C}\{f_1, \ldots, f_n\} \subset A$$

be isomorphic to $\mathbb{C}\{y_1, \ldots, y_n\}$. Then A is a free R-module.

Proof. We prove the proposition by induction on n. For $n = 0$ we have $R = \mathbb{C}$ and A is free as a $\mathbb{C}$-vector space. Now let $n > 0$.

We consider the analytic algebra $\bar{A} = A/(f_n)$. Then $\dim \bar{A} = n - 1$ and $(\bar{f}_1, \ldots, \bar{f}_{n-1})$ is a prime sequence in $\bar{A}$. So $\bar{A}$ is also a CM-ring. Let $\bar{R} = \mathbb{C}\{f_1, \ldots, f_n\}/(f_n)$. Then $\bar{R}$ is isomorphic to $\mathbb{C}\{y_1, \ldots, y_{n-1}\}$. By the induction hypothesis, $\bar{A}$ is a free $\bar{R}$-module. Let $e_1, \ldots, e_q \in A$ be chosen so that their residue classes $\bar{e}_1, \ldots, \bar{e}_q$ in $\bar{A}$ form an $\bar{R}$-basis of $\bar{A}$. From

$$A \subset \sum R \cdot e_j + A \cdot f_n \subset \sum R \cdot e_j + A\mathfrak{m}_R$$

it follows with the help of Nakayama's lemma that

$$A \subset \sum R \cdot e_j.$$

This means that $e_1, \ldots, e_q$ generate the R-module A. We show that $e_1, \ldots, e_q$ are also R-linearly independent. Suppose $\sum a_j e_j = 0$ for elements $a_j \in R$. We write $\bar{a}_j$ for the residue classes of a_j in $\bar{R}$. Then $\sum \bar{a}_j \bar{e}_j = 0$. Since the $\bar{e}_1, \ldots, \bar{e}_q$ are linearly independent in $\bar{A}$ over $\bar{R}$, it follows that $\bar{a}_j = 0$ for $j = 1, \ldots, q$. Thus $a_j = f_n a_j^{(1)}$ with $a_j^{(1)} \in R$. Hence

$$f_n\left(\sum a_j^{(1)} e_j\right) = \sum a_j e_j = 0.$$

Since, by hypothesis and Proposition 2.49, f_n is a nonzerodivisor of A, it follows that

$$\sum a_j^{(1)} e_j = 0.$$

In this way one obtains successively for each $k = 1, 2, \ldots$ an $a_j^{(k)} \in R$ with $a_j^{(k)} = f_n a_j^{(k+1)}$. Thus

$$a_j = f_n^k a_j^{(k)} \in \bigcap_{k=1}^{\infty} \mathfrak{m}_R^k = 0$$

for $j = 1, \ldots, q$. This demonstrates Proposition 2.52. □

2.11. Elimination theory for analytic sets

We now want to show that in certain circumstances the image of an analytic set under a proper holomorphic map is again an analytic set. That this holds in general is the content of the Remmert mapping theorem. This is a deep result, usually proved by sheaf theoretic methods. We consider here only a special case, where one can apply elimination theory. We follow [**Mum76**, Proposition (4.11)].

The aim of this section is the proof of the following proposition.

Proposition 2.53. *Let $p : \mathbb{C}^{n+k} \to \mathbb{C}^n$ be a linear projection, let $U \subset \mathbb{C}^{n+k}, V \subset \mathbb{C}^n$ be open sets with $p(U) \subset V$, and let $X \subset U$ be a proper analytic subset. If the restriction $p|_X : X \to V$ is proper, then $p(X)$ is an analytic subset of V and $p|_X$ is finite (i.e., proper with finite fibers).*

Proof. a) We show first by induction on k that it is enough to prove it for the case $k = 1$.

To do this we factorize $p : \mathbb{C}^{n+k} \to \mathbb{C}^n$ as

$$\mathbb{C}^{n+k} \xrightarrow{p_1} \mathbb{C}^{n+k-1} \xrightarrow{p_2} \mathbb{C}^n$$

and put $U_1 = p_1(U) \subset \mathbb{C}^{n+k+1}$. Since $p|_X$ is proper, so too is $p_1|_X : X \to U_1$ proper. From the result for the case $k = 1$ it follows that $X_1 = p_1(X)$ is an analytic subset of U_1 and $p_1|_X$ is finite. Then it follows that $p_2|_{X_1}$ is proper. By the induction hypothesis, $p_2(X_1) = p(X)$ is then an analytic subset of V and $p_2|_{X_1}$ is finite. Since the composition of finite maps is again finite, $p|_X$ is also finite.

b) So suppose $k = 1$. Let $y \in V$. By a suitable choice of coordinates we can assume that $p : \mathbb{C}^{n+1} \to \mathbb{C}^n$ is the projection given by $(z_1, \ldots, z_{n+1}) \mapsto (z_1, \ldots, z_n)$. We now show that it is enough to prove the result for the case where the preimage of y under $p|_X$ consists of exactly one point.

Since $p|_X$ is proper, $X \cap p^{-1}(y)$ is a compact proper analytic subset of $U \cap p^{-1}(y)$. The set $U \cap p^{-1}(y)$ is, however, an open subset of $\{y\} \times \mathbb{C}$. Now, by Proposition 2.24, the codimension of $X \cap p^{-1}(y)$ in $\{y\} \times \mathbb{C}$ is greater than or equal to 1, so $X \cap p^{-1}(y)$ is discrete. Besides, since $X \cap p^{-1}(y)$ is compact, $X \cap p^{-1}(y)$ consists of finitely many points and $p|_X$ is finite. Let $X \cap p^{-1}(y) = \{(y, x_1), \ldots, (y, x_k)\}$. We choose pairwise disjoint neighborhoods $U_1, \ldots, U_k$ of the points $(y, x_1), \ldots, (y, x_k)$ in $\mathbb{C}^{n+1}$. Let $W \subset V$ be a neighborhood of y such that

$$X \cap p^{-1}(W) \subset U_1 \cup \cdots \cup U_k.$$

If one puts $X_j := X \cap p^{-1}(W) \cap U_j$, then $X \cap p^{-1}(W)$ is the disjoint union of analytic subsets X_j of U_j and

$$p(X) \cap W = p(X_1) \cup \cdots \cup p(X_k).$$

Since a finite union of analytic subsets is again analytic (Proposition 2.30), it is enough to show that $p(X_j)$ is an analytic subset of W. But the subset X_j contains only one point of $p^{-1}(y)$.

c) We now assume additionally that $X \cap p^{-1}(y)$ consists of only one point. By a suitable choice of coordinates we can further assume that $y = 0$, $X \cap$

$p^{-1}(y) = \{0\}$. Let $f_1, \ldots, f_k \in \mathbb{C}\{z_1, \ldots, z_{n+1}\}$ be generators of the ideal $I(X)$ of $(X, 0)$. By Lemma 2.1 there are coordinates $z_1, \ldots, z_{n+1}$ around 0 such that f_1 is z_{n+1}-regular of order d. By the Weierstrass preparation theorem we can assume that f_1 has the following form:

$$f_1(z) = z_{n+1}^d + a_1 z_{n+1}^{d-1} + \ldots + a_d,$$

where $a_j \in \mathbb{C}\{z_1, \ldots, z_n\}$, $a_j(0) = 0$. By the Weierstrass division theorem we can further assume that for $f_2, \ldots, f_k$,

$$\begin{aligned} f_2(z) &= b_{21} z_{n+1}^{d-1} + \ldots + b_{2d} \\ \vdots \ &\ \vdots \ \ \vdots \\ f_k(z) &= b_{k1} z_{n+1}^{d-1} + \ldots + b_{kd} \end{aligned}$$

with $b_{jl} \in \mathbb{C}\{z_1, \ldots, z_n\}$. We now define polynomials $f_j(z_1, \ldots, z_n)(z_{n+1}) \in \mathbb{C}[z_{n+1}]$ by $f_j(z_1, \ldots, z_n)(z_{n+1}) = f_j(z_1, \ldots, z_{n+1})$. We consider the polynomials $f_1(z_1, \ldots, z_n)(z_{n+1})$ and $\sum_{j=2}^k t_j f_j(z_1, \ldots, z_n)(z_{n+1})$ for $(t_2, \ldots, t_k) \in \mathbb{C}^{k-1}$. For their resultant R we have

$$R = \sum_{|\nu|=d} t^\nu R_\nu, \quad R_\nu \in \mathbb{C}\{z_1, \ldots, z_n\}.$$

Now let $U_0 \subset U$ be a neighborhood of $0 \in \mathbb{C}^{n+1}$ such that $f_1, \ldots, f_k$ are defined on U_0, have the above representation, and

$$X \cap U_0 = \{z \in U_0 \mid f_1(z) = \ldots = f_k(z) = 0\}.$$

Let $V' \subset V$ be a neighborhood of $0 \in \mathbb{C}^n$ such that all a_j, b_{jl}, and so all R_ν, are defined on V' and all roots of $f_1(z_1, \ldots, z_n)(z_{n+1}) = 0$ lie in U_0 if $(z_1, \ldots, z_n) \in V'$. We put $U' := U_0 \cap p^{-1}(V')$. We now show that

$$p(X \cap U') = \{(z_1, \ldots, z_n) \in V' \mid R_\nu(z_1, \ldots, z_n) = 0 \text{ for all } \nu\}.$$

First let $(z_1, \ldots, z_n) \in p(X \cap U')$. Then there is a $z_{n+1} \in \mathbb{C}$ with $(z_1, \ldots, z_{n+1}) \in X \cap U'$, and z_{n+1} is a common zero of $f_1(z_1, \ldots, z_n)(z_{n+1})$, and

$$\sum_{j=2}^k t_j f_j(z_1, \ldots, z_n)(z_{n+1})$$

for all $(t_2, \ldots, t_k) \in \mathbb{C}^{k-1}$. From Proposition 1.23 it follows that

$$\sum_{|\nu|=d} t^\nu R_\nu(z_1, \ldots, z_n) = 0$$

for all $(t_2, \ldots, t_k) \in \mathbb{C}^{k-1}$ and hence $R_\nu(z_1, \ldots, z_n) = 0$ for all ν.

Conversely, suppose $(z_1, \ldots, z_n) \in V'$ and $R_\nu(z_1, \ldots, z_n) = 0$ for all ν. Then by Proposition 1.23 for all $(t_2, \ldots, t_k) \in \mathbb{C}^{k-1}$ both the polynomials

$f_1(z_1, \ldots, z_n)(z_{n+1})$ and $\sum_{j=2}^{k} t_j f_j(z_1, \ldots, z_n)(z_{n+1})$ must have a common zero. Let $\lambda_1, \ldots, \lambda_d$ be the zeros of $f_1(z_1, \ldots, z_n)(z_{n+1})$. We put

$$T_l := \left\{ (t_2, \ldots, t_k) \in \mathbb{C}^{k-1} \;\middle|\; \sum_{j=2}^{k} t_j f_j(z_1, \ldots, z_n)(\lambda_l) = 0 \right\}$$

for $l = 1, \ldots, d$. Then every $(k-1)$-tuple $(t_2, \ldots, t_k)$ is contained in a T_l. Hence

$$T_1 \cup \cdots \cup T_d = \mathbb{C}^{k-1}.$$

But this is a union of subspaces of the vector space $\mathbb{C}^{k-1}$. There must therefore be an l_0 with $T_{l_0} = \mathbb{C}^{k-1}$. This means that λ_{l_0} is a zero of all the polynomials $f_j(z_1, \ldots, z_n)(z_{n+1})$. Thus $(z_1, \ldots, z_n)$ lies in $p(X \cap U')$.

We have now shown that $p(X \cap U')$ is an analytic subset of V'. Proposition 2.53 follows from a), b) and c). □

We now prove another rider to Proposition 2.53. The definition of an irreducible analytic set germ can be taken word-for-word from that of an analytic set:

Definition. Let $U \subset \mathbb{C}^n$ be an open subset. An analytic subset $X \subset U$ is called *irreducible* if from $X = X_1 \cup X_2$ for analytic subsets $X_1, X_2 \subset U$ it follows that $X_1 = X$ or $X_2 = X$. Otherwise X is called *reducible.*

Proposition 2.54. *If, under the hypotheses of Proposition 2.53, X is irreducible, then $p(X)$ is also irreducible.*

Proof. Let $p(X) = Y_1 \cup Y_2$, where Y_1 and Y_2 are analytic subsets of V. We put $X_1 = X \cap p^{-1}(Y_1)$, $X_2 = X \cap p^{-1}(Y_2)$. Then X_1 and X_2 are analytic subsets of U and $X = X_1 \cup X_2$. Since X is irreducible, it follows that $X = X_1$ or $X = X_2$. Thus X must be contained in one of the two subsets $p^{-1}(Y_1)$ or $p^{-1}(Y_2)$; i.e., $X \subset p^{-1}(Y_1)$ or $X \subset p^{-1}(Y_2)$. It follows that $p(X) = Y_1$ or $p(X) = Y_2$, as was to be shown. □

Chapter 3

Isolated singularities of holomorphic functions

3.1. Differentiable manifolds

For the study of isolated singularities of holomorphic functions we need some of the fundamentals of differential topology as will be presented in this and the next three sections. We begin with the definition of a differentiable manifold.

Definition. Let M be an n-dimensional topological manifold (see §1.1).

An atlas of a manifold M is called differentiable if all its transition functions are *differentiable*. (Here differentiable means C^∞-differentiable.)

Two differentiable atlases $\mathfrak{A}$ and $\mathfrak{W}$ are called *equivalent* if and only if $\mathfrak{A} \cup \mathfrak{W}$ is also a differentiable atlas.

A *differentiable structure* on M is an equivalence class of differentiable atlases on M.

A *differentiable* manifold is a topological manifold together with a differentiable structure.

Remark 3.1. Each differentiable structure contains a uniquely determined maximal atlas $\mathfrak{A}^*$: If $\mathfrak{A}$ is an arbitrary atlas from the corresponding equivalence class, then

$$\mathfrak{A}^* := \left\{ \varphi : U \longrightarrow V \text{ chart} \;\middle|\; \begin{array}{l}\text{transition function from } \varphi \text{ to all charts of } \mathfrak{A} \\ \text{is differentiable}\end{array} \right\}$$

is a maximal atlas.

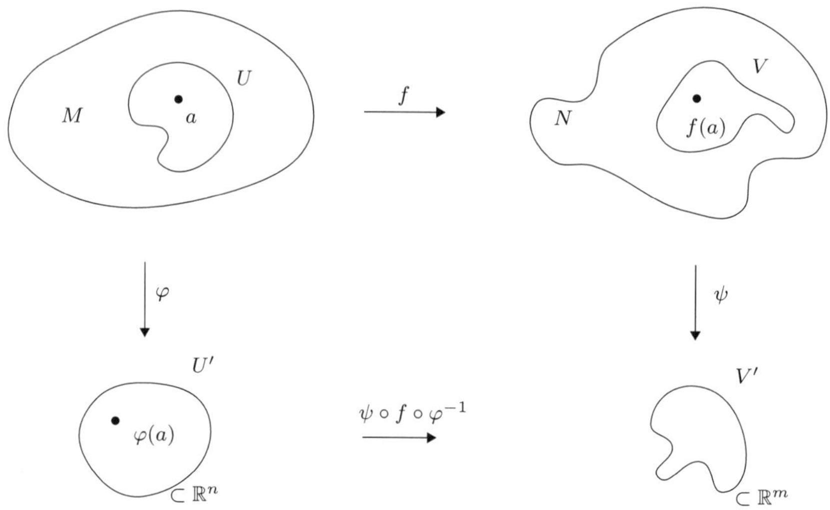

Figure 3.1. Definition of a differentiable map

Convention. If M is a differentiable manifold, then a chart of M is always a chart of the maximal atlas of the differentiable structure.

Definition. Let M, N be differentiable manifolds. A continuous map $f : M \to N$ is called *differentiable at* $a \in M$ if and only if for each chart $\varphi : U \to U'$, $a \in U$, and $\psi : V \to V'$, $f(a) \in V$, of M, resp. N, the composition $\psi \circ f \circ \varphi^{-1}$ is differentiable at $\varphi(a) \in U'$ (see Figure 3.1). (The domain of definition of $\psi \circ f \circ \varphi^{-1}$ is $\varphi(f^{-1}(V) \cap U)$.) The map f is called *differentiable* if it is differentiable at every $a \in M$.

A differentiable map $f : M \to N$ is called a *diffeomorphism* if and only if it is invertible and its inverse map is differentiable.

Remark 3.2. Let M be a differentiable manifold, $\varphi : U \to U'$ a chart of M. Then U and U' are also differentiable manifolds as open subsets of M and $\mathbb{R}^n$, and $\varphi : U \to U'$ is a diffeomorphism between U and U'. A chart $\varphi : U \to U'$ around a point $a \in M$, $a \in U$, is often also called a local coordinate system: Let $\varphi_1, \ldots, \varphi_n$ be the component functions of φ. These are differentiable functions. By a translation in $\mathbb{R}^n$ one can ensure that $\varphi(a) = 0$. After the choice of coordinates $x_1, \ldots, x_n$ of $\mathbb{R}^n$ we call

$$x_1 = \varphi_1(x), \ldots, x_n = \varphi_n(x)$$

the coordinates of the point $x \in U$. The point $a \in M$ then has the coordinates $(0, \ldots, 0)$. Thus every point of U can be described by unique coordinates. A function on U is differentiable if and only if it is differentiable in the usual sense as a function of the coordinates.

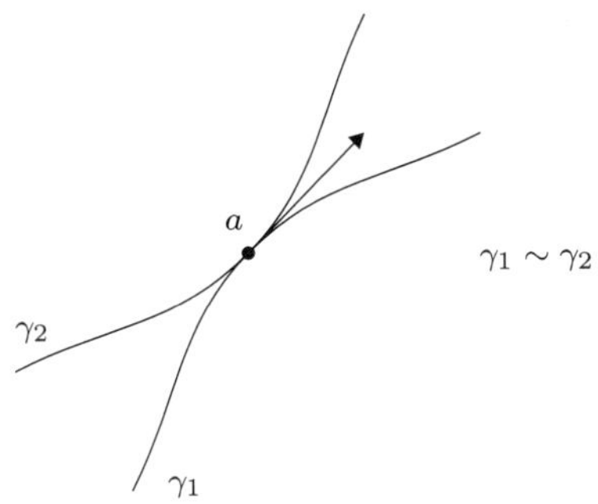

Figure 3.2. Tangent vector

We now want to introduce the concept of a tangent space. Let M be an n-dimensional differentiable manifold, a a point of M.

A *(differentiable) path through* a is a differentiable map $\gamma : (-\varepsilon, \varepsilon) \to M$ with $\varepsilon > 0$ and $\gamma(0) = a$. We define an equivalence relation on the set of all paths through a: $\gamma_1 \sim \gamma_2 :\Leftrightarrow$ there is a chart $\varphi : U \to U'$, $a \in U$, with

$$\frac{d}{dt}\varphi \circ \gamma_1(0) = \frac{d}{dt}\varphi \circ \gamma_2(0).$$

(If this equation holds for one chart, then it holds for every chart since the transition functions are differentiable.) The equivalence class of a path $\gamma : (-\varepsilon, \varepsilon) \to M$ with $\gamma(0) = a$ is denoted by $[\gamma]$ (see Figure 3.2).

Definition. An equivalence class $[\gamma]$ is called a *tangent vector to* M *at* a. The set of all tangent vectors is called the *tangent space to* M *at* a, symbolically, T_aM.

We now derive an algebraic description of the tangent space to a differentiable manifold.

Definition. Let M be a differentiable manifold, $a \in M$. We denote by $\bar{f}$ the germ at a of a differentiable function $f : U \to \mathbb{R}$, U an open neighborhood of a. Let $\mathcal{E}_{M,a}$ be the set of all germs at a of differentiable functions $f : U \to \mathbb{R}$, U an open neighborhood of a. In particular we put

$$\mathcal{E}_{n,0} := \mathcal{E}_{\mathbb{R}^n,0}.$$

Definition. Let M be a differentiable manifold, $a \in M$. A *derivation* of $\mathcal{E}_{M,a}$ is an $\mathbb{R}$-linear map $\delta : \mathcal{E}_{M,a} \to \mathbb{R}$ that obeys the product rule

$$\delta(\bar{f} \cdot \bar{g}) = \bar{f}(a)\delta(\bar{g}) + \bar{g}(a)\delta(\bar{f})$$

for all $\bar{f}, \bar{g} \in \mathcal{E}_{M,a}$.

We denote the set of all derivations of $\mathcal{E}_{M,a}$ by $\operatorname{Der}\mathcal{E}_{M,a}$. They form a vector space. We wish to show that

$$T_aM \cong \operatorname{Der}\mathcal{E}_{M,a}.$$

We consider first the derivations of $\mathcal{E}_{n,0}$. Let $x_1, \dots, x_n$ be the coordinates of $\mathbb{R}^n$. Then we define derivations

$$\frac{\partial}{\partial x_j} : \mathcal{E}_{n,0} \longrightarrow \mathbb{R}$$

by $\bar{f} \mapsto (\partial/\partial x_j)f(0)$ (partial derivatives with respect to the coordinate x_j).

Lemma 3.1. *Let U be an open ball around $0 \in \mathbb{R}^n$ and $f : U \to \mathbb{R}$ a differentiable function. Then f can be written in the form*

$$f(x) = f(0) + \sum_{j=1}^{n} x_j \cdot f_j(x)$$

with differentiable functions $f_j : U \to \mathbb{R}$, $j = 1, \dots, n$.

Proof. We have

$$\begin{aligned} f(x) - f(0) &= \int_0^1 \frac{d}{dt} f(tx_1, \dots, tx_n)dt \\ &= \sum_{j=1}^{n} x_j \int_0^1 \frac{\partial}{\partial x_j} f(tx_1, \dots, tx_n)dt\,. \end{aligned}$$

The desired representation follows on putting

$$f_j(x) := \int_0^1 (\partial/\partial x_j) f(tx_1, \dots, tx_n)dt.$$

□

Proposition 3.1. *The derivations $\partial/\partial x_j$, $j = 1, \dots, n$, form a basis of the vector space* $\operatorname{Der}\mathcal{E}_{n,0}$.

Proof. We put $V = \operatorname{Der}\mathcal{E}_{n,0}$.

We show first the linear independence of the $\partial/\partial x_j$: Let

$$\sum_{j=1}^{n} a_j \frac{\partial}{\partial x_j} = 0, \quad a_j \in \mathbb{R}.$$

Then

$$a_k = \sum_{j=1}^{n} a_j \frac{\partial \bar{x}_k}{\partial x_j} = 0 \text{ for all } k.$$

Now let $\delta \in V$, $a_k := \delta(\bar{x}_k)$. We show

$$\delta = \sum_{j=1}^{n} a_j \frac{\partial}{\partial x_j}.$$

Let $\bar{f} \in \mathcal{E}_{n,0}$ and let $f : U \to \mathbb{R}$ be a representative of $\bar{f}$. By Lemma 3.1, f can be written as $f = f(0) + \sum_{j=1}^{n} x_j f_j$ with $\bar{f}_j \in \mathcal{E}_{n,0}$.

Then

$$\begin{aligned} \delta(\bar{f}) &= \delta(f(0)) + \sum_{j=1}^{n} \delta(\bar{x}_j) \cdot f_j(0) \\ &= \sum_{j=1}^{n} a_j \cdot f_j(0) \\ &= \sum_{j=1}^{n} a_j \frac{\partial \bar{f}}{\partial x_j}, \end{aligned}$$

as was to be shown. □

Now let M again be a differentiable manifold, $a \in M$. Given a $[\gamma] \in T_a M$, we can specify a derivation $\delta_\gamma : \mathcal{E}_{M,a} \to \mathbb{R}$ by

$$\delta_\gamma(\bar{f}) = \frac{d}{dt} f \circ \gamma(0).$$

Proposition 3.2. *The map*

$$\begin{array}{ccc} T_a M & \longrightarrow & \operatorname{Der} \mathcal{E}_{M,a} \\ [\gamma] & \longmapsto & \delta_\gamma \end{array}$$

is bijective.

Proof. a) The map is injective: Let γ, γ' be paths through $a \in M$ with $\delta_\gamma = \delta_{\gamma'}$. Let $\varphi : U \to U' \subset \mathbb{R}^n$, $a \in U$, be a chart and $\varphi_j : U \to \mathbb{R}$, $j = 1, \ldots, n$, the component functions of φ. Since $\delta_\gamma = \delta_{\gamma'}$, we have

$$\frac{d}{dt} \varphi_j \circ \gamma(0) = \frac{d}{dt} \varphi_j \circ \gamma'(0)$$

for all $j = 1, \ldots, n$. Thus $\gamma \sim \gamma'$.

b) The map is also surjective: Let $\delta \in \operatorname{Der} \mathcal{E}_{M,a}$. By Proposition 3.1, δ can be written in local coordinates $x_1, \ldots, x_n$ around a as

$$\delta = \sum_{j=1}^{n} a_j \frac{\partial}{\partial x_j} \quad \text{with } a_j \in \mathbb{R}.$$

Put

$$\begin{array}{rccc} \gamma : & (-\varepsilon, \varepsilon) & \longrightarrow & M \\ & t & \longmapsto & (a_1 t, \ldots, a_n t) \end{array}$$

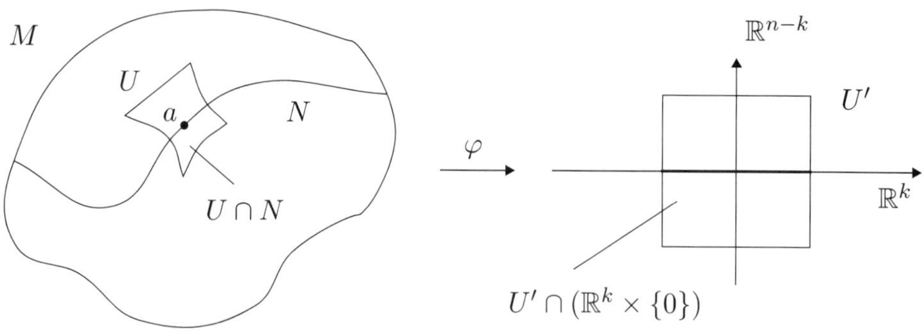

Figure 3.3. Chart of a submanifold

for $\varepsilon > 0$ small enough. It is easy to calculate that $\delta = \delta_\gamma$. □

Corollary 3.1. *T_aM is a real vector space of dimension n.*

Proof. This follows from Proposition 3.1 and Proposition 3.2. □

Let M, N be differentiable manifolds and $f : M \to N$ a differentiable map. If $a \in M$, then f induces an algebra homomorphism $f^* : \mathcal{E}_{N,f(a)} \to \mathcal{E}_{M,a}$.

Definition. The map

$$\begin{array}{rccc} T_af : & T_aM & \longrightarrow & T_{f(a)}N \\ & \delta & \longmapsto & \delta \circ f^* \end{array}$$

is called the *tangent map* (or the *differential*) of f at a.

Remark 3.3. T_af is a linear map.

We now define the concept of a submanifold of a differentiable manifold.

Definition. Let M be a differentiable manifold of dimension n. Furthermore, let $0 \le k \le n$. A subset $N \subset M$ is called a *k-dimensional differentiable submanifold* of M if for every $a \in N$ there is a chart $\varphi : U \to U' \subset \mathbb{R}^n = \mathbb{R}^k \times \mathbb{R}^{n-k}$ with $a \in U$ such that

$$\varphi(U \cap N) = U' \cap (\mathbb{R}^k \times \{0\})$$

(see Figure 3.3).

The number $n - k$ is called the *codimension* of the submanifold N.

A differentiable submanifold N is itself again a differentiable manifold: From a chart φ (as in the definition) one obtains a chart $\varphi' = \varphi|_{U \cap N} : U \cap N \to U' \cap \mathbb{R}^k$, where we identify $\mathbb{R}^k$ with $\mathbb{R}^k \times \{0\} \subset \mathbb{R}^n$, and the set of all these charts forms a differentiable atlas. The tangent space to N at a point $a \in N$ is in the natural way a subspace of T_aM.

3.2. Tangent bundles and vector fields

When M_1 and M_2 are differentiable manifolds of dimension n_1, resp. n_2, we can equip the Cartesian product $M_1 \times M_2$ in a natural way with the structure of an $(n_1 + n_2)$-dimensional differentiable manifold: If $\varphi_\nu : U_\nu \to U'_\nu$ are charts of M_ν, $\nu = 1, 2$, then

$$\varphi_1 \times \varphi_2 : U_1 \times U_2 \longrightarrow U'_1 \times U'_2 \subset \mathbb{R}^{n_1} \times \mathbb{R}^{n_2} = \mathbb{R}^{n_1+n_2}$$

is a chart of $M_1 \times M_2$, and the atlas of all these charts defines the differentiable structure of $M_1 \times M_2$. The canonical projections

$$\mathrm{pr}_1 : M_1 \times M_2 \longrightarrow M_1, \quad \mathrm{pr}_2 : M_1 \times M_2 \longrightarrow M_2$$

are then differentiable maps.

The concept of a (locally trivial) differentiable fiber bundle is a generalization of the Cartesian product of two differentiable manifolds.

Definition. A *(locally trivial) differentiable fiber bundle* is a quadruple (E, π, B, F), where E, B, F are differentiable manifolds and $\pi : E \to B$ is a surjective differentiable map satisfying the

Axiom of local triviality: Each point $b \in B$ has a neighborhood U for which there is a diffeomorphism

$$\psi : \pi^{-1}(U) \longrightarrow U \times F$$

such that the following diagram commutes:

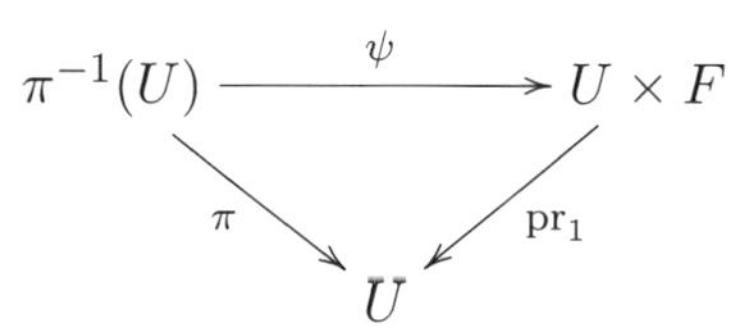

where pr_1 is the projection onto the first component.

Notation. (E, π, B, F) is called a differentiable fiber bundle *over B with fiber F*. The space E is called its *total space*, π the *projection*, B the *basis* and F the *fiber* of the bundle.

Example 3.1. If M_1, M_2 are differentiable manifolds, then $(M_1 \times M_2, \mathrm{pr}_1, M_1, M_2)$ is a differentiable fiber bundle. Such a bundle is called *trivial*. A differentiable fiber bundle (E, π, B, F) is trivial if one can take all of B as the neighborhood in the axiom of local triviality.

Definition. $\psi : \pi^{-1}(U) \to U \times F$ in the axiom of local triviality is also called a "*bundle chart*".

Vector bundles form a special class of differentiable fiber bundles.

Definition. A(n n-dimensional differentiable) *vector bundle* is a differentiable fiber bundle (E, π, B, F), where $F = \mathbb{R}^n$, where each fiber $E_b = \pi^{-1}(b)$ is equipped with the structure of an n-dimensional real vector space, and for each bundle chart $\psi : \pi^{-1}(U) \to U \times \mathbb{R}^n$ we have

$$\psi_b := \psi|_{E_b} : E_b \longrightarrow \{b\} \times \mathbb{R}^n$$

is a vector space isomorphism for all $b \in U$.

Example 3.2 (Tangent bundle). Let M be an n-dimensional manifold. We put

$$TM := \bigcup_{a \in M} T_a M$$
$$\text{and } \pi : TM \longrightarrow M$$

the canonical projection ($\pi(v) = a$ for $v \in T_a M$). Then $(TM, \pi, M, \mathbb{R}^n)$ is an n-dimensional vector bundle:

Let $\{\varphi_\alpha : U_\alpha \to U'_\alpha\}$ be an atlas of M. Given a chart $\varphi_\alpha : U_\alpha \to U'_\alpha$, one obtains a bundle chart

$$\psi_\alpha : \pi^{-1}(U_\alpha) \longrightarrow U_\alpha \times \mathbb{R}^n$$

as follows: Let $p \in U_\alpha$ and $v \in T_p M \subset \pi^{-1}(U_\alpha)$. Then $T_p\varphi_\alpha(v) \in T_{\varphi_\alpha(p)}\mathbb{R}^n$. If $x_1, \dots, x_n$ are coordinates of $\mathbb{R}^n$, then, by Proposition 3.1,

$$T_p\varphi_\alpha(v) = \sum_{j=1}^{n} a_j \left.\frac{\partial}{\partial x_j}\right|_{\varphi_\alpha(p)}.$$

We define

$$\psi_\alpha(v) := (p, a_1, \dots, a_n) \in U_\alpha \times \mathbb{R}^n.$$

The map ψ_α is bijective, the diagram

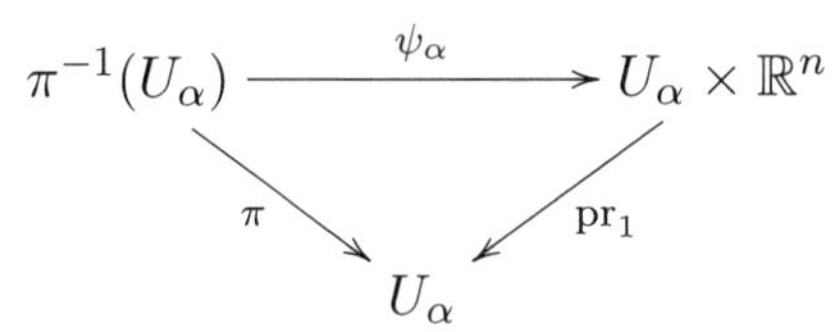

commutes and

$$\psi_{\alpha,b} := \psi_\alpha|_{T_b M} : T_b M \longrightarrow \{b\} \times \mathbb{R}^n$$

is an $\mathbb{R}$-linear isomorphism.

The total space TM has so far been defined only as a set. We must equip TM with the structure of a differentiable manifold. First we define a topology on TM.

We call a subset $U \subset TM$ *open* if and only if $\psi_\alpha(U \cap \pi^{-1}(U_\alpha))$ is open in $U_\alpha \times \mathbb{R}^n$ for all α.

To show that this is a good definition, we consider another chart $\varphi_\beta : U_\beta \to U'_\beta$ with $p \in U_\beta$. Let $q \in U_\beta$ and $\varphi_\beta(q) = (y_1, \ldots, y_n)$. The map $\psi_\beta : \pi^{-1}(U_\beta) \to U_\beta \times \mathbb{R}^n$ is defined by $\psi_\beta(w) := (q, b_1, \ldots, b_n)$ where $w \in T_qM$ and

$$T_q\varphi_\beta(w) = \sum_{j=1}^{n} b_j \left.\frac{\partial}{\partial y_j}\right|_{\varphi_\beta(q)}.$$

Then for a function germ $\bar{g} \in \mathcal{E}_{M,p}$

$$\begin{aligned}
\left.\frac{\partial}{\partial y_k}\right|_{\varphi_\beta(p)} (\overline{g \circ \varphi_\beta^{-1}}) &= \left.\frac{\partial}{\partial y_k}\right|_{\varphi_\beta(p)} (\overline{g \circ \varphi_\alpha^{-1} \circ \varphi_\alpha \circ \varphi_\beta^{-1}}) \\
&= \sum_{j=1}^{n} \left.\frac{\partial}{\partial x_j}\right|_{\varphi_\alpha(p)} (\overline{g \circ \varphi_\alpha^{-1}}) \frac{\partial(\varphi_\alpha \circ \varphi_\beta^{-1})_j}{\partial y_k}(\varphi_\beta(p)) \\
&= \sum_{j=1}^{n} \left.\frac{\partial}{\partial x_j}\right|_{\varphi_\alpha(p)} (\overline{g \circ \varphi_\alpha^{-1}}) c_{jk}(p),
\end{aligned}$$

where

$$c_{jk}(p) = \frac{\partial(\varphi_\alpha \circ \varphi_\beta^{-1})_j}{\partial y_k}(\varphi_\beta(p)), \quad \varphi_\alpha \circ \varphi_\beta^{-1} = ((\varphi_\alpha \circ \varphi_\beta^{-1})_1, \ldots, (\varphi_\alpha \circ \varphi_\beta^{-1})_n).$$

Thus

$$\frac{\partial}{\partial y_k} = \sum_{j=1}^{n} c_{jk} \frac{\partial}{\partial x_j}.$$

The map

$$\psi_\alpha \circ \psi_\beta^{-1} : (U_\alpha \cap U_\beta) \times \mathbb{R}^n \longrightarrow (U_\alpha \cap U_\beta) \times \mathbb{R}^n$$

can therefore be written as follows:

$$(\psi_\alpha \circ \psi_\beta^{-1})((p, b_1, \ldots, b_n)) = (p, \sum_{k=1}^{n} c_{1k} b_k, \ldots, \sum_{k=1}^{n} c_{nk} b_k)$$

for all $p \in U_\alpha \cap U_\beta$. Hence this map is a diffeomorphism.

In particular the sets $\pi^{-1}(U_\alpha)$ are open in the above topology and the maps

$$(\varphi_\alpha \times \mathrm{id}) \circ \psi_\alpha : \pi^{-1}(U_\alpha) \longrightarrow U'_\alpha \times \mathbb{R}^n \subset \mathbb{R}^{2n}$$

are charts of a differentiable atlas. In this way we have a differentiable structure on TM so that the projection π and the bundle charts ψ_α are differentiable. Then $(TM, \pi, M, \mathbb{R}^n)$ is an n-dimensional differentiable vector bundle. This vector bundle is called the *tangent bundle* of M.

Definition. Let M, N be differentiable manifolds and $f : M \to N$ a differentiable map. Let

$$Tf : TM \longrightarrow TN$$

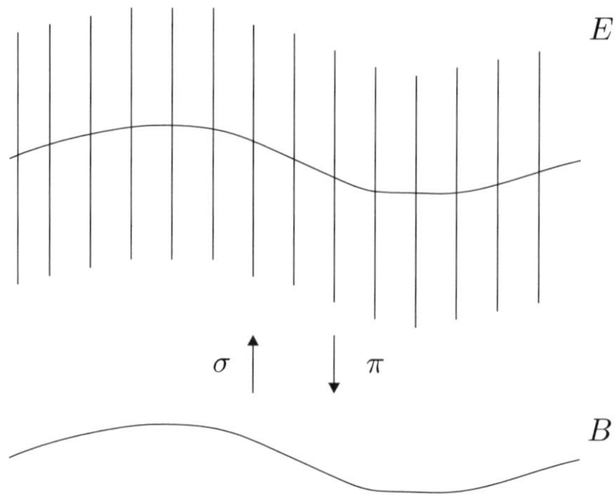

Figure 3.4. Section of a differentiable fiber bundle

be defined by the tangent maps

$$T_p f : T_p M \longrightarrow T_{f(p)} N$$

for $p \in M$. This is a differentiable map. It is called the *tangent map* (or the *differential*) of f.

Remark 3.4. Tf is a *linear map of vector bundles over* f, i.e., a differentiable map $Tf : TM \to TN$ such that the diagram

$$\begin{array}{ccc} TM & \xrightarrow{Tf} & TN \\ {\scriptstyle\pi}\downarrow & & \downarrow{\scriptstyle\pi} \\ M & \xrightarrow[f]{} & N \end{array}$$

commutes and each fiber $T_p M$ is mapped linearly into $T_{f(p)} N$.

Definition. A *section* of a differentiable fiber bundle (E, π, B, F) is a differentiable map $\sigma : B \to E$ with $\pi \circ \sigma = \mathrm{id}_B$, i.e., $\sigma(b) \in E_b$ for all $b \in B$ (see Figure 3.4).

Example 3.3. Every vector bundle has, for example, its *zero section*

$$\begin{array}{rccc} \sigma_0 : & B & \longrightarrow & E \\ & b & \longmapsto & 0 \in E_b\,. \end{array}$$

Definition. Let M be a differentiable manifold. A (differentiable) *vector field* on M is a section

$$X : M \longrightarrow TM$$

of its tangent bundle.

In what follows let M be an n-dimensional differentiable manifold.

Definition. A *one parameter group of diffeomorphisms* of M is a differentiable map

$$g : \mathbb{R} \times M \longrightarrow M$$

with the properties

(i) for each $t \in \mathbb{R}$ the map $g_t : M \to M$, $a \mapsto g(t,a)$, is a diffeomorphism of M onto itself,

(ii) $g_{t+s} = g_t \circ g_s$ for all $t, s \in \mathbb{R}$.

Remark 3.5. It follows from (ii) that $g_0 = \mathrm{id}$ (how?).

Definition. Let $U \subset M$ be open, $\varepsilon > 0$, $I = (-\varepsilon, \varepsilon) \subset \mathbb{R}$. A *local one parameter group of diffeomorphisms* of U to M is a differentiable map

$$g : I \times U \longrightarrow M$$

with the properties

(i) $g_t : U \to M$, $a \mapsto g(t,a)$, is a diffeomorphism of U onto an open subset of M for each $t \in I$,

(ii) $g_{s+t}(a) = g_s \circ g_t(a)$ for all $s, t \in I$ with $s + t \in I$ and all $a \in U$ with $g_t(a) \in U$,

(iii) $g_0 = \mathrm{id}$.

Let $g : I \times U \to M$ be a local one parameter group of diffeomorphisms. Then we define a vector field X on U as follows: For $a \in U$ and $\bar{f} \in \mathcal{E}_{U,a}$ we put

$$X_a(\bar{f}) = \left.\frac{d}{dt}\right|_{t=0} f \circ g_t(a).$$

We say that X is *induced* by the group g.

For $a \in U$

$$\begin{array}{rccc} \varphi_a : & I & \longrightarrow & M \\ & t & \longmapsto & g_t(a) \end{array}$$

is a differentiable path in M. This path is called the *phase curve* of g through a (see Figure 3.5). The vector $X_a \in T_aM$ is then precisely the tangent vector at the point a to the phase curve φ_a of g through a.

Proposition 3.3. *Let X be a vector field on M and let $a \in M$. Then there are an open neighborhood U of a, an interval $I = (-\varepsilon, \varepsilon)$, $\varepsilon > 0$, and a uniquely determined local one parameter group of diffeomorphisms of U to M that induces X on U.*

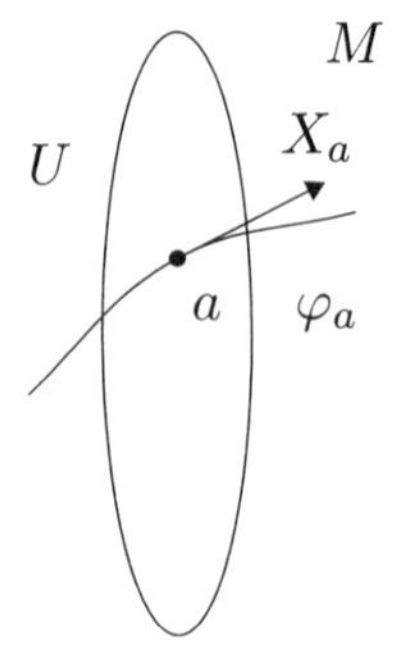

Figure 3.5. Tangent vector to a phase curve

Proof. Since we are dealing with a local assertion, we can assume that M is an open subset of $\mathbb{R}^n$. Let $x_1, \ldots, x_n$ be the coordinates of $\mathbb{R}^n$.

By Proposition 3.1 we can write the vector field X as

$$X_a = \sum_{j=1}^{n} f_j(a) \left. \frac{\partial}{\partial x_j} \right|_a \quad \text{for } a \in M,$$

where the $f_j : M \to \mathbb{R}$ are differentiable functions. We put

$$f = (f_1, \ldots, f_n) : M \longrightarrow \mathbb{R}^n.$$

If now X is induced by a local one parameter group $g : I \times U \to M$ of diffeomorphisms and φ is the phase curve of g through a, then

$$\begin{aligned} \varphi'(t) &= f(\varphi(t)) \text{ for all } t \in I, \\ \varphi(0) &= a. \end{aligned}$$

Thus $\varphi : I \to M$ is a solution of the differential equation

$$y' = f(y)$$

with initial value $\varphi(0) = a$.

By the Picard-Lindelöf Theorem, for each $x \in M$ there are a $\delta > 0$ and a solution $\varphi : (-\delta, \delta) \to M$ of the differential equation $y' = f(y)$ with $\varphi(0) = x$. Hence there are a neighborhood U_0 of a and a $\delta > 0$ such that, with $I_0 = (-\delta, \delta)$, a map

$$g : I_0 \times U_0 \longrightarrow M$$

exists such that

$$\begin{aligned} \frac{\partial}{\partial t} g(t, x) &= f(g(t, x)), \\ g(0, x) &= x. \end{aligned}$$

The map g is differentiable. This is the theorem on the differentiable dependence of the solution of a differential equation on the initial value, as can be found in any textbook on differential equations (see e.g. [**Arn06**]).

Again we put $g_t(x) := g(t,x)$. Choose $I = (-\varepsilon, \varepsilon)$, $\varepsilon > 0$, and a neighborhood U of a such that

$$s + t \in I_0,\ g_{s+t}(U) \subset U_0 \text{ for all } s, t \in I.$$

We shall show that $g : I \times U \to M$ is a local one parameter group of diffeomorphisms. Fix $x \in U$ and $s \in I$. Put $\varphi : I \to M$, $t \mapsto g_{t+s}(x)$, $\psi : I \to M$, $t \mapsto g_t \circ g_s(x)$. Then φ and ψ are both solutions of the differential equation $y' = f(y)$ with the initial value $\varphi(0) = \psi(0) = g_s(x)$. By the Picard-Lindelöf uniqueness theorem φ and ψ must agree everywhere. Since this holds for all $x \in U$, it follows that

$$g_{t+s} = g_t \circ g_s \text{ for all } s, t \in I.$$

In particular $g_t \circ g_{-t} = g_0 = \mathrm{id}$ for each $t \in I$, so each map g_t is a diffeomorphism.

By construction, X is induced by the group g. By the uniqueness theorem $g : I \times U \to M$ is uniquely determined. □

Remark 3.6. The construction of vector fields on a differentiable manifold M and the application of Proposition 3.3 is an essential method of differential topology in the construction of diffeomorphisms. We shall see an application of Proposition 3.3 in §3.7.

3.3. Transversality

In this section we shall introduce the concept of transversality.

Let M, N be differentiable manifolds and $f : M \to N$ a differentiable map.

Definition. The *rank* of $f : M \to N$ at $a \in M$ is the number

$$\mathrm{rank}_a f := \mathrm{rank}\, T_a f.$$

Proposition 3.4 (Rank theorem). *Let $f : M \to N$ be a differentiable map, $\dim M = m$, $\dim N = n$, and $\mathrm{rank}_a f = r$ for all $a \in M$. Then for all $a \in M$ there are a chart $\varphi : U \to U' \subset \mathbb{R}^m$ around a and a chart $\psi : V \to V' \subset \mathbb{R}^n$ around $f(a)$ such that the map $\psi \circ f \circ \varphi^{-1} : U' \to V'$ is given by*

$$(x_1, \ldots, x_m) \longmapsto (x_1, \ldots, x_r, 0, \ldots, 0).$$

Proof. We may assume that $M \subset \mathbb{R}^m$ and $N \subset \mathbb{R}^n$ are open. The proof is then identical to the proof of the rank theorem for holomorphic maps, cf. Proposition 2.34. □

Let M, X, Y be differentiable manifolds.

Figure 3.6. Transversal – not transversal

Definition. Let $f : X \to M$ and $g : Y \to M$ be differentiable maps, and let $x \in X$, $y \in Y$ with $f(x) = g(y) = a \in M$. Then f and g are called *transversal* at x and y if

$$T_xf(T_xX) + T_yg(T_yY) = T_aM.$$

(Note that the sum need not be direct!) The maps f and g are called *transversal* to one another if f and g are transversal at all $x \in X$, $y \in Y$ with $f(x) = g(y)$.

If X, Y are submanifolds of M, one says that X and Y *cut transversally* if the natural inclusions $i : X \to M$ and $j : Y \to M$ are transversal to one another (see Figure 3.6).

Definition. A differentiable map $f : X \to M$ is called an *embedding* if $f(X) \subset M$ is a differentiable submanifold and $f : X \to f(X)$ is a diffeomorphism.

Proposition 3.5. *If $f : X \to M$ and $g : Y \to M$ are transversal to one another and g is an embedding, then $f^{-1}(f(X) \cap g(Y))$ is a differentiable submanifold of X of dimension* $\dim X + \dim Y - \dim M$.

Proof. This follows from the rank theorem. Exercise (cf. [**BJ82**, (5.12)]). □

Definition. Let M, N be differentiable manifolds. A differentiable map $f : M \to N$ is called an *immersion* (resp. *submersion*) if

$$\text{rank}_a f = \dim M \quad (\text{resp. } \text{rank}_a f = \dim N)$$

for all $a \in M$.

Proposition 3.6. *Let M, N be differentiable manifolds and $f : M \to N$ an immersion. Then for each $a \in M$ there is an open neighborhood U of a in M such that*

(i) *$f|_U : U \to f(U)$ is a homeomorphism, where $f(U)$ is endowed with the topology induced from N,*

(ii) *$f(U)$ is a submanifold of N.*

Proof. This too follows from the rank theorem (Proposition 3.4). □

Definition. Let M, N be differentiable manifolds and $f : M \to N$ a differentiable map. A point $a \in M$ is called a *critical point* of f if the differential $T_a f$ is not surjective. A point $b \in N$ is called a *critical value* of f if $b = f(a)$ for a critical point $a \in M$.

We shall need the following fundamental result in §3.8.

Proposition 3.7 (Sard's theorem). *Let M, N be differentiable manifolds and $f : M \to N$ a differentiable map. Then the set of critical values D of f is a Lebesgue null set in N ($:\Leftrightarrow \varphi(D \cap U)$ is a Lebesgue null set in $\mathbb{R}^n$ for each chart $\varphi : U \to U' \subset \mathbb{R}^n$ of N).*

We refer to Bröcker and Jänich [**BJ82**, §6] for a proof.

3.4. Lie groups

We now introduce Lie groups and present a few facts about them that we will need in §3.9.

Definition. A *Lie group* is a set G with the following properties:

(i) G is a group.

(ii) G is a differentiable manifold.

(iii) The map $G \times G \to G$, $(a, b) \mapsto ab^{-1}$, is differentiable.

Example 3.4. a) $\mathbb{R}^n$ is a Lie group under addition.

b) The group $\mathrm{GL}(n, \mathbb{R})$ of invertible $n \times n$-matrices with real entries is a Lie group under matrix multiplication. Since the vector space $M(n \times n, \mathbb{R})$ of all $n \times n$-matrices with real entries is isomorphic to $\mathbb{R}^{n^2}$,

$$\mathrm{GL}(n, \mathbb{R}) = \{A \in M(n \times n, \mathbb{R}) \mid \det A \neq 0\}$$

can be regarded as an open subset of $\mathbb{R}^{n^2}$. Then $\mathrm{GL}(n, \mathbb{R})$ is a submanifold of $\mathbb{R}^{n^2}$. It is clear that matrix multiplication is a differentiable map. Similarly $\mathrm{GL}(n, \mathbb{C})$, the group of invertible $n \times n$-matrices with complex entries, is a Lie group. Here we consider $\mathrm{GL}(n, \mathbb{C})$ as a submanifold of $\mathbb{R}^{n^2} \oplus \mathbb{R}^{n^2}$.

Let G be a Lie group with neutral element $e \in G$. We consider the tangent space $T_e G$ to G at e. Let $v \in T_e G$. Using v and the group structure of G, we define a vector field on G. Given $a \in G$, we consider the left translation $l_a : G \to G$, defined by $l_a(g) = ag$. By the definition of a Lie group, l_a is a diffeomorphism. We define

$$X_a^v := T_e l_a(v).$$

Then $X^v : G \to TG$, $a \mapsto X_a^v$, is a vector field on G. We have

$$T_b l_a(X_b^v) = T_b l_a T_e l_b(v) = T_e l_{ab}(v) = X_{ab}^v$$

for all $a, b \in G$. This means that the vector field X^v is left invariant.

Definition. A vector field X on G is called *left invariant* if $(Tl_a)(X) = X$ for all $a \in G$.

If $h : \mathbb{R} \times G \to G$ is a one parameter group of diffeomorphisms of G and $a \in G$, we define $(d/dt)|_{t=0}\, h_t(a)$ by

$$\left(\left.\frac{d}{dt}\right|_{t=0} h_t(a) \right)(\bar{f}) = \left.\frac{d}{dt}\right|_{t=0} f \circ h_t(a)$$

for all $\bar{f} \in \mathcal{E}_{G,a}$.

Lemma 3.2. *Let X be a left invariant vector field on G. Then there is a one parameter group $h : \mathbb{R} \times G \to G$ of diffeomorphisms that induces X.*

Proof. By Proposition 3.3 there are an open neighborhood U of e in G, an interval $I = (-\varepsilon, \varepsilon)$, $\varepsilon > 0$, and a uniquely determined local one parameter group $h : I \times U \to G$. We show first that we can extend h_t for $t \in I$ to all of G.

Since X is left invariant, $T_b l_a(X_b) = X_{ab}$ for all $a, b \in G$. This means that, for all $a, b \in U$,

$$\left.\frac{d}{dt}\right|_{t=0} a h_t(b) = \left.\frac{d}{dt}\right|_{t=0} h_t(ab).$$

It follows that

$$a h_t(b) = h_t(ab) \tag{3.1}$$

for all $a, b \in U$. In particular

$$h_t(a) = a h_t(e)$$

for all $a \in U$. We use this formula to define h_t on all G.

Finally we define $h : \mathbb{R} \times G \to G$ by $h_t = (h_{(t/n)})^n$ for $t \in \mathbb{R}$, where n is a positive integer, large enough that $|t/n| < \varepsilon$. This is well defined. For, if m is another such number, then

$$(h_{t/m})^m = ((h_{t/mn})^n)^m = (h_{t/mn})^{nm} = ((h_{t/mn})^m)^n = (h_{t/n})^n.$$

By the left invariance of X it follows that the map $h : \mathbb{R} \times G \to G$ defined in this way is a one parameter group of diffeomorphisms that induces X. $\square$

Proposition 3.8. *Let $v \in T_eG$, let X^v be the corresponding left invariant vector field and let $h^v : \mathbb{R} \times G \to G$ be the one parameter group of diffeomorphisms that induces X^v according to Lemma 3.2. Then the map*

$$\begin{array}{rccc} h: & \mathbb{R} \times G \times T_eG & \longrightarrow & G \\ & (t, g, v) & \longmapsto & h_t^v(g) \end{array}$$

is differentiable and

(i) $h_t^v(ab) = ah_t^v(b)$,

(ii) $h_s^{tv} = h_{ts}^v$.

Proof. For v fixed, h^v is a one parameter group of diffeomorphisms and so it is differentiable. Varying v means changing the initial conditions of the differential equation that defines h^v as in the proof of Proposition 3.3. Since the solutions of a differential equation depend differentiably on the initial conditions, h is differentiable.

The equation (i) is just the formula (3.1) from the proof of Lemma 3.2.

As to (ii), note that h_s^{tv} as well as h_{st}^v are, for fixed t and v, one parameter groups of diffeomorphisms of G. Also, h_s^{tv} induces the vector field X^{tv}. For all $g \in G$

$$\left.\frac{d}{ds}\right|_{s=0} h_{st}^v(g) = t\left.\frac{d}{dr}\right|_{r=0} h_r^v(g) = tX_g^v = X_g^{tv}.$$

Thus h_{st}^v also induces the vector field X^{tv}. By the uniqueness of the one parameter groups of diffeomorphisms it follows that

$$h_s^{tv} = h_{st}^v.$$

This completes the proof of Proposition 3.8. □

Definition. We define the *exponential map*

$$\exp : T_eG \to G$$

by $\exp(v) := h_1^v(e)$ for all $v \in T_eG$.

Proposition 3.9. *The map* $\exp : T_eG \to G$ *is differentiable and is a diffeomorphism of an open neighborhood of* $0 \in T_eG$ *onto an open neighborhood of* $e \in G$. *We have* $T_0 \exp = \mathrm{id}_{T_eG}$, *where we identify* $T_0(T_eG)$ *with* T_eG.

Proof. The map exp is differentiable by Proposition 3.8. By Proposition 3.8(ii)

$$(T_0 \exp)(v) = \left.\frac{d}{dt}\right|_{t=0} \exp tv = \left.\frac{d}{dt}\right|_{t=0} h_1^{tv}(e) = \left.\frac{d}{dt}\right|_{t=0} h_t^v(e) = X_e^v = v$$

for all $v \in T_eG$. Hence $T_0 \exp = \mathrm{id}_{T_eG}$. □

Corollary 3.2. *Let V and W be subspaces of T_eG such that $V \oplus W = T_eG$. Define $\gamma : T_eG \to G$ by $\gamma(v, w) = \exp(v)\exp(w)$ for $v \in V$ and $w \in W$. Then γ is a diffeomorphism of an open neighborhood of $0 \in T_eG$ onto an open neighborhood of $e \in G$.*

Proof. It is clear that γ is differentiable. Now

$$(T_0\gamma)|_V = (T_0 \exp)|_V = \mathrm{id}_V$$

by Proposition 3.9. Similarly $(T_0\gamma)|_W = \mathrm{id}_W$. It follows that

$$T_0\gamma = \mathrm{id}_{T_eG} .$$

This proves Corollary 3.2. □

Definition. Let G be a Lie group. A subset $H \subset G$ is a *Lie subgroup* if

(i) H is a subgroup of G,

(ii) H is a submanifold of G,

(iii) H is a Lie group with the induced group structure and the induced differentiable structure.

We now want to show

Proposition 3.10. *A closed subgroup H of a Lie group G is a Lie subgroup.*

We need two lemmas for this.

Lemma 3.3. *Let H be a closed subgroup of G. Let $\| \ \|$ be a norm on T_eG. Let $v_1, v_2, \dots$ be a sequence of vectors in T_eG with $v_i \neq 0$, $\exp v_i \in H$ for all i, such that*

$$\lim_{i\to\infty} v_i = 0 \quad \textit{and also} \quad \lim_{i\to\infty} \frac{v_i}{\|v_i\|} = \tilde{v}.$$

Then $\exp t\tilde{v} \in H$ for all $t \in \mathbb{R}$.

Proof. Let $t \in \mathbb{R}$. Let $k_i(t)$ be the largest integer smaller than or equal to $t/\|v_i\|$. Then

$$\frac{t}{\|v_i\|} - 1 < k_i(t) \leq \frac{t}{\|v_i\|},$$

and so

$$\lim_{i\to\infty} k_i(t)\|v_i\| = t.$$

Since exp is differentiable, and so in particular it is continuous,

$$\lim_{i\to\infty} \exp(k_i(t)v_i) = \lim_{i\to\infty} \exp\left(k_i(t)\|v_i\|\frac{v}{\|v_i\|}\right) = \exp(t\tilde{v}).$$

But now by Proposition 3.8(ii)

$$\exp(k_i(t)v_i) = h_1^{k_i(t)v_i}(e) = h_{k_i(t)}^{v_i}(e) = (h_1^{v_i}(e))^{k_i(t)}$$

and this element lies in H. Since H is closed, we have $\exp(t\tilde{v}) \in H$. □

Lemma 3.4. *Let H be a closed subgroup of G and*

$$V = \{v \in T_eG \mid \exp(tv) \in H \text{ for all } t \in \mathbb{R}\}.$$

Then V is a subspace of T_eG.

Proof. It is clear that V is closed under scalar multiplication. Thus it is enough to show that V is closed with respect to addition. So consider $v, w \in V$ with $v + w \neq 0$. Then

$$\exp(tv)\exp(tw) \in H \text{ for all } t \in \mathbb{R}.$$

Since by Proposition 3.9 exp is a diffeomorphism of an open neighborhood of $0 \in T_eG$ onto an open neighborhood of $e \in G$, for every small t there is a uniquely determined element $f(t) \in T_eG$ with

$$\exp(tv)\exp(tw) = \exp f(t).$$

The element $f(t)$ depends differentiably on t so it defines a differentiable curve in T_eG. By Proposition 3.8

$$\exp(tv)\exp(tw) = h_1^{tv}(e)h_1^{tw}(e) = h_t^v(e)h_t^w(e).$$

Hence

$$\left.\frac{d}{dt}\right|_{t=0} \exp(tv)\exp(tw) = \left.\frac{d}{dt}\right|_{t-0} h_t^v(e) + \left.\frac{d}{dt}\right|_{t-0} h_t^w(e) = v + w.$$

Similarly

$$\left.\frac{d}{dt}\right|_{t=0} \exp t(v + w) = v + w.$$

It follows that

$$\lim_{t\to 0} \frac{1}{t}(\exp(f(t)) - \exp t(v + w)) = 0.$$

Since exp is a local diffeomorphism in the neighborhood of 0, we have

$$\lim_{t\to 0} \frac{f(t)}{t} = v + w.$$

We now apply Lemma 3.3 with $v_i = f(1/i)$. Then $\lim_{i\to\infty} v_i = 0$ and $\exp v_i \in H$ for all i, and

$$\lim_{i\to\infty} \frac{f\left(\frac{1}{i}\right)}{\left(\left\|f\left(\frac{1}{i}\right)\right\|\right)} = \lim_{i\to\infty} \left\|\frac{\frac{1}{i}}{f\left(\frac{1}{i}\right)}\right\| \frac{f\left(\frac{1}{i}\right)}{\frac{1}{i}} = \frac{v + w}{\|v + w\|}.$$

By Lemma 3.3, $\exp t(v + w) \in H$ for all $t \in \mathbb{R}$. Hence $v + w \in V$. □

Proof of Proposition 3.10. Let V be the subspace of T_eG as in Lemma 3.4 and let W be a subspace complementary to it. We show first that $\exp(V)$ is a neighborhood of e in H.

Suppose that this is not the case. Then there is a sequence $a_1, a_2, \ldots$ in H with $\lim_{i\to\infty} a_i = e$ but $a_i \notin \exp(V)$. Since $\gamma : T_eG \to G$, $(v, w) \mapsto$

$\exp(v)\exp(w)$, is, by Corollary 3.2, a local diffeomorphism in the neighborhood of 0, there are vectors $v_i \in V$ and $w_i \in W$ with $\exp(v_i)\exp(w_i) = a_i$. From this it follows that $\exp w_i = (\exp v_i)^{-1} a_i \in H$ for all i. By passing to a subsequence, if necessary, we can assume that

$$\lim_{i\to\infty} \frac{w_i}{\|w_i\|} = w \in W, \quad \|w\| = 1.$$

By Lemma 3.3 it follows that $w \in V$. But then it follows that

$$w \in V \cap W = \{0\},$$

in contradiction to $\|w\| = 1$.

Thus $\exp(V)$ is a neighborhood of e in H. Hence there are an open neighborhood U' of $0 \in T_eG$ and an open neighborhood U of e in G such that $\exp|_{U'} : U' \to U$ is a diffeomorphism and maps $U' \cap V$ diffeomorphically onto $U \cap H$. The map $\varphi := (\exp|_{U'})^{-1} : U \to U'$ is therefore a chart of G around e that maps $U \cap H$ onto $U' \cap V$. On translating by the left translation l_g, we obtain a differentiable atlas for G that exhibits H as a submanifold. □

Now let H be a closed subgroup of G. We endow the quotient space G/H with the quotient topology; cf. §1.3. Since H is also a Lie subgroup, we can say more.

Proposition 3.11. *Let H be a closed subgroup of a Lie group G. Then G/H is a differentiable manifold of dimension*

$$\dim G/H = \dim G - \dim H.$$

Moreover, the quotient map $\pi : G \to G/H$ is differentiable and for each $q \in G/H$ there exist an open neighborhood Q of q and a differentiable map $\tau : Q \to G$ with $\pi \circ \tau = \mathrm{id}_Q$, i.e., τ is a local section of π.

Proof. Let W be a complement of T_eH in T_eG. Let U be an open neighborhood of 0 in T_eG such that $\exp|_U : U \to \exp(U)$ is a diffeomorphism. Then

$$\pi \circ \exp|_{U\cap W} : U \cap W \to \pi(\exp(U \cap W)) \subset G/H$$

is a homeomorphism, for $\exp(U \cap W)$ is open in G so $Q := \pi(\exp(U \cap W))$ is also open in G/H. It is easy to see that $(\pi \circ \exp|_{U\cap W})^{-1}$ is continuous. Thus $\varphi := (\pi \circ \exp|_{U\cap W})^{-1} : Q \to U \cap W$ is a chart of G/H around the residue class of e. With the help of the left translation l_g we obtain a chart around the residue class gH. Hence G/H is a differentiable manifold and

$$\dim G/H = \dim W = \dim G - \dim H.$$

Furthermore, π is differentiable and the differentiable map $\tau := \exp \circ \varphi : Q \to G$ is the desired local section around the residue class of e. □

In §1.3, we have already considered group actions on topological spaces. Now we consider the action of a Lie group G on a differentiable manifold X.

Definition. Let M be a differentiable manifold and G a Lie group. An *action of* G *on* M is a differentiable map $M \times G \to M$, $(p, g) \mapsto pg$ with the following properties:

(i) $(pg_1)g_2 = p(g_1g_2)$ for all $p \in M$ and $g_1, g_2 \in G$,

(ii) $pe = p$ for all $p \in M$.

Let $p \in M$. The set $pG := \{pg \mid g \in G\}$ is called the *orbit* (of the action of G on M) through p.

The subgroup $G_p := \{g \in G \mid pg = p\}$ is called the *isotropy group* of p.

Remark 3.7. The isotropy group G_p of a point $p \in M$ is a closed subgroup of G and thus a Lie subgroup of G.

Proposition 3.12. *Let G be a Lie group that acts on a differentiable manifold M, and let $p \in M$. Then the natural map $\lambda : G/G_p \to M$, $gG_p \mapsto pg$, is an immersion and maps G/G_p bijectively onto the orbit pG through p. In particular there is an open neighborhood U of the residue class of e in G/G_p such that $pG \cap \lambda(U)$ is a differentiable submanifold of M of dimension* $\dim G/G_p$.

Proof. It is clear that λ is differentiable and maps G/G_p bijectively onto pG. We show that λ is an immersion. To do this, it is enough to show that $T_{\bar{e}}\lambda$ is injective, where $\bar{e} = eG_p$ in G/G_p. For let $\mathrm{Diff}(M)$ be the group of all diffeomorphisms of M and $\rho : G \to \mathrm{Diff}(M)$ the map given by $\rho(g)(q) = qg$ for $g \in G$ and $q \in M$. The left translation $l_g : G \to G$ induces a differentiable map $\bar{l}_g : G/G_p \to G/G_p$ and $\lambda(\bar{g}) = (\rho(g) \circ \lambda \circ (\bar{l}_g)^{-1})(\bar{g})$, whence

$$T_{\bar{g}}\lambda = T_p(\rho(g)) \circ T_{\bar{e}}\lambda \circ (T_{\bar{g}}\bar{l}_g)^{-1}.$$

Now let $w \in T_{\bar{e}}(G/G_p)$ with $T_{\bar{e}}\lambda(w) = 0$. Let Q be a suitable neighborhood of $\bar{e}$ in G/G_p, let $\sigma : G \to M$ be the map given by $\sigma(g) = pg$, and let $\tau : Q \to G$ be the local section of $\pi : G \to G/G_p$, which exists by Proposition 3.11. Then $\lambda|_Q = \sigma \circ \tau$. If then $v := T_{\bar{e}}\tau(w) \in T_eG$, it follows that $T_e\sigma(v) = 0$. Let $\tilde{h} : \mathbb{R} \times M \to M$ be the map given by $\tilde{h}(t, q) = qh_t^v(e)$ for $t \in \mathbb{R}$, $q \in M$. This is a one parameter group of diffeomorphisms of M for which $h_{t+s}^v(e) = h_t^v(e)h_s^v(e)$, by Proposition 3.8. Let X be the vector field induced by $\tilde{h}$. Since $\tilde{h}_t(p) = \sigma(h_t^v(e))$, it follows that

$$X_p = \left.\frac{d}{dt}\right|_{t=0} \tilde{h}_t(p) = \left.\frac{d}{dt}\right|_{t=0} \sigma(h_t^v(e)) = T_e\sigma(v) = 0.$$

Since, by the proof of Proposition 3.3, $\tilde{h}_t(p)$ is the solution of a differential equation $y' = f(y)$ with $y'(0) = 0$ (for $X_p = 0$) and initial condition $\tilde{h}_0(p) = p$, the uniqueness of such a solution shows that $\tilde{h}_t(p) = p$ for all t in a small interval around 0. The usual argument shows that $\tilde{h}_t(p) = p$ for all

$t \in \mathbb{R}$. This means that $h_t^v(e)$ must lie in G_p for all $t \in \mathbb{R}$. It follows that $v = T_{\bar{e}}\tau(w) \in T_e G_p$. Hence $w = T_e\pi T_{\bar{e}}\tau(w) = 0$. Therefore $T_{\bar{e}}\lambda$ is injective.

The other assertions follow from Proposition 3.6. □

3.5. Complex manifolds

Let us now introduce complex manifolds. The concepts are analogous to those for differentiable manifolds.

Definition. Let M be a $2n$-dimensional topological manifold. An atlas of M is called *complex* (or *holomorphic*) if all its transition functions are holomorphic. (Here we identify $\mathbb{R}^{2n}$ in the natural way with $\mathbb{C}^n$.)

Two complex atlases $\mathfrak{A}$ and $\mathfrak{W}$ are called *equivalent* if and only if $\mathfrak{A} \cup \mathfrak{W}$ is also a complex atlas.

A *complex structure* on M is an equivalence class of complex atlases on M.

A *complex manifold* is a topological manifold together with a complex structure. If M is a complex manifold, then n is called the (complex) *dimension* of the complex manifold M.

Remark 3.8. Each complex structure contains a uniquely determined maximal atlas $\mathfrak{A}^*$: If $\mathfrak{A}$ is an arbitrary atlas from the corresponding equivalence class, then

$$\mathfrak{A}^* := \left\{ \varphi : U \longrightarrow V \text{ chart} \;\middle|\; \begin{array}{l} \text{transition function from } \varphi \text{ to all charts of } \mathfrak{A} \\ \text{is holomorphic} \end{array} \right\}$$

is a maximal atlas.

Convention. If M is a complex manifold, then a chart of M is always a chart of the maximal atlas of the complex structure.

Example 3.5. The Riemann surfaces considered in §1.1 are connected complex manifolds of real dimension 2 and complex dimension 1.

Definition. Let M, N be complex manifolds. A continuous map $f : M \to N$ is called *holomorphic at* $a \in M$ if and only if the composition $\psi \circ f \circ \varphi^{-1}$ is holomorphic at $\varphi(a) \in U'$ for each chart $\varphi : U \to U'$, $a \in U$, and $\psi : V \to V'$, $f(a) \in V$, of M, resp. N (see Figure 3.1) (the domain of definition of $\psi \circ f \circ \varphi^{-1}$ being $\varphi(f^{-1}(V) \cap U)$.) The map f is called *holomorphic* if it is holomorphic at every $a \in M$.

A holomorphic map $f : M \to N$ is called *biholomorphic* if and only if it is invertible and its inverse map is holomorphic.

Remark 3.9. Let M be a complex manifold, $\varphi : U \to U'$ a chart of M. Then U and U' are also complex manifolds as open subsets of M and $\mathbb{C}^n$, and

$\varphi : U \to U'$ is a biholomorphic map between U and U'. A chart $\varphi : U \to U'$ around a point $a \in M$, $a \in U$, is often also called a local coordinate system: Let $\varphi_1, \ldots, \varphi_n$ be the component functions of φ. They are holomorphic functions. By translation in $\mathbb{C}^n$ one can ensure that $\varphi(a) = 0$. After the choice of coordinates $x_1, \ldots, x_n$ of $\mathbb{C}^n$ we call

$$x_1 = \varphi_1(x), \ldots, x_n = \varphi_n(x)$$

the coordinates of the point $x \in U$. The point $a \in M$ then has the coordinates $(0, \ldots, 0)$. In this way every point of U can be described uniquely in terms of coordinates. A function on U is holomorphic if and only if it is holomorphic in the usual sense as a function of the coordinates.

Definition. Let M be a complex manifold, $a \in M$. The set of all germs at a of holomorphic functions $f : U \to \mathbb{C}$, U an open neighborhood of a, is denoted by $\mathcal{O}_{M,a}$. In particular we put

$$\mathcal{O}_{n,0} := \mathcal{O}_{\mathbb{C}^n,0}.$$

Definition. Let M be a complex manifold and $a \in M$. A *derivation* of $\mathcal{O}_{M,a}$ is a $\mathbb{C}$-linear map $\delta : \mathcal{O}_{M,a} \to \mathbb{C}$ that satisfies the product rule

$$\delta(\bar{f} \cdot \bar{g}) = \bar{f}(a)\delta(\bar{g}) + \bar{g}(a)\delta(\bar{f})$$

for all $\bar{f}, \bar{g} \in \mathcal{O}_{M,a}$.

Let us denote by $\operatorname{Der} \mathcal{O}_{M,a}$ the set of all derivations of $\mathcal{O}_{M,a}$. These form a vector space.

Let $x_1, \ldots, x_n$ be the coordinates of $\mathbb{C}^n$. Then we define derivations

$$\frac{\partial}{\partial x_j} : \mathcal{O}_{n,0} \longrightarrow \mathbb{C}$$

as in §3.1 by $f \mapsto (\partial/\partial x_j)f(0)$ (partial derivative with respect to the coordinate x_j).

Proposition 3.13. *The derivations $\partial/\partial x_j$, $j = 1, \ldots, n$, form a basis of the vector space* $\operatorname{Der} \mathcal{O}_{n,0}$.

Proof. We put $V = \operatorname{Der} \mathcal{O}_{n,0}$.

We show first that the $\partial/\partial x_j$ are linearly independent: Let

$$\sum_{j=1}^{n} a_j \frac{\partial}{\partial x_j} = 0, \quad a_j \in \mathbb{C}.$$

Then it follows that

$$a_k = \sum_{j=1}^{n} a_j \frac{\partial \bar{x}_k}{\partial x_j} = 0 \text{ for all } k.$$

Now let $\delta \in V$, $a_k := \delta(\bar{x}_k)$. We show

$$\delta = \sum_{j=1}^{n} a_j \frac{\partial}{\partial x_j}.$$

Let $\bar{f} \in \mathcal{O}_{n,0}$ and let $f : U \to \mathbb{C}$ be a representative of $\bar{f}$. The existence of holomorphic functions $f_j : U \to \mathbb{C}$, $j = 1, \ldots, n$, with $f = f(0) + \sum_{j=1}^{n} x_j f_j$ follows from Proposition 2.10.

Then

$$\begin{aligned} \delta(\bar{f}) &= \delta(f(0)) + \sum_{j=1}^{n} \delta(\bar{x}_j) \cdot f_j(0) \\ &= \sum_{j=1}^{n} a_j \cdot f_j(0) \\ &= \sum_{j=1}^{n} a_j \frac{\partial \bar{f}}{\partial x_j}, \end{aligned}$$

as was to be shown. □

Definition. Let M be a complex manifold, $a \in M$. The *tangent space of M at a*, symbolically, T_aM, is the complex vector space $\operatorname{Der} \mathcal{O}_{M,a}$.

Remark 3.10. If M has complex dimension n, then T_aM has complex dimension n.

Remark 3.11. Let (X, x) be an analytic set germ with corresponding analytic algebra $\mathcal{O}_{X,x}$ and let $\mathfrak{m}$ be the maximal ideal of $\mathcal{O}_{X,x}$. Then the vector space

$$T_xX := \operatorname{Hom}_{\mathbb{C}}(\mathfrak{m}/\mathfrak{m}^2, \mathbb{C}),$$

where $\operatorname{Hom}_{\mathbb{C}}(\mathfrak{m}/\mathfrak{m}^2, \mathbb{C})$ denotes the set of $\mathbb{C}$-linear maps $\varphi : \mathfrak{m}/\mathfrak{m}^2 \to \mathbb{C}$, is called the *Zariski tangent space* of (X, x). One should note that this tangent space is always defined, independently of whether x is a regular or singular point of X. The space

$$\mathfrak{m}/\mathfrak{m}^2$$

is correspondingly called the *cotangent space* of (X, x). We have

$$T_xX \cong \operatorname{Der} \mathcal{O}_{X,x}$$

(Proof: Exercise).

Let M, N be complex manifolds and $f : M \to N$ a holomorphic map. If $a \in M$, then f induces an algebra homomorphism $f^* : \mathcal{O}_{N,f(a)} \to \mathcal{O}_{M,a}$.

Definition. The map

$$\begin{array}{rccl} T_a f: & T_a M & \longrightarrow & T_{f(a)} N \\ & \delta & \longmapsto & \delta \circ f^* \end{array}$$

is called the *tangent map* (or the *differential*) of f at a.

Remark 3.12. $T_a f$ is a complex linear map.

Let X be a complex manifold and $a \in X$. We wish to study the connection between the tangent space $T_a X$ to X at a and the tangent space to the underlying differentiable manifold M at a.

First we need a little linear algebra: Let V be a vector space over $\mathbb{R}$. A *complex structure* on V is an $\mathbb{R}$-linear isomorphism $J : V \to V$ with $J^2 = -\operatorname{id}$. Let V be an $\mathbb{R}$-vector space with a complex structure J. Then we can make V into a complex vector space in the following way: We define a scalar multiplication by $\alpha + i\beta \in \mathbb{C}$, $\alpha, \beta \in \mathbb{R}$, $i = \sqrt{-1}$, as follows:

$$(\alpha + i\beta)v := \alpha v + \beta J v.$$

It is easy to check that this makes V a complex vector space.

Conversely, suppose V is a complex vector space. Then V can also be considered as an $\mathbb{R}$-vector space $V_{\mathbb{R}}$. Let $J : V_{\mathbb{R}} \to V_{\mathbb{R}}$ be multiplication by i. Then J is a complex structure on $V_{\mathbb{R}}$. If $\{v_1, \dots, v_n\}$ is a basis of V over $\mathbb{C}$, then $\{v_1, Jv_1, \dots, v_n, Jv_n\}$ is a basis of $V_{\mathbb{R}}$ over $\mathbb{R}$.

In particular, let

$$V = \mathbb{C}^n = \{(z_1, \dots, z_n) \mid z_j \in \mathbb{C}\},$$

and let $z_j = x_j + iy_j$ be the decomposition into real and imaginary parts. Then

$$V_{\mathbb{R}} = \mathbb{R}^{2n} = \{(x_1, y_1, \dots, x_n, y_n) \mid x_j, y_j \in \mathbb{R}\}.$$

Scalar multiplication by i in $\mathbb{C}^n$ furnishes a map

$$\begin{array}{rccc} J: & \mathbb{R}^{2n} & \longrightarrow & \mathbb{R}^{2n} \\ & (x_1, y_1, \dots, x_n, y_n) & \longmapsto & (-y_1, x_1, \dots, -y_n, x_n). \end{array}$$

This is an $\mathbb{R}$-linear map, specified by a matrix of the form

$$\begin{pmatrix} 0 & 1 & & & \\ -1 & 0 & & & \\ & & \ddots & & \\ & & & 0 & 1 \\ & & & -1 & 0 \end{pmatrix},$$

and $J^2 = -\operatorname{id}$. We call this complex structure the *standard complex structure* on $\mathbb{R}^{2n}$. Note that there are other complex structures on $\mathbb{R}^{2n}$!

Next we consider $\mathbb{C}^n$ with the above coordinates. By Proposition 3.13, $\{\partial/\partial z_1, \ldots, \partial/\partial z_n\}$ is a basis of $T_0\mathbb{C}^n$, and, by Proposition 3.1, $\{\partial/\partial x_1, \partial/\partial y_1, \ldots, \partial/\partial x_n, \partial/\partial y_n\}$ forms a basis of $T_0\mathbb{R}^{2n}$. If we identify

$$T_0\mathbb{C}^n \cong \mathbb{C}^n \quad \text{by} \quad \left\{\frac{\partial}{\partial z_1}, \ldots, \frac{\partial}{\partial z_n}\right\},$$

$$T_0\mathbb{R}^{2n} \cong \mathbb{R}^{2n} \quad \text{by} \quad \left\{\frac{\partial}{\partial x_1}, \frac{\partial}{\partial y_1}, \ldots, \frac{\partial}{\partial x_n}, \frac{\partial}{\partial y_n}\right\},$$

it follows from

$$\frac{\partial}{\partial z_j} = \frac{\partial}{\partial x_j} + i\frac{\partial}{\partial y_j}$$

that $T_0\mathbb{C}^n$ induces the standard complex structure on $T_0\mathbb{R}^{2n}$.

Now let X be a complex manifold of complex dimension n and M the underlying $2n$-dimensional differentiable manifold. For a point $a \in X$ let T_aX be the (complex) tangent space to X at a and T_aM the (real) tangent space to M at a.

Proposition 3.14. *T_aM is in canonical manner isomorphic to the underlying real vector space of T_aX, and T_aX induces a complex structure J_a on T_aM.*

Proof. a) Let $\varphi : U \to U' \subset \mathbb{C}^n$, $a \in U$, be a chart of X around a. We obtain a chart $\tilde{\varphi} : U \to U' \subset \mathbb{R}^{2n}$ of M by

$$\tilde{\varphi}(x) = (\operatorname{Re}\varphi_1(x), \operatorname{Im}\varphi_1(x), \ldots, \operatorname{Re}\varphi_n(x), \operatorname{Im}\varphi_n(x)).$$

Then $T_a\varphi$ determines a $\mathbb{C}$-linear isomorphism between T_aX and $T_0\mathbb{C}^n$ and $T_a\tilde{\varphi}$ an $\mathbb{R}$-linear isomorphism between T_aM and $T_0\mathbb{R}^{2n}$:

$$T_aX \xrightarrow[T_a\varphi]{\cong} T_0\mathbb{C}^n \cong \mathbb{C}^n,$$

$$T_aM \xrightarrow[T_a\tilde{\varphi}]{\cong} T_0\mathbb{R}^{2n} \cong \mathbb{R}^{2n}.$$

It follows that T_aM is isomorphic to T_aX and with respect to this isomorphism T_aX induces a complex structure J_a on T_aM.

b) We show that the complex structure J_a on T_aM is independent of the choice of the holomorphic chart φ. Suppose that $\psi : V \to V'$, $a \in V$, is another chart of X around a. Let $W = \varphi(U \cap V) \subset \mathbb{C}^n$, and assume that $\varphi(a) = \psi(a) = 0$. Then $f = \psi \circ \varphi^{-1} : W \to W$ is the corresponding transition function, f is biholomorphic, and $f(0) = 0$. We put

$$u = \operatorname{Re} f, \quad v = \operatorname{Im} f.$$

By the Cauchy-Riemann differential equations the functional matrix A of the real map f is as follows:

$$A = \begin{pmatrix} \frac{\partial u_1}{\partial x_1} & \frac{\partial u_1}{\partial y_1} & \cdots & \frac{\partial u_1}{\partial x_n} & \frac{\partial u_1}{\partial y_n} \\ -\frac{\partial u_1}{\partial y_1} & \frac{\partial u_1}{\partial x_1} & & -\frac{\partial u_1}{\partial y_n} & \frac{\partial u_1}{\partial x_n} \\ \vdots & & \ddots & \vdots & \\ \frac{\partial u_n}{\partial x_1} & \frac{\partial u_n}{\partial y_1} & \cdots & \frac{\partial u_n}{\partial x_n} & \frac{\partial u_n}{\partial y_n} \\ -\frac{\partial u_n}{\partial y_1} & \frac{\partial u_n}{\partial x_1} & & -\frac{\partial u_n}{\partial y_n} & \frac{\partial u_n}{\partial x_n} \end{pmatrix}.$$

If J is the matrix of the canonical complex structure of $\mathbb{R}^{2n}$,

$$\begin{pmatrix} \begin{matrix} 0 & 1 \\ -1 & 0 \end{matrix} & & 0 \\ & \ddots & \\ 0 & & \begin{matrix} 0 & 1 \\ -1 & 0 \end{matrix} \end{pmatrix},$$

then, as is easy to see,

$$AJ = JA.$$

It follows that J induces the same complex structure on T_aM for each choice of holomorphic coordinate system at a. □

The concept of a complex submanifold of an open subset of the $\mathbb{C}^n$ was introduced in §2.7. We now define in general the concept of a submanifold of a complex manifold.

Definition. Let M be a complex manifold of dimension n. Furthermore, let $0 \leq k \leq n$. A subset $N \subset M$ is called a *k-dimensional complex submanifold* of M if for every $a \in N$ there is a chart $\varphi : U \to U' \subset \mathbb{C}^n = \mathbb{C}^k \times \mathbb{C}^{n-k}$ with $a \in U$ such that

$$\varphi(U \cap N) = U' \cap (\mathbb{C}^k \times \{0\})$$

(see Figure 3.3, p. 118).

The number $n - k$ is called the *codimension* of the submanifold N.

A complex submanifold N is itself again a complex manifold: From a chart φ as in the definition one obtains a chart $\varphi' = \varphi|_{U\cap N} : U\cap N \to U'\cap\mathbb{C}^k$, where we identify $\mathbb{C}^k$ with $\mathbb{C}^k \times \{0\} \subset \mathbb{C}^n$; and the set of all these charts forms a complex atlas. The tangent space to N at a point $a \in N$ is a subspace of T_aM in the natural way.

Remark 3.13. The results about vector fields and one parameter groups of diffeomorphisms were formulated in §3.2 only for differentiable manifolds, but they hold analogously also for complex manifolds, holomorphic vector fields and holomorphic one parameter groups. If M is a complex manifold, one can correspondingly define a complex tangent bundle and *holomorphic* *vector fields* as holomorphic sections in the complex tangent bundle. A *holomorphic one parameter group* is a holomorphic map $g : \mathbb{C} \times M \to M$ with the obvious properties. In particular, for each $t \in \mathbb{C}$ the map $g_t : M \to M$, $a \mapsto g(t, a)$, is a biholomorphic transformation of M. Correspondingly one also defines the concept of a *holomorphic local one parameter group*: This is a holomorphic map $g : I \times U \to M$, $U \subset M$ open, with the corresponding properties. Then, as in the differentiable case, holomorphic vector fields and holomorphic local one parameter groups correspond. The proofs are identical and will not be presented here.

Definition. A *complex Lie group* is a set G with the following properties:

(i) G is a group.

(ii) G is a complex manifold.

(iii) The map $G \times G \to G$, $(a, b) \mapsto ab^{-1}$, is holomorphic.

Example 3.6. The Lie group $\mathrm{GL}(n, \mathbb{C})$ is also a complex Lie group of complex dimension n^2.

Definition. Let M be a complex manifold and G a complex Lie group. An *action of G on M* is a holomorphic map $M \times G \to M$, $(p, g) \mapsto pg$ with the following properties:

(i) $(pg_1)g_2 = p(g_1 g_2)$ for all $p \in M$ and $g_1, g_2 \in G$,

(ii) $pe = p$ for all $p \in M$.

Remark 3.14. All the results of §3.4 hold analogously for complex Lie groups and actions of complex Lie groups on complex manifolds. The proofs can be taken over almost word-for-word to this situation, replacing the concept differentiable by complex or holomorphic throughout.

3.6. Isolated critical points

We now come to the main theme of this chapter, namely the study of isolated singularities of holomorphic functions.

Let M be an $(n+1)$-dimensional complex manifold and $f : M \to \mathbb{C}$ a holomorphic function.

Definition. A point $p \in M$ is called a *critical point* or *singularity* of f if its differential $T_p f$ is the null map. If a is a critical point of f, then $f(a)$ is called a *critical value* of f.

Remark 3.15. If $(z_1, \ldots, z_{n+1})$ is a local coordinate system around a (with $z_j(a) = 0$), then a is a critical point of f if and only if

$$\frac{\partial f}{\partial z_1}(0) = \ldots = \frac{\partial f}{\partial z_{n+1}}(0) = 0.$$

Definition. A point $p \in M$ is called an *isolated critical point* or an *isolated singularity* of f if there is a neighborhood U of p in M such that no point of $U \setminus \{p\}$ is critical.

Remark 3.16. One should note that this definition also includes the case when p is not a critical point at all. This may appear illogical but is, however, the usual definition.

Let $f : (\mathbb{C}^{n+1}, 0) \to (\mathbb{C}, 0)$ be the germ of a holomorphic function with an isolated critical point at 0. If $M \subset \mathbb{C}^{n+1}$ is a suitable open neighborhood of $0 \in \mathbb{C}^{n+1}$, if $\tilde{f} : M \to \mathbb{C}$ is a representative of f, and $X := \tilde{f}^{-1}(0)$, then X has a regular or an isolated singular point at 0. We also then say simply that f has an isolated singularity at 0. One often calls f or the analytic set germ $(X, 0)$ an isolated singularity. In the sequel we shall not distinguish notationally between a germ f and a representative $f : M \to \mathbb{C}$, where M is an open neighborhood of $0 \in \mathbb{C}^{n+1}$.

By the implicit function theorem the level surface $f^{-1}(w)$ for $w \in \mathbb{C}$, $w \neq 0$, $|w|$ suitably small, is a complex submanifold of $\mathbb{C}^{n+1}$ in a neighborhood of $0 \in \mathbb{C}^{n+1}$. The zero set $f^{-1}(0)$ has a singularity at $0 \in \mathbb{C}^{n+1}$, but it is a complex submanifold of $\mathbb{C}^{n+1}$ in a neighborhood of 0, apart from 0.

Lemma 3.5. *There exists an $\varepsilon > 0$ such that the sphere $S_\rho \subset \mathbb{C}^{n+1}$, center 0, of radius $\rho \leq \varepsilon$, intersects the zero set $f^{-1}(0)$ transversally.*

For the proof of this auxiliary result we need the following lemma, which we cite without proof.

Lemma 3.6 (Curve selection lemma). *Let $V \subset \mathbb{R}^m$ be an open neighborhood of $p \in \mathbb{R}^m$, let $f_1, \ldots, f_k, g_1, \ldots, g_l : V \to \mathbb{R}$ be real analytic functions, and let*

$$Z := \{x \in V \mid f_1(x) = \ldots = f_k(x) = 0, g_1(x) > 0, \ldots, g_l(x) > 0\}.$$

If $p \in \bar{Z}$, there exists a real analytic curve $\gamma : [0, \delta) \to V$, $0 < \delta$, with $\gamma(0) = p$ and $\gamma(t) \in Z$ for all $t \in (0, \delta)$.

For the proof see [**Mil68**, §3]. It is assumed there that $V = \mathbb{R}^m$ and that the $f_1, \ldots, f_k, g_1, \ldots, g_l$ are polynomials. The proof can be adapted immediately to the analytic case.

Proof of Lemma 3.5. We consider the function $r|_{f^{-1}(0)} : f^{-1}(0) \to \mathbb{R}$, $r(z) = |z|^2$. The critical points of r on $f^{-1}(0)$ are exactly those points of $f^{-1}(0)$ at which $S_{\sqrt{r}}$ and $f^{-1}(0)$ do not intersect transversally. Put

$$Z = \{z \in f^{-1}(0) \mid z \text{ a critical point of } r|_{f^{-1}(0)\setminus\{0\}}\}.$$

Then Z is specified by real analytic equations and inequalities, so it satisfies the hypothesis of the curve selection lemma. We must show that 0 is not an accumulation point of Z.

Suppose $0 \in \bar{Z}$. Then, by the curve selection lemma, there are a $\delta > 0$ and a curve $\gamma : [0, \delta) \to f^{-1}(0)$ with $\gamma(0) = 0$ and $\gamma(t) \in Z$ for $t \in (0, \delta)$. Thus

$$(r \circ \gamma)'(t) = \langle \operatorname{grad} r(\gamma(t)), \gamma'(t) \rangle = 0$$

for all $t \in (0, \delta)$. Hence $r \circ \gamma$ is constant, and, since $r \circ \gamma(0) = 0$, we have $r \circ \gamma \equiv 0$ on $[0, \delta)$. Since $r^{-1}(0) = 0$, it follows that $\gamma(t) = 0$ for all $t \in [0, \delta)$. This, however, contradicts $\gamma(t) \in Z$ for $t \in (0, \delta)$. □

Now let M be an open neighborhood of $0 \in \mathbb{C}^{n+1}$ and let $f : M \to \mathbb{C}$ be a representative of f. Let $\varepsilon > 0$ be as in Lemma 3.5, $B_\varepsilon \subset M$ the open ball around $0 \in \mathbb{C}^{n+1}$ of radius ε. This is a complex submanifold of M. It follows from Lemma 3.5 that there is an $\eta_0 > 0$, $\eta_0 \ll \varepsilon$, such that $f^{-1}(w)$ intersects the sphere S_ε transversally for $w \in \mathbb{C}$, $|w| \leq \eta_0$.

Now again let M be an $(n+1)$-dimensional complex manifold and $f : M \to \mathbb{C}$ a holomorphic function.

Definition. A critical point $p \in M$ of f is called *nondegenerate* if there is a local coordinate system $(z_1, \ldots, z_{n+1})$ around p (with $p = 0$) such that $\det\big((\partial^2 f / \partial z_i \partial z_j)(0)\big) \neq 0$. Here $\det\big((\partial^2 f / \partial z_i \partial z_j)(0)\big)$ is the determinant of the Hessian matrix of f in the coordinates $(z_1, \ldots, z_{n+1})$ at 0.

Remark 3.17. One can show that the condition for a nondegenerate critical point is independent of the choice of the coordinate system.

There is a complex Morse lemma:

Proposition 3.15 (Complex Morse lemma). *Let M be an $(n+1)$-dimensional complex manifold, $f : M \to \mathbb{C}$ a holomorphic function and p a nondegenerate critical point of f. Then there is a local coordinate system $(z_1, \ldots, z_{n+1})$ in a neighborhood V of p with $z_i(p) = 0$, $i = 1, \ldots, n+1$, such that on V,*

$$f(z_1, \ldots, z_{n+1}) = f(p) + z_1^2 + \ldots + z_{n+1}^2.$$

To prove Proposition 3.15, we need the following lemma.

Lemma 3.7. *Let $f : (\mathbb{C}^{n+1}, 0) \to (\mathbb{C}, 0)$ be the germ of a holomorphic function with a nondegenerate critical point in 0. Then the partial derivatives $\partial f / \partial z_1, \ldots, \partial f / \partial z_{n+1}$ generate the maximal ideal $\mathfrak{m}_{n+1}$ of $\mathcal{O}_{n+1}$.*

Proof. Let $U \subset \mathbb{C}^{n+1}$ be an open neighborhood of 0 and $f : U \to \mathbb{C}$ a representative of f. We consider the map

$$\begin{array}{rccl} \operatorname{grad} f : & U & \longrightarrow & \mathbb{C}^{n+1} \\ & z & \longmapsto & \left(\frac{\partial f}{\partial z_1}(z), \ldots, \frac{\partial f}{\partial z_{n+1}}(z)\right). \end{array}$$

Since 0 is a nondegenerate critical point of f, we have

$$\operatorname{rank}\left(J_{\operatorname{grad} f}(0)\right) = n+1.$$

From the rank theorem (Proposition 2.34) follows: There are open neighborhoods V, V' of 0 in U and a biholomorphic map $\varphi : V \to V'$ such that for all $z \in V$,

$$\operatorname{grad} f \circ \varphi(z_1, \ldots, z_{n+1}) = (z_1, \ldots, z_{n+1}).$$

Since $z_1, \ldots, z_{n+1}$ generate the maximal ideal $\mathfrak{m}_{n+1}$ of $\mathcal{O}_{n+1}$, the result follows. □

Proof of Proposition 3.15. Since we are dealing with a local assertion, we can assume that M is an open neighborhood of 0 in $\mathbb{C}^{n+1}$, $p = 0$ and $f(p) = 0$.

By a linear change of coordinates we can ensure that in suitable coordinates f has the form

$$f(u) = u_1^2 + \ldots + u_{n+1}^2 + \varphi(u)$$

with $\varphi \subset \mathfrak{m}_{n+1}^3$. We consider the function

$$F(u,t) := u_1^2 + \ldots + u_{n+1}^2 + t\varphi(u)$$

for u in a neighborhood U of 0 in $\mathbb{C}^{n+1}$ and $t \in [0,1]$. We now search for a holomorphic one parameter group $g : [0,1] \times U' \to U'$ for a suitable neighborhood U' of $0 \in \mathbb{C}^{n+1}$ such that

(3.2)
$$F(g_t(u), t) = u_1^2 + \ldots + u_{n+1}^2, \; g_0(u) = u, \; g_t(0) = 0 \text{ for } u \in U',\, t \in [0,1].$$

A holomorphic vector field

$$X = \frac{\partial}{\partial t} + \sum_{j=1}^{n+1} a_{t,j} \frac{\partial}{\partial u_j}$$

is induced on $[0,1] \times U'$ by g. From the equation (3.2) it follows that $XF = 0$ for this vector field. Hence one obtains the following equation for the functions $a_{t,j}$ sought:

$$\frac{\partial F}{\partial t} + \sum_{j=1}^{n+1} a_{t,j} \frac{\partial F}{\partial u_j} = \varphi + \sum_{j=1}^{n+1} a_{t,j} \frac{\partial F}{\partial u_j} = 0.$$

We can rewrite this equation as

$$\sum_{j=1}^{n+1} a_{t,j} \frac{\partial F}{\partial u_j} = -\varphi. \tag{3.3}$$

However, the function F_t with $F_t(u) = F(u,t)$ has a nondegenerate critical point in 0 for each $t \in [0,1]$. By Lemma 3.7 the partial derivatives $\partial F_t/\partial u_1, \ldots, \partial F_t/\partial u_{n+1}$ generate the maximal ideal $\mathfrak{m}_{n+1}$ of $\mathcal{O}_{n+1}$. Hence in a suitable neighborhood W of $\{0\} \times [0,1]$ in $U \times \mathbb{C}$ we can take $w_1 := \partial F/\partial u_1, \ldots, w_{n+1} := \partial F/\partial u_{n+1}$ and t as new coordinates.

Since by hypothesis φ lies in $\mathfrak{m}_{n+1}^3$, we can write φ as

$$\varphi = \sum_{j=1}^{n+1} w_j \psi_j,$$

where the ψ_j are holomorphic functions on W with $\psi_j(0,t) = 0$ for $t \in [0,1]$. We can thus solve the equation (3.3) as follows:

$$a_{t,j} := -\psi_j.$$

By Proposition 3.3 and Remark 3.13 there is now a local holomorphic one parameter group $g' : I \times U'' \to \mathbb{C}^{n+1}$, where $U'' \subset U$ is a suitable neighborhood of $0 \in \mathbb{C}^{n+1}$, that induces the vector field X. The vector field X in this case also depends on t, but Proposition 3.3 can be generalized immediately to time dependent vector fields too. Since $X_{(0,t)} = 0$ for $t \in [0,1]$, we can extend g' to a holomorphic one parameter group $g : [0,1] \times U' \to U'$, where, if necessary, we shrink U'' to a suitable neighborhood U' of $0 \in \mathbb{C}^{n+1}$. Then $g_t(0) = 0$ for $t \in [0,1]$. The coordinate transformation sought is thus g_1. □

Corollary 3.3. *Nondegenerate critical points are isolated.*

Definition. A holomorphic function $f : M \to \mathbb{C}$ is called a *Morse function* if all its critical points are nondegenerate and all its critical values are distinct.

3.7. The universal unfolding

In this section we introduce the concept of an unfolding of a holomorphic function germ.

Let $f : (\mathbb{C}^{n+1}, 0) \to (\mathbb{C}, 0)$ be a holomorphic function germ.

Definition. An *unfolding* of f is a holomorphic function germ $F : (\mathbb{C}^{n+1} \times \mathbb{C}^k, 0) \to (\mathbb{C}, 0)$ with $F(z, 0) = f(z)$.

Definition. Two unfoldings $F : (\mathbb{C}^{n+1} \times \mathbb{C}^k, 0) \to (\mathbb{C}, 0)$, $G : (\mathbb{C}^{n+1} \times \mathbb{C}^k, 0) \to (\mathbb{C}, 0)$ of f are called *equivalent* if there is a holomorphic map germ

$$\psi : (\mathbb{C}^{n+1} \times \mathbb{C}^k, 0) \longrightarrow (\mathbb{C}^{n+1}, 0) \text{ with } \psi(z, 0) = z$$

such that

$$G(z,u) = F(\psi(z,u),u).$$

Definition. Let $F : (\mathbb{C}^{n+1} \times \mathbb{C}^k, 0) \to (\mathbb{C}, 0)$ be an unfolding of f and let $\varphi : (\mathbb{C}^l, 0) \to (\mathbb{C}^k, 0)$ be a holomorphic map germ. The unfolding $G : (\mathbb{C}^{n+1} \times \mathbb{C}^l, 0) \to (\mathbb{C}, 0)$ with

$$G(z,t) = F(z, \varphi(t))$$

is called the *unfolding of f induced from F by φ*.

Definition. An unfolding $F : (\mathbb{C}^{n+1} \times \mathbb{C}^k, 0) \to (\mathbb{C}, 0)$ of f is called *versal* if every unfolding of f is equivalent to an unfolding induced from F.

A versal unfolding $F : (\mathbb{C}^{n+1} \times \mathbb{C}^k, 0) \to (\mathbb{C}, 0)$ of f is called *universal* (or *miniversal*) if k is minimal.

Proposition 3.16. *Let $f : (\mathbb{C}^{n+1}, 0) \to (\mathbb{C}, 0)$ be a holomorphic function germ with an isolated singularity at* 0*. Then*

$$\mathcal{O}_{n+1}/\mathcal{O}_{n+1}\left(\frac{\partial f}{\partial z_1}, \ldots, \frac{\partial f}{\partial z_{n+1}}\right)$$

is a finite-dimensional $\mathbb{C}$-vector space.

Proof. Let $f_j := \partial f / \partial z_j$, $j = 1, \ldots, n+1$, be the partial derivatives of f and let J_f be the ideal spanned by the $f_1, \ldots, f_{n+1}$. Now f has an isolated singularity at 0, so 0 is the unique solution of

$$f_1(z) = \ldots = f_{n+1}(z) = 0$$

in an open neighborhood M of 0. Then

$$V(J_f) = V(f_1, \ldots, f_{n+1}) = (\{0\}, 0).$$

It follows that

$$I(V(J_f)) = \mathfrak{m},$$

where $\mathfrak{m}$ is the maximal ideal of $\mathcal{O}_{n+1}$. By Rückert's Nullstellensatz (Proposition 2.27)

$$I(V(J_f)) = \operatorname{rad} J_f.$$

Hence there is a natural number $q > 0$ such that $\mathfrak{m}^q \subset J_f$. But $\mathcal{O}_{n+1}/\mathfrak{m}^q$ is a finite-dimensional $\mathbb{C}$-vector space (cf. the proof of Proposition 2.42). Hence $\mathcal{O}_{n+1}/J_f$ is also a finite-dimensional $\mathbb{C}$-vector space. This completes the proof of Proposition 3.16. □

If f has an isolated singularity at 0, one can construct a universal unfolding of f as follows.

Proposition 3.17. *Let $f : (\mathbb{C}^{n+1}, 0) \to (\mathbb{C}, 0)$ be a holomorphic function germ with an isolated singularity at* 0. *Then one obtains a universal unfolding F of f as follows: Let $g_0 = -1, g_1, \ldots, g_{p-1}$ be representatives of a basis of the $\mathbb{C}$-vector space*

$$\mathcal{O}_{n+1}/\mathcal{O}_{n+1}\left(\frac{\partial f}{\partial z_1}, \ldots, \frac{\partial f}{\partial z_{n+1}}\right).$$

Then put

$$\begin{array}{rccc} F: & (\mathbb{C}^{n+1} \times \mathbb{C}^p, 0) & \longrightarrow & (\mathbb{C}, 0) \\ & (z, u) & \longmapsto & f(z) + \sum\limits_{j=0}^{p-1} g_j(z) u_j. \end{array}$$

We present a proof of this proposition modelled on [**AGV85, Mar82**]. We need a few preparations.

Let $F : (\mathbb{C}^{n+1} \times \mathbb{C}^k, 0) \to (\mathbb{C}, 0)$, $(z, u) \mapsto F(z, u)$ be an unfolding of f. We put

$$\dot{F}_j(z) = \left.\frac{\partial F(z, u)}{\partial u_j}\right|_{u=0}, \; j = 1, \ldots, k.$$

This defines function germs $\dot{F}_j \in \mathcal{O}_{n+1}$. We denote the set of all linear combinations of $\dot{F}_1, \ldots, \dot{F}_k$ with complex coefficients by

$$\mathbb{C}(\dot{F}_1, \ldots, \dot{F}_k).$$

Furthermore, for brevity, we put

$$\mathcal{O}_{n+1}\left(\frac{\partial f}{\partial z_i}\right) = \mathcal{O}_{n+1}\left(\frac{\partial f}{\partial z_1}, \ldots, \frac{\partial f}{\partial z_{n+1}}\right)$$

for the ideal of $\mathcal{O}_{n+1}$ spanned by the partial derivatives $\partial f/\partial z_1, \ldots, \partial f/\partial z_{n+1}$.

Definition. An unfolding $F : (\mathbb{C}^{n+1} \times \mathbb{C}^k, 0) \to (\mathbb{C}, 0)$ of f is called *infinitesimally versal* if

$$\mathcal{O}_{n+1} = \mathcal{O}_{n+1}\left(\frac{\partial f}{\partial z_i}\right) + \mathbb{C}(\dot{F}_1, \ldots, \dot{F}_k).$$

Example 3.7. The unfolding F as in Proposition 3.17 is infinitesimally versal.

Now let $F : (\mathbb{C}^{n+1} \times \mathbb{C}^k, 0) \to (\mathbb{C}, 0)$, $(z, u) \mapsto F(z, u)$, be an infinitesimally versal unfolding of f. Let

$$\begin{array}{rccc} G: & (\mathbb{C}^{n+1} \times \mathbb{C}^k \times \mathbb{C}, 0) & \longrightarrow & (\mathbb{C}, 0) \\ & (z, u, v) & \longmapsto & G(z, u, v) \end{array}$$

be a one-parameter unfolding of F, i.e., $G(z, u, 0) = F(z, u)$. Since $G(z, 0, 0) = F(z, 0) = f(z)$, we can also regard G as an unfolding of f.

Lemma 3.8. *The unfolding G of f is equivalent to an unfolding of f induced from F.*

Proof. Since F is infinitesimally versal,

$$\mathcal{O}_{n+1} = \mathcal{O}_{n+1}\left(\frac{\partial f}{\partial z_i}\right) + \mathbb{C}(\dot{F}_1, \ldots, \dot{F}_k).$$

Since

$$\begin{aligned} \frac{\partial G}{\partial z_i}(z,0,0) &= \frac{\partial f}{\partial z_i},\ i = 1, \ldots, n+1, \\ \frac{\partial G}{\partial u_j}(z,0,0) &= \dot{F}_j,\ j = 1, \ldots, k, \end{aligned}$$

it follows that

(3.4)

$$\mathcal{O}_{n+1+k+1} = \mathcal{O}_{n+1+k+1}\left(\frac{\partial G}{\partial z_i}\right) + \mathcal{O}_{n+1+k+1}\mathfrak{m}_{k+1} + \mathbb{C}\left(\frac{\partial G}{\partial u_1}, \ldots, \frac{\partial G}{\partial u_k}\right),$$

where $\mathfrak{m}_{k+1}$ is the maximal ideal of $\mathcal{O}_{k+1} = \mathbb{C}\{u_1, \ldots, u_k, v\}$. Now let

$$\begin{array}{rccc} \pi: & \mathbb{C}^{n+1} \times \mathbb{C}^k \times \mathbb{C} & \longrightarrow & \mathbb{C}^k \times \mathbb{C} \\ & (z,u,v) & \longmapsto & (u,v) \end{array}$$

be the canonical projection and $\pi^* : \mathcal{O}_{k+1} \to \mathcal{O}_{n+1+k+1}$ the algebra homomorphism induced by it. Let

$$M := \mathcal{O}_{n+1+k+1}/\mathcal{O}_{n+1+k+1}\left(\frac{\partial G}{\partial z_i}\right).$$

Then M is obviously finite over $\mathcal{O}_{n+1+k+1}$. We apply the Weierstrass preparation theorem for modules (Corollary 2.5) to π^*, M and $\partial G/\partial u_1, \ldots, \partial G/\partial u_k$. From this proposition and the equation (3.4) it follows that

$$\mathcal{O}_{n+1+k+1} = \mathcal{O}_{n+1+k+1}\left(\frac{\partial G}{\partial z_i}\right) + \mathcal{O}_{k+1}\left(\frac{\partial G}{\partial u_1}, \ldots, \frac{\partial G}{\partial u_k}\right).$$

Now we consider the element $\partial G/\partial v \in \mathcal{O}_{n+1+k+1}$. Then $\partial G/\partial v$ has a representation

$$\frac{\partial G}{\partial v} = \sum_{i=1}^{n+1} \xi_i(z,u,v)\frac{\partial G}{\partial z_i} + \sum_{j=1}^{k} \eta_j(u,v)\frac{\partial G}{\partial u_j}$$

with $\xi_i \in \mathcal{O}_{n+1+k+1}$ and $\eta_j \in \mathcal{O}_{k+1}$. This equation can be formulated as follows: For the germ of the holomorphic vector field

$$X = \frac{\partial}{\partial v} - \sum_{j=1}^{k} \eta_j(u,v)\frac{\partial}{\partial u_j} - \sum_{i=1}^{n+1} \xi_i(z,u,v)\frac{\partial}{\partial z_i}$$

we have

$$X(G) = 0.$$

This vector field defines, by Proposition 3.3 and Remark 3.13, a holomorphic local one parameter group and a system of phase curves. For a point (z, u, v) near 0 let $(\psi(z, u, v), \varphi(u, v), 0)$ be the point of intersection of the phase curve through (z, u, v) with the hyperplane $\mathbb{C}^{n+1} \times \mathbb{C}^k \times \{0\}$. This defines holomorphic map germs

$$\psi : (\mathbb{C}^{n+1} \times \mathbb{C}^k \times \mathbb{C}, 0) \longrightarrow (\mathbb{C}^{n+1}, 0),$$

$$\varphi : (\mathbb{C}^k \times \mathbb{C}, 0) \longrightarrow (\mathbb{C}^k, 0)$$

with $\psi(z, 0, 0) = z$. From the condition $X(G) = 0$ it follows that G is constant along the phase curves. Thus

$$G(z, u, v) = G(\psi(z, u, v), \varphi(u, v), 0) = F(\psi(z, u, v), \varphi(u, v)).$$

This means that ψ is an equivalence of G with the unfolding F induced by φ. This proves Lemma 3.8. □

Proof of Proposition 3.17. Let $F : (\mathbb{C}^{n+1} \times \mathbb{C}^p, 0) \to (\mathbb{C}, 0)$ be the unfolding of f from Proposition 3.17 and let $F' : (\mathbb{C}^{n+1} \times \mathbb{C}^l, 0) \to (\mathbb{C}, 0)$, $(z, u') \mapsto F'(z, u')$, be an arbitrary unfolding of f. We form the "sum" of F and F':

$$\begin{array}{rlcl} H: & (\mathbb{C}^{n+1} \times \mathbb{C}^p \times \mathbb{C}^l, 0) & \longrightarrow & (\mathbb{C}, 0) \\ & (z, u, u') & \longmapsto & F(z, u) + F'(z, u') - f(z). \end{array}$$

We denote by $H_0, H_1, \ldots, H_l$ the restrictions of H to the subspaces

$$\mathbb{C}^{n+1} \times \mathbb{C}^p \subset \mathbb{C}^{n+1} \times \mathbb{C}^p \times \mathbb{C} \subset \ldots \subset \mathbb{C}^{n+1} \times \mathbb{C}^p \times \mathbb{C}^l.$$

In particular $H_0 = F$ and $H_l = H$. Since F is infinitesimally versal, so too the unfoldings $H_0, H_1, \ldots, H_l$ are infinitesimally versal, as is easy to see. We can therefore apply Lemma 3.8 successively to show that the unfolding H is equivalent to an unfolding of f induced from F. Since the unfolding F' of f is induced from H, it follows that the unfolding F' is equivalent to an unfolding of f induced from F. Thus the unfolding F is universal and Proposition 3.17 is proved. □

We record yet another corollary of Proposition 3.17.

Corollary 3.4. *Let $f : (\mathbb{C}^{n+1}, 0) \to (\mathbb{C}, 0)$ be a holomorphic function germ with a nondegenerate critical point at* 0. *Then*

$$\begin{array}{rlcl} F: & (\mathbb{C}^{n+1} \times \mathbb{C}, 0) & \longrightarrow & (\mathbb{C}, 0) \\ & (z, t) & \longmapsto & f(z) - t \end{array}$$

is a universal unfolding of f.

Proof. This follows from Proposition 3.17 and Lemma 3.7. □

3.8. Morsifications

Let $f : (\mathbb{C}^{n+1}, 0) \to (\mathbb{C}, 0)$ be a holomorphic function germ.

Definition. A *morsification* of $f : (\mathbb{C}^{n+1}, 0) \to (\mathbb{C}, 0)$ is a representative $F : M \times U \to \mathbb{C}$ of an unfolding

$$\begin{array}{rlcl} F: & (\mathbb{C}^{n+1} \times \mathbb{C}, 0) & \longrightarrow & (\mathbb{C}, 0) \\ & (z, \lambda) & \longmapsto & f_\lambda(z) \end{array}$$

of f such that for almost all $\lambda \in U \setminus \{0\}$ (everywhere except on a Lebesgue null set) the function $f_\lambda : M \to \mathbb{C}$ is a Morse function. The Morse function f_λ is itself often called a morsification of f.

We next want to show that every isolated singularity of f has a morsification.

Proposition 3.18. *Let $f : (\mathbb{C}^{n+1}, 0) \to (\mathbb{C}, 0)$ be a holomorphic function germ with an isolated singularity at* 0. *Then f has a morsification.*

Proof. Let $f : M \to \mathbb{C}$ be a representative of f, where M is an open neighborhood of 0 in $\mathbb{C}^{n+1}$. We consider the map

$$\begin{array}{rlcl} \operatorname{grad} f: & M & \longrightarrow & \mathbb{C}^{n+1} \\ & z & \longmapsto & \operatorname{grad} f(z) = \left(\frac{\partial f}{\partial z_1}, \dots, \frac{\partial f}{\partial z_{n+1}}\right). \end{array}$$

By Proposition 3.7 the critical values of $\operatorname{grad} f$ form a Lebesgue null set in $\mathbb{C}^{n+1}$. Let $\lambda(\tilde{a}_1, \dots, \tilde{a}_{n+1})$ for $\lambda \in \mathbb{C}$ be a regular value near $0 \in \mathbb{C}^{n+1}$. Put

$$\tilde{f}_\lambda(z) := f(z) - \lambda \sum_{i=1}^{n+1} \tilde{a}_i z_i.$$

A point $p \in \mathbb{C}^{n+1}$ is then a critical point of $\tilde{f}_\lambda$ if and only if $\operatorname{grad} f(p) = \lambda(\tilde{a}_1, \dots, \tilde{a}_{n+1})$. But since $\lambda(\tilde{a}_1, \dots, \tilde{a}_{n+1})$ is not a critical value of $\operatorname{grad} f$, we see that $\operatorname{grad} f$ is biholomorphic at p, whence

$$\det\left(\frac{\partial^2 \tilde{f}_\lambda}{\partial z_i \partial z_j}(p)\right) \neq 0,$$

i.e., p is a nondegenerate critical point of $\tilde{f}_\lambda$. Therefore $\tilde{f}_\lambda$ has only nondegenerate critical points. Since the set of regular values of the map $\operatorname{grad} f$ is open, we may replace $\lambda(\tilde{a}_1, \dots, \tilde{a}_{n+1})$ by a neighboring regular value $\lambda(a_1, \dots, a_{n+1})$ so that the function

$$f_\lambda(z) := f(z) - \lambda \sum_{i=1}^{n+1} a_i z_i$$

has only nondegenerate points with distinct critical values. □

Definition. The number $\mu = \mu(f)$ of nondegenerate critical points of a morsification f_λ of f is called the *Milnor number* (or *multiplicity*) of the singularity f.

In the following proposition we give another description of this number. It follows from this proposition that the Milnor number is independent of the morsification chosen. Our proof follows [**Orl76**, I.5].

Proposition 3.19. *Let $f : (\mathbb{C}^{n+1}, 0) \to (\mathbb{C}, 0)$ be a holomorphic function germ with an isolated singularity at 0. Then*

$$\mu(f) = \dim_{\mathbb{C}} \mathcal{O}_{n+1} \Big/ \mathcal{O}_{n+1}\left(\frac{\partial f}{\partial z_1}, \ldots, \frac{\partial f}{\partial z_{n+1}}\right).$$

Proof. By Proposition 3.16 the number on the right side is finite. We put

$$d := \dim_{\mathbb{C}} \mathcal{O}_{n+1} \Big/ \mathcal{O}_{n+1}\left(\frac{\partial f}{\partial z_1}, \ldots, \frac{\partial f}{\partial z_{n+1}}\right).$$

By definition, $\mu(f)$ is equal to the number of nondegenerate critical points of a morsification f_λ of f. By the proof of Proposition 3.18 this last number is equal to the number of preimage points (in a neighborhood of 0) of a regular value near 0 of the finite holomorphic map

$$\operatorname{grad} f : M \longrightarrow \mathbb{C}^{n+1},$$

where M is a suitable open neighborhood of 0 in $\mathbb{C}^{n+1}$. We now show that this number is equal to d.

We denote the component functions of the map $\operatorname{grad} f$ by $f_1 := \partial f/\partial z_1, \ldots, f_{n+1} := \partial f/\partial z_{n+1}$. Since f has an isolated singularity at 0, by hypothesis, 0 is an isolated point of $(\operatorname{grad} f)^{-1}(0)$. By Proposition 2.42 the map $(\operatorname{grad} f)^* : \mathcal{O}_{n+1} \to \mathcal{O}_{n+1}$ is finite. Since $\mathcal{O}_{n+1}$ has dimension $n+1$ (Corollary 2.6), it follows from the proof of Proposition 2.44 that the homomorphism $(\operatorname{grad} f)^*$ is injective. To distinguish, we denote the coordinates of the image space of $\operatorname{grad} f$ by $y_1, \ldots, y_{n+1}$. Furthermore, we put $A = \mathbb{C}\{z_1, \ldots, z_{n+1}\}$, $R = \mathbb{C}\{f_1, \ldots, f_{n+1}\} \subset A$. Since $(\operatorname{grad} f)^* : \mathbb{C}\{y_1, \ldots, y_{n+1}\} \to A$ is injective, R is isomorphic to $\mathbb{C}\{y_1, \ldots, y_{n+1}\}$.

Now $d < \infty$ so $(f_1, \ldots, f_{n+1})$ is a parameter system for A and by Corollary 2.9 it is also a prime sequence in A. From Proposition 2.52 it follows that A is a free R-module. By the Weierstrass preparation theorem for modules (Corollary 2.5), A is generated as an R-module by d elements and d is the minimal number of generators. Thus A has an R-basis of d elements. Let K_R and K_A be the respective quotient fields of R and A. Then K_A is a field extension of K_R of degree d. Since all elements of K_A are algebraic over K_R, we have $K_A = K_R(z_1, \ldots, z_{n+1})$. By the primitive element theorem (see e.g. [**Lan02**, V, §4]) there is a $\xi \in K_A$, $\xi = \sum_{i=1}^{n+1} c_i z_i$ with $c_i \in K_R$, with $K_A = K_R(\xi)$. By multiplying the minimal polynomial of ξ over K_R by the common denominator of the coefficients, we can assume that ξ satisfies an equation

$$b_0\xi^d + b_1\xi^{d-1} + \ldots + b_d = 0$$

with $b_j \in R$, $j = 0, \ldots, d$, $b_0 \neq 0$. On multiplying this equation by b_0^{d-1}, one obtains the equation

$$(b_0\xi)^d + b_1(b_0\xi)^{d-1} + \ldots + b_0^{d-1}b_d = 0.$$

It follows from this equation that $b_0\xi$ is integral over R. By changing coordinates we can assume that $b_0\xi = z_1$. Then z_1 satisfies an equation $p(f_1, \ldots, f_{n+1})(z_1) = 0$, where

$$p(f_1, \ldots, f_{n+1})(t) = t^d + a_1(f_1, \ldots, f_{n+1})t^{d-1} + \ldots + a_d(f_1, \ldots, f_{n+1})$$

and $a_i(f_1, \ldots, f_{n+1})$, for $i = 1, \ldots, d$, is a power series in $f_1, \ldots, f_{n+1}$ that converges in a neighborhood $U \subset M$ of 0. By Proposition 2.43, $a_i(0) = 0$ for $i = 1, \ldots, d$.

Next we consider the hypersurface

$$V = \{(f_1, \ldots, f_{n+1}, z_1) \in U \times \mathbb{C} \mid p(f_1, \ldots, f_{n+1})(z_1) = 0\} \subset U \times \mathbb{C}$$

with the projection

$$\begin{array}{rccc} \pi: & V & \longrightarrow & U \\ & (f_1, \ldots, f_{n+1}, z_1) & \longmapsto & (f_1, \ldots, f_{n+1}). \end{array}$$

Let $\Delta(f_1, \ldots, f_{n+1})$ be the discriminant of the polynomial $p(f_1, \ldots, f_{n+1})(t)$ (cf. §1.6). This is a polynomial in the coefficients $a_i(f_1, \ldots, f_{n+1})$ of the polynomial $p(f_1, \ldots, f_{n+1})(t)$. Let

$$D = \{(f_1, \ldots, f_{n+1}) \in U \mid \Delta(f_1, \ldots, f_{n+1}) = 0\} \subset U.$$

The set D is a hypersurface in U. Outside D the equation

$$p(f_1, \ldots, f_{n+1})(z_1) = 0$$

has exactly d distinct solutions in the variable z_1. Let $C = \pi^{-1}(D)$.

We consider further the map

$$\begin{array}{rccc} \tilde{\sigma}: & M & \longrightarrow & M \times \mathbb{C} \\ & (z_1, \ldots, z_{n+1}) & \longmapsto & (f_1(z), \ldots, f_{n+1}(z), z_1). \end{array}$$

Let $W = \tilde{\sigma}^{-1}(V)$. Then W is a neighborhood of 0 in M. We put $\sigma = \tilde{\sigma}|_W$, $B = \sigma^{-1}(C)$. By shrinking U and V, if necessary, we can ensure that σ maps the set $W \setminus B$ homeomorphically onto $V \setminus C$. The composition $\sigma \circ \pi : W \to U$ agrees with the map $\operatorname{grad} f : W \to U$. Since $\sigma \circ \pi$ has exactly d preimage points outside D, this is true for $\operatorname{grad} f$ too. By the remark at the beginning of the proof it follows that

$$\mu(f) = d.$$

This completes the proof of Proposition 3.19. □

Let $f : (\mathbb{C}^{n+1}, 0) \to (\mathbb{C}, 0)$ be a holomorphic function germ with an isolated singularity at 0, $\operatorname{grad} f(0) = 0$. To be able to survey all possible morsifications, we consider a universal unfolding F of f. By Proposition 3.16 one obtains a universal unfolding F of f as follows: Let $g_0 = -1, g_1, \ldots, g_{\mu-1}$ be representatives of a basis of the $\mathbb{C}$-vector space

$$\mathcal{O}_{n+1} \Big/ \left(\frac{\partial f}{\partial z_1}, \ldots, \frac{\partial f}{\partial z_n} \right) \mathcal{O}_{n+1},$$

which has dimension μ; cf. Proposition 3.19. Then put

$$\begin{array}{rccl} F: & (\mathbb{C}^{n+1} \times \mathbb{C}^{\mu}, 0) & \longrightarrow & (\mathbb{C}, 0) \\ & (z, u) & \longmapsto & f(z) + \sum_{j=0}^{\mu-1} g_j(z) u_j. \end{array}$$

Let

$$F : M \times U \to \mathbb{C}$$

be a representative of the unfolding F, where M is an open neighborhood of 0 in $\mathbb{C}^{n+1}$ and U is an open neighborhood of 0 in $\mathbb{C}^{\mu}$. We put

$$\begin{aligned} \mathcal{Y} &:= \{(z, u) \in M \times U \mid F(z, u) = 0\}, \\ \mathcal{Y}_u &:= \{z \in M \mid F(z, u) = 0\}. \end{aligned}$$

Since $F(z, 0) = f(z)$, there is, by Lemma 3.5, an $\varepsilon > 0$ such that each sphere $S_\rho \subset M$ around 0 of radius $\rho \le \varepsilon$ intersects the set $\mathcal{Y}_0$ transversally. Let $\varepsilon > 0$ be so chosen. Then there is also an $\eta > 0$ such that for $|u| \le \eta$ the set $\{u \in \mathbb{C}^{\mu} \mid |u| \le \eta\}$ lies entirely in U and $\mathcal{Y}_u$ intersects the sphere S_ε transversally. Let η be so chosen. We put

$$\begin{aligned} \mathcal{X} &:= \{(z, u) \in \mathcal{Y} \mid |z| < \varepsilon, |u| < \eta\}, \\ \bar{\mathcal{X}} &:= \{(z, u) \in \mathcal{Y} \mid |z| \le \varepsilon, |u| < \eta\}, \\ \partial \bar{\mathcal{X}} &:= \{(z, u) \in \mathcal{Y} \mid |z| = \varepsilon, |u| < \eta\}, \\ S &:= \{u \in U \mid |u| < \eta\}, \end{aligned}$$

$$\begin{array}{rccc} p: & \bar{\mathcal{X}} & \longrightarrow & S \\ & (z, u) & \longmapsto & u. \end{array}$$

Let C be the set of critical points of p. If $p_0, \ldots, p_{\mu-1}$ are the component functions of p, i.e., $p_i(z, u) = u_i$ for $(z, u) \in \bar{\mathcal{X}}$, then the critical points of p

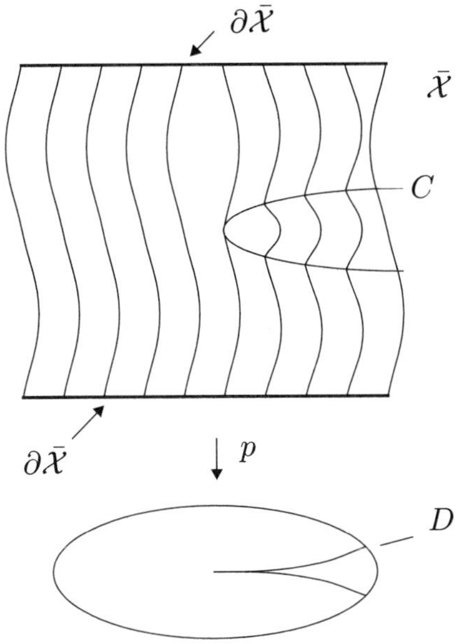

Figure 3.7. Critical set C and discriminant D

are determined by the equations and inequalities

$$F(z,u) = 0,$$
$$\operatorname{grad} F = \lambda_0 \operatorname{grad} p_0 + \ldots + \lambda_{\mu-1} \operatorname{grad} p_{\mu-1},$$
$$|z| \leq \varepsilon, \ |u| < \eta,$$

in the unknowns $z_1, \ldots, z_{n+1}, u_0, \ldots, u_{\mu-1}, \lambda_0, \ldots, \lambda_{\mu-1} \in \mathbb{C}$. It follows that

$$\begin{aligned} C &= \left\{ (z,u) \in M \times U \;\middle|\; \begin{array}{c} |z| \leq \varepsilon, \ |u| < \eta, \\ F(z,u) = 0, \\ \frac{\partial F}{\partial z_1}(z,u) = \ldots = \frac{\partial F}{\partial z_{n+1}}(z,u) = 0 \end{array} \right\} \\ &= \left\{ (z,u) \in \bar{\mathcal{X}} \mid z \text{ is a critical point of } F(\cdot,u) \right\}. \end{aligned}$$

Let $D = p(C) \subset S$ be the discriminant of p (see Figure 3.7). The map

$$p' := p|_{\bar{\mathcal{X}} - p^{-1}(D)} : \bar{\mathcal{X}} - p^{-1}(D) \longrightarrow S - D$$

is then a submersion.

For the proof of the following proposition we need a generalization of the Morse lemma (Proposition 3.15).

Proposition 3.20 (Generalized Morse lemma)**.** *Let $f : (\mathbb{C}^{n+1}, 0) \to (\mathbb{C}, 0)$ be a holomorphic function germ with*

$$\operatorname{grad} f(0) = 0 \quad \textit{and} \quad \operatorname{rank}\left((\partial^2 f/\partial z_i \partial z_j)(0)\right) = k.$$

Then there are coordinates $(z_1, \ldots, z_{n+1})$ *of* $\mathbb{C}^{n+1}$ *around* 0 *such that* f *is represented in a neighborhood of* 0 *as*

$$(z_1, \ldots, z_{n+1}) \longmapsto z_1^2 + \ldots + z_k^2 + g(z_{k+1}, \ldots, z_{n+1}),$$

where $g \in \mathfrak{m}_{n+1}^3$.

Proof. By a linear coordinate transformation one can ensure that f is of the form

$$f(y_1, \ldots, y_{n+1}) = y_1^2 + \ldots + y_k^2 + h(y_1, \ldots, y_{n+1})$$

with $h \in \mathfrak{m}^3$. Let $f_0 : (\mathbb{C}^k, 0) \to (\mathbb{C}, 0)$ be defined by

$$f_0(y_1, \ldots, y_k) = f(y_1, \ldots, y_k, 0, \ldots, 0).$$

Then one can regard f as an unfolding of f_0. The function germ f_0 has a nondegenerate critical point at 0. By Corollary 3.4, f is equivalent as an unfolding to an unfolding of f_0 induced from the universal unfolding

$$\begin{array}{rccc} F: & (\mathbb{C}^k \times \mathbb{C}, 0) & \longrightarrow & (\mathbb{C}, 0) \\ & (y, t) & \longmapsto & f_0(y) - t. \end{array}$$

Thus there are holomorphic map germs

$$g : (\mathbb{C}^{n+1-k}, 0) \longrightarrow (\mathbb{C}, 0),$$
$$\psi : (\mathbb{C}^k \times \mathbb{C}^{n+1-k}, 0) \longrightarrow (\mathbb{C}^k, 0)$$

with $\psi(y_1, \ldots, y_k, 0, \ldots, 0) = (y_1, \ldots, y_k)$ such that

$$f(y_1, \ldots, y_{n+1}) = f_0(\psi(y_1, \ldots, y_{n+1})) + g(y_{k+1}, \ldots, y_{n+1}).$$

By the hypothesis on ψ the map germ

$$\begin{array}{rccc} \tilde{\psi} : & (\mathbb{C}^{n+1}, 0) & \longrightarrow & (\mathbb{C}^{n+1}, 0) \\ & y = (y_1, \ldots, y_{n+1}) & \longmapsto & (\psi_1(y), \ldots, \psi_k(y), y_{k+1}, \ldots, y_{n+1}) \end{array}$$

is biholomorphic. We put

$$(z_1, \ldots, z_{n+1}) := (\psi_1(y), \ldots, \psi_k(y), y_{k+1}, \ldots, y_{n+1}).$$

It follows that

$$(f \circ \tilde{\psi}^{-1})(z_1, \ldots, z_{n+1}) = f_0(z_1, \ldots, z_k) + g(z_{k+1}, \ldots, z_{n+1}).$$

We now apply the Morse lemma (Proposition 3.15) to f_0, proving the result. □

Proposition 3.21. *For a suitable* $\eta > 0$ *we have*

(i) *the map* $p : \bar{\mathcal{X}} \to S$ *is proper,*

(ii) C *is a nonsingular analytic subset of* $\mathcal{X}$ *and is closed in* $\bar{\mathcal{X}}$,

(iii) *the restriction* $p|_C : C \to S$ *is finite (i.e., proper with finite fibers),*

(iv) *the discriminant* D *is an irreducible hypersurface in* S.

Proof. (i) If $K \subset S$ is compact, so too is $\bar{\mathcal{X}} \cap p^{-1}(K)$ compact. Thus $p : \bar{\mathcal{X}} \to S$ is proper.

(ii) By the choice of ε and η the map $p|_{\partial\bar{\mathcal{X}}}$ is a submersion. It follows that $C \cap \partial\mathcal{X} = \emptyset$, i.e., $C \subset \mathcal{X}$. From the description above, C is an analytic subset of $\mathcal{X}$ and so is closed in $\bar{\mathcal{X}}$.

We must next show that C is nonsingular for some suitable $\eta > 0$. By Proposition 3.20 we can write f in suitable coordinates as

$$f(z_1, \ldots, z_{n+1}) = h(z_1, \ldots, z_r) + z_{r+1}^2 + \ldots + z_{n+1}^2$$

where $h \in \mathfrak{m}_{n+1}^3$. Then we can choose the coordinate functions $z_1, \ldots, z_r$ as representatives $g_1, \ldots, g_r$ of elements of

$$\mathcal{O}_{n+1} \Big/ \left(\frac{\partial f}{\partial z_1}, \ldots, \frac{\partial f}{\partial z_{n+1}}\right) \mathcal{O}_{n+1} \ .$$

Then a representative of the universal unfolding $F : (\mathbb{C}^{n+1} \times \mathbb{C}^{\mu}, 0) \to (\mathbb{C}, 0)$ of f has the form

$$(3.5)\quad F(z, u) = -u_0 + u_1 z_1 + u_2 z_2 + \ldots + u_r z_r + z_{r+1}^2 + \ldots + z_{n+1}^2 + a(z, u),$$

where $a(z, u)$ has terms only of third or higher order.

The critical set C can now be specified as

$$C = \{(z, u) \in M \times U \mid |z| < \varepsilon, |u| < \eta, \sigma(z, u) = 0\},$$

where

$$\sigma : M \times U \to \mathbb{C}^{n+2}$$

is the map given by $\sigma = (F, \partial F/\partial z_1, \ldots, \partial F/\partial z_{n+1})$. In view of the above form for F we have

$$\frac{\partial F}{\partial z_j}(0) = 0, \quad \frac{\partial^2 F}{\partial z_i \partial z_j}(0) = 2\delta_{ij} \text{ for } 1 \le i \le n+1, r+1 \le j \le n+1,$$
$$\frac{\partial F}{\partial u_0}(0) = -1, \quad \frac{\partial F}{\partial u_i}(0) = 0 \text{ for } 1 \le i \le r,$$
$$\frac{\partial^2 F}{\partial z_i \partial u_j}(0) = \delta_{ij} \text{ for } 1 \le i \le n+1, 0 \le j \le r.$$

The matrix

$$\left(\begin{array}{ccc|ccc}
\frac{\partial F}{\partial z_{r+1}}(0) & \cdots & \frac{\partial F}{\partial z_{n+1}}(0) & \frac{\partial F}{\partial u_0}(0) & \cdots & \frac{\partial F}{\partial u_r}(0) \\
\frac{\partial^2 F}{\partial z_1 \partial z_{r+1}}(0) & \cdots & \frac{\partial^2 F}{\partial z_1 \partial z_{n+1}}(0) & \frac{\partial^2 F}{\partial z_1 \partial u_0}(0) & \cdots & \frac{\partial^2 F}{\partial z_1 \partial u_r}(0) \\
\vdots & \ddots & \vdots & \vdots & \ddots & \vdots \\
\frac{\partial^2 F}{\partial z_{n+1} \partial z_{r+1}}(0) & \cdots & \frac{\partial^2 F}{\partial z_{n+1} \partial z_{n+1}}(0) & \frac{\partial^2 F}{\partial z_{n+1} \partial u_0}(0) & \cdots & \frac{\partial^2 F}{\partial z_{n+1} \partial u_r}(0)
\end{array}\right)$$

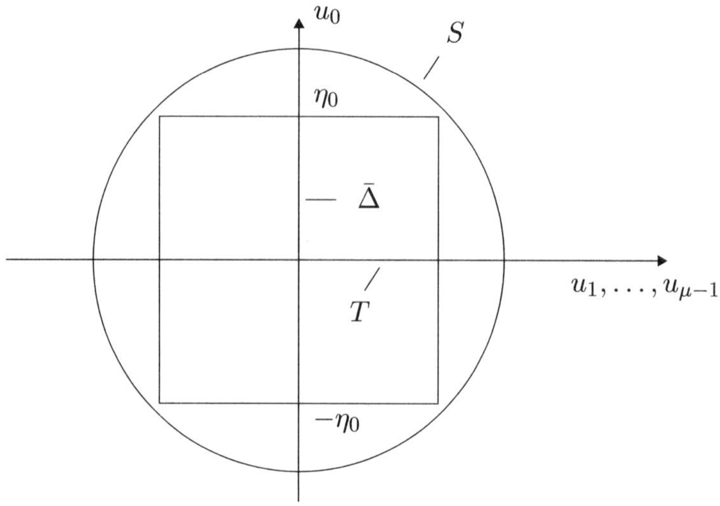

Figure 3.8. $\bar{\Delta} \times T \subset S$

is therefore an $(n+2) \times (n+2)$-submatrix of rank $n+2$ of the functional matrix of σ at 0. By the implicit function theorem C is therefore a $(\mu-1)$-dimensional submanifold of $\mathcal{X}$ in a neighborhood of 0. Therefore C is not singular for $\eta > 0$ sufficiently small.

(iii), (iv) By (ii), C is an irreducible analytic subset of $\mathcal{X}$. The map $p : \bar{\mathcal{X}} \to S$ is the restriction of a linear projection and, by (i), is proper. Since, by (ii), C is closed in $\bar{\mathcal{X}}$, so too $p|_C : C \to S$ is proper. By Proposition 2.53 the image $D = p(C)$ is thus an analytic subset of S and the map $p|_C : C \to S$ is finite. By Proposition 2.54, D is also irreducible. Since the dimension of the analytic set germ (C, x) is equal to $\mu - 1$ (Corollary 2.7) for each point $x \in C$, and since $p|_C : C \to D$ is finite, we have $\dim_u D = \mu - 1$ for all $u \in D$. It follows from Proposition 2.47 that D is a hypersurface in S. □

Let $T := \{t = (u_1, \ldots, u_{\mu-1}) \in \mathbb{C}^{\mu-1} \mid |t| < \eta_1\}$ be an open ball of radius $\eta_1 > 0$ around $0 \in \mathbb{C}^{\mu-1}$ and let $\Delta \subset \mathbb{C}$ be a disc of radius $\eta_0 > 0$ around 0 such that

$$\bar{\Delta} \times T = \{(u_0, u_1, \ldots, u_{\mu-1}) \in \mathbb{C}^\mu \mid u_0 \in \bar{\Delta}, (u_1, \ldots, u_{\mu-1}) \in T\} \subset S$$

(see Figure 3.8). We replace S by $\bar{\Delta} \times T$ and $\bar{\mathcal{X}}$ by $\bar{\mathcal{X}} \cap p^{-1}(\bar{\Delta} \times T)$, retaining, however, the same symbols. For $t \in T$ we put

$$\begin{aligned} S_t &:= \bar{\Delta} \times \{t\} \\ &= \{(u_0, u_1, \ldots, u_{\mu-1}) \in \bar{\Delta} \times T \mid (u_1, \ldots, u_{\mu-1}) = t\}. \end{aligned}$$

We consider a suitable representative $\tilde{F} : \bar{\mathcal{X}} \to S = \bar{\Delta} \times T$ of

$$\begin{array}{rccc} \tilde{F}: & (\mathbb{C}^{n+1} \times \mathbb{C}^{\mu}, 0) & \longrightarrow & (\mathbb{C} \times \mathbb{C}^{\mu-1}, 0) \\ & (z, u) & \longmapsto & (F(z,u), u_1, \ldots, u_{\mu-1}) \end{array}$$

where F is of the form

$$F(z, u) = f(z) - u_0 + \sum_{j=1}^{\mu-1} g_j(z) u_j \,.$$

Lemma 3.9. *Let $t \in T$, $t \neq 0$. The function*

$$f_{\lambda t} : M \longrightarrow \Delta \times \{\lambda t\}$$

with $f_{\lambda t}(z) = \tilde{F}(z, 0, \lambda t)$ represents a morsification of f for $\lambda \neq 0$.

Proof. We again consider a representative of the universal unfolding of f having the form (3.5) as in the proof of Proposition 3.21. Then it follows as in the proof of Proposition 3.18 that $F(\cdot, u)$, for $(u_1, \ldots, u_{\mu-1}) = \lambda t$ outside a Lebesgue null set in a neighborhood of 0, has only nondegenerate critical points with pairwise distinct critical values. Hence $f_{\lambda t}$ for $\lambda \neq 0$ is a morsification of f. $\square$

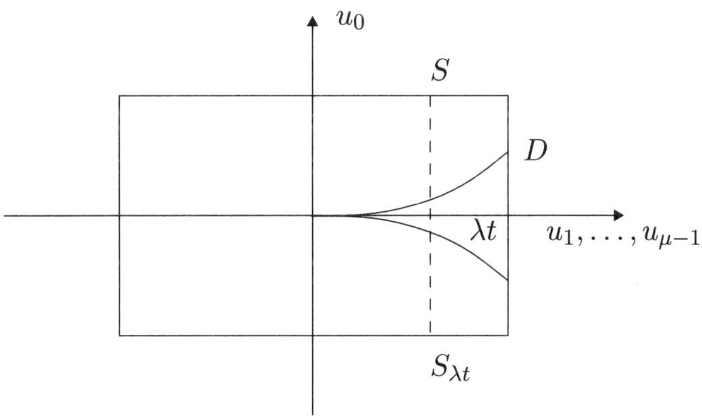

Figure 3.9. The line $\mathbb{C} \times \{\lambda t\}$ intersects the discriminant D transversally

Remark 3.18. That $f_{\lambda t}$ is a Morse function is tantamount to the line $\mathbb{C} \times \{\lambda t\}$ being in general position in the discriminant D, i.e., intersecting the discriminant D transversally at regular points (see Figure 3.9). This follows from the description

$$C = \{(z, u) \in \bar{\mathcal{X}} \mid z \text{ is a critical point of } F(\cdot, u)\}.$$

If the representative F of the universal unfolding of f has the form (3.5), it follows that the line $\mathbb{C} \times \{0\}$ does not lie in the tangent space (tangent cone) of D at 0. It follows that for λt outside a Lebesgue null set the line $\mathbb{C} \times \{\lambda t\}$ is in general position to the discriminant D, whence $f_{\lambda t}$ is a Morse function.

3.9. Finitely determined function germs

Next we consider an equivalence relation on the set of all holomorphic function germs $f : (\mathbb{C}^{n+1}, p) \to (\mathbb{C}, s)$.

Definition. Two holomorphic function germs $f_1 : (\mathbb{C}^{n+1}, p_1) \to (\mathbb{C}, s_1)$, $f_2 : (\mathbb{C}^{n+1}, p_2) \to (\mathbb{C}, s_2)$ are called *right equivalent* if there are representatives $\tilde{f}_1 : U_1 \to \mathbb{C}$ of f_1 and $\tilde{f}_2 : U_2 \to \mathbb{C}$ of f_2 and a biholomorphic map $\varphi : U_1 \to U_2$ with $\varphi(p_1) = p_2$ such that $\psi \circ \tilde{f}_1 = \tilde{f}_2 \circ \varphi$, where $\psi : \mathbb{C} \to \mathbb{C}$ is the translation $z \mapsto z + (s_2 - s_1)$, i.e., if the following diagram commutes:

$$\begin{array}{ccc} U_1 & \xrightarrow{\varphi} & U_2 \\ {\scriptstyle \tilde{f}_1}\downarrow & & \downarrow{\scriptstyle \tilde{f}_2} \\ \mathbb{C} & \xrightarrow[\psi]{} & \mathbb{C} \end{array}$$

The right equivalence class of a holomorphic function germ $f : (\mathbb{C}^{n+1}, p) \to (\mathbb{C}, s)$ is denoted by $[f]$. If f is an isolated singularity, we also call $[f]$ an isolated singularity.

Definition. Let $f : (\mathbb{C}^{n+1}, 0) \to (\mathbb{C}, 0)$ be a holomorphic function germ with an isolated singularity at 0. The *modality* (or *module number*) of f, symbolically, $\operatorname{mod}(f)$, is the smallest number m for which there exists a representative $p : \mathcal{X} \to S$ of the universal unfolding $F : (\mathbb{C}^{n+1} \times \mathbb{C}^{\mu}, 0) \to (\mathbb{C}, 0)$ of f such that for all $(z, u) \in \mathcal{X}$ the function germs $F_u : (\mathbb{C}^{n+1}, z) \to (\mathbb{C}, F(z, u))$ given by $F_u(z') = F(z', u)$ fall into finitely many families of right equivalence classes depending on at most m (complex) parameters. The modality depends only on the right equivalence class $[f]$ of f, so we may write $\operatorname{mod}[f]$. If $\operatorname{mod}[f] = m$, we call $[f]$ *m-modal* (or *m-modular*). For $m = 0, 1, 2$ one says *simple*, *unimodal* (or *unimodular*), and *bimodal* (*bimodular*), respectively.

The fact that f is simple means that there are finitely many function germs $f_i : (\mathbb{C}^{n+1}, 0) \to (\mathbb{C}, 0)$ $(i = 1, \ldots, k)$ such that for each unfolding $F : (\mathbb{C}^{n+1} \times \mathbb{C}^r, 0) \to (\mathbb{C}, 0)$ of f there exists a representative $p : \mathcal{X} \to S$ such that for all $(z, u) \in \mathcal{X}$ the function germ $F_u : (\mathbb{C}^{n+1}, z) \to (\mathbb{C}, F(z, u))$ is right equivalent to an f_i.

Example 3.8. If $f : (\mathbb{C}^{n+1}, 0) \to (\mathbb{C}, 0)$ is a holomorphic function germ with a nondegenerate critical point at 0, then f is simple. This follows from Corollary 3.4.

Example 3.9. Let $f : (\mathbb{C}^2, 0) \to (\mathbb{C}, 0)$ be given by $f(z_1, z_2) = z_1 z_2 (z_1 + z_2)(z_1 - z_2)$. We show that f is not simple. To do this, we consider the unfolding

$$F(z_1, z_2, u) := (z_1 + u z_2) z_2 (z_1 + z_2)(z_1 - z_2).$$

Then the germs of the functions

$$f_u : (\mathbb{C}^2, 0) \to (\mathbb{C}, 0), \quad f_u(z_1, z_2) = F(z_1, z_2, u),$$

are in general not right equivalent to one another, for, if this were the case, for two values $u_1, u_2 \in \mathbb{C}$ there would have to be a biholomorphic map $\varphi : U_1 \to U_2$, $U_1, U_2 \subset \mathbb{C}^2$ open, with $\varphi(0) = 0$, such that $f_{u_1} = f_{u_2} \circ \varphi$. Since f_{u_1} and f_{u_2} are homogeneous polynomials of fourth degree, one can assume that φ is linear. The space of all lines in $\mathbb{C}^2$ through the origin is the 1-dimensional complex projective space $\mathbb{P}_1\mathbb{C}$. This space is isomorphic to the Riemann sphere $\hat{\mathbb{C}}$. The four lines

$$z_1 + u z_2 = 0, \ z_2 = 0, \ z_1 + z_2 = 0, \ z_1 - z_2 = 0$$

determine four points on the projective line $\mathbb{P}_1\mathbb{C}$. The map φ induces a fractional linear transformation $\hat{\varphi}$ of $\hat{\mathbb{C}}$ onto itself. By Proposition 1.26 a fractional linear transformation leaves the cross ratio of these four points invariant. The cross ratio of these four points depends, however, on u.

In what follows we shall classify the simple function germs. To do this, we need a means of deciding whether two function germs are right equivalent. We now address this matter.

Let $f : (\mathbb{C}^{n+1}, 0) \to (\mathbb{C}, 0)$ be a holomorphic function germ.

Definition. The Taylor polynomial of f around 0 of degree r is called the *r-jet* of f and is denoted by $j^r f$.

Definition. The function germ f is called *r-determined* if every holomorphic function germ $g : (\mathbb{C}^{n+1}, 0) \to (\mathbb{C}, 0)$ with $j^r f = j^r g$ is right equivalent to f.

In particular an r-determined function germ f is right equivalent to the polynomial $j^r f$.

Let us again write $\mathfrak{m}$ for the maximal ideal of $\mathcal{O}_{n+1}$. Then

$$\mathfrak{m}^r = \{ f \in \mathcal{O}_{n+1} \,|\, j^{r-1} f = 0 \}.$$

Theorem 3.1 (Mather). *Let $f : (\mathbb{C}^{n+1}, 0) \to (\mathbb{C}, 0)$ be a holomorphic function germ. If*

$$\mathfrak{m}^{r+1} \subset \mathfrak{m}^2 \left(\frac{\partial f}{\partial z_1}, \ldots, \frac{\partial f}{\partial z_{n+1}} \right) + \mathfrak{m}^{r+2},$$

then f is r-determined.

Proof. We must show that for each element $h \in \mathfrak{m}^{r+1}$ the function germ $f+h$ is right equivalent to f. We proceed in a way similar to that of the proof of the complex Morse lemma (Proposition 3.15).

So we consider the function

$$F(z,t) = f(z) + th(z)$$

for z in a neighborhood U of 0 in $\mathbb{C}^{n+1}$ and $t \in [0,1]$. We seek a holomorphic local one parameter group $g : [0,1] \times U' \to U'$ for a suitable neighborhood U' of $0 \in \mathbb{C}^{n+1}$ such that

$$(3.6) \qquad F(g_t(z),t) = f(z), \quad g_0(z) = z, \quad g_t(0) = 0 \text{ for } z \in U', t \in [0,1].$$

A holomorphic vector field

$$X = \frac{\partial}{\partial t} + \sum_{j=1}^{n+1} a_{t,j} \frac{\partial}{\partial z_j}$$

is induced on $[0,1] \times U'$ by g. From the equation (3.6) it follows that $XF = 0$ for this vector field. Hence we obtain the following equation for the functions $a_{t,j}$ that we seek:

$$\frac{\partial F}{\partial t} + \sum_{j=1}^{n+1} a_{t,j} \frac{\partial F}{\partial z_j} = h + \sum_{j=1}^{n+1} a_{t,j} \frac{\partial F}{\partial z_j} = 0.$$

We can rewrite this equation as follows:

$$(3.7) \qquad \sum_{j=1}^{n+1} a_{t,j} \frac{\partial F}{\partial z_j} = -h.$$

Claim 3.1. Under the hypotheses of Theorem 3.1 the equation (3.7) is solvable for the coefficients $a_{t,j}$.

Proof. Let $t_0 \in [0,1]$, let $\mathcal{E}$ be the ring of holomorphic function germs

$$(\mathbb{R} \times \mathbb{C}^{n+1}, (t_0, 0)) \to (\mathbb{C}, 0)$$

and let $\mathfrak{n}$ be the corresponding maximal ideal. Let

$$J = \mathcal{E}\left(\frac{\partial F}{\partial z_1}, \dots, \frac{\partial F}{\partial z_{n+1}}\right).$$

By hypothesis, $h \in \mathfrak{m}^{r+1}$. We can solve the equation (3.7) for $a_{t,j}$ if

$$\mathfrak{m}^{r+1} \subset J\mathcal{E}\mathfrak{m}.$$

From the definition of F

$$\frac{\partial F}{\partial z_j} = \frac{\partial f}{\partial z_j} + t\frac{\partial h}{\partial z_j},$$

and so

$$\frac{\partial f}{\partial z_j} = \frac{\partial F}{\partial z_j} - t\frac{\partial h}{\partial z_j}.$$

Hence

$$\left(\frac{\partial f}{\partial z_1}, \ldots, \frac{\partial f}{\partial z_{n+1}}\right) \subset J + \mathfrak{n}\mathfrak{m}^{r-1}.$$

Since $\mathfrak{m}^{r+1} \subset \mathfrak{m}^2 (\partial f/\partial z_1, \ldots, \partial f/\partial z_{n+1}) + \mathfrak{m}^{r+2}$, by hypothesis, we obtain

$$\begin{aligned} \mathcal{E}\mathfrak{m}^{r+1} &\subset \mathcal{E}\mathfrak{m}^2 \left(\frac{\partial f}{\partial z_1}, \ldots, \frac{\partial f}{\partial z_{n+1}}\right) + \mathcal{E}\mathfrak{m}^{r+2} \\ &\subset \mathfrak{m}^2 J + \mathfrak{n}\mathcal{E}\mathfrak{m}^{r+1}. \end{aligned}$$

The claim follows from Nakayama's lemma (Corollary 2.3). □

As in the proof of the complex Morse lemma it now follows that there exists a holomorphic local one parameter group $g : [0,1] \times U' \to U'$ satisfying the equation (3.6). Then the map germ $g_1 : (U', 0) \to (U', 0)$ satisfies

$$(f + h) \circ g_1 = f$$

so f is right equivalent to $f + h$. □

Corollary 3.5. *Let $f : (\mathbb{C}^{n+1}, 0) \to (\mathbb{C}, 0)$ be a holomorphic function germ. If*

$$\mathfrak{m}^{r-1} \subset \left(\frac{\partial f}{\partial z_1}, \ldots, \frac{\partial f}{\partial z_{n+1}}\right),$$

for $r \geq 1$, then f is r-determined.

Proof. This follows from Theorem 3.1 since $\mathfrak{m}^{r+1} = \mathfrak{m}^2 \cdot \mathfrak{m}^{r-1}$ for $r \geq 1$. □

Corollary 3.6. *Let $f : (\mathbb{C}^{n+1}, 0) \to (\mathbb{C}, 0)$ be a holomorphic function germ with an isolated singularity at 0. Then f is right equivalent to a polynomial.*

Proof. Again let $f_j := \partial f/\partial z_j$, $j = 1, \ldots, n+1$, be the partial derivatives of f, and let J_f of $f_1, \ldots, f_{n+1}$ be the ideal they span. From the proof of Proposition 3.16 it follows that $\mathfrak{m}^{r-1} \subset J_f$ for some $r \geq 1$. By Corollary 3.5 f is right equivalent to $j^r f$, so to a polynomial. □

We next apply Proposition 3.12 and Remark 3.14 to the study of holomorphic function germs.

Proposition 3.22. *Suppose $k \geq 3$ and let $f(z_1, \ldots, z_k)$ be a homogeneous polynomial of degree 3 having an isolated singularity at 0. Then the corresponding holomorphic function germ $f : (\mathbb{C}^k, 0) \to (\mathbb{C}, 0)$ is not simple.*

Proof. The homogeneous polynomials in the variables $z_1, \ldots, z_k$ of degree 3 form a complex vector space V. The complex dimension of this vector space is the number of distinct monomials in the variables $z_1, \dots, z_k$, i.e.,

$$\binom{k+3-1}{3} = \frac{1}{6}k(k+1)(k+2).$$

We consider the unfolding

$$F(z,u) = f(z) + \sum_{|\nu|=3} u_\nu z^\nu$$

of f, where ν runs through all multiindices $\nu = (\nu_1, \dots, \nu_k)$ with $|\nu| = \nu_1 + \dots + \nu_k = 3$. All the function germs $f_u = F(\cdot, u)$ are homogeneous polynomials of degree 3. Two homogeneous polynomials g and $\tilde{g}$ of degree 3 are right equivalent if and only if there is a $\varphi \in \mathrm{GL}(k, \mathbb{C})$ with $\tilde{g} = g \circ \varphi$.

Now $\mathrm{GL}(k, \mathbb{C})$ is a complex Lie group of complex dimension k^2. The orbit of the natural action of this Lie group on V under $g \in V$ corresponds to the right equivalence class of g. By Proposition 3.12 and Remark 3.14 the orbit is locally a complex submanifold of V of complex dimension $\leq k^2$. But

$$k^2 < \frac{1}{6}k(k+1)(k+2)$$

for $k \geq 3$. On applying this to the unfolding F of f, we see that in every neighborhood of f we can find infinitely many function germs not right equivalent to one another. Therefore f is not simple. □

Corollary 3.7. *Let $f : (\mathbb{C}^{n+1}, 0) \to (\mathbb{C}, 0)$ be a simple holomorphic function germ with an isolated singularity at 0. Then*

$$\operatorname{rank}\left(\frac{\partial^2 f}{\partial z_i \partial z_j}(0)\right) \geq n-1.$$

Proof. Suppose that

$$\operatorname{rank}\left(\frac{\partial^2 f}{\partial z_i \partial z_j}(0)\right) = k < n-1.$$

By the generalized Morse lemma (Proposition 3.20) there are coordinates $(z_1, \dots, z_{n+1})$ such that in these coordinates f can be written as

$$f(z_1, \dots, z_{n+1}) = z_1^2 + \dots + z_k^2 + g(z_{k+1}, \dots, z_{n+1}),$$

where $g \in \mathfrak{m}_{n+1}^3$. But the 3-jet of g is a homogeneous polynomial of degree 3 in $n+1-k \geq 3$ variables having an isolated singularity at 0. By Proposition 3.22, $j^3(g)$, hence g also, is not simple. By the following proposition f is then also not simple. □

Proposition 3.23. *Let $f_i : (\mathbb{C}^k \times \mathbb{C}^l, 0) \to (\mathbb{C}, 0)$, $(x,y) \mapsto f_i(x,y)$, $i = 1, 2$, be two function germs of the following form:*

$$\begin{aligned} f_1(x,y) &= x_1^2 + \dots + x_k^2 + g_1(y) \quad \textit{with } g_1 \in \mathfrak{m}^3, \\ f_2(x,y) &= x_1^2 + \dots + x_k^2 + g_2(y) \quad \textit{with } g_2 \in \mathfrak{m}^3. \end{aligned}$$

If f_1 and f_2 are right equivalent, then so too are g_1 and g_2 right equivalent.

Proof. Let

$$\begin{array}{rccl}\varphi: & (\mathbb{C}^k \times \mathbb{C}^l, 0) & \longrightarrow & (\mathbb{C}^k \times \mathbb{C}^l, 0) \\ & (x, y) & \longmapsto & (x', y')\end{array}$$

be the germ of a biholomorphic map with

$$f_1(x', y') = f_1 \circ \varphi(x, y) = f_2(x, y). \tag{3.8}$$

We consider the partial derivative of f_2 with respect to the coordinate x_i at a point $(0, y)$. Let $\varphi(0, y) = (x', y')$. Then

$$0 = \frac{\partial f_2}{\partial x_i}(0, y) = \sum_{j=1}^{k} \frac{\partial f_1}{\partial x'_j}(x', y')\frac{\partial x'_j}{\partial x_i}(0, y) + \sum_{j=1}^{l} \frac{\partial f_1}{\partial y'_j}(x', y')\frac{\partial y'_j}{\partial x_i}(0, y).$$

Let $A(0, y)$ be the matrix

$$A(0, y) = \left(\left(\frac{\partial x'_j}{\partial x_i}(0, y)\right)\right)_{j=1,\ldots,k}^{i=1,\ldots,k}.$$

Since

$$\frac{\partial f_1}{\partial x'_j}(x', y') = 2x'_j$$

and

$$\frac{\partial f_1}{\partial y'_j}(x', y') = \frac{\partial g_1}{\partial y'_j}(x', y'),$$

we obtain following system of equations

$$-2A(0, y)x' = \left(\sum_{j=1}^{l} \frac{\partial g_1}{\partial y'_j}(x', y')\frac{\partial y'_j}{\partial x_1}(0, y), \ldots, \sum_{j=1}^{l} \frac{\partial g_1}{\partial y'_j}(x', y')\frac{\partial y'_j}{\partial x_k}(0, y)\right)^t.$$

The matrix $A(0, y)$ is invertible for y in a neighborhood of 0 in $\mathbb{C}^l$. Hence we can solve the system of equations for x' in a neighborhood of $(0, 0)$ in $\varphi(\{0\} \times \mathbb{C}^l)$, i.e., write the coordinate x'_i as a function $x'_i = \xi_i(y)$ of y, where

$$\xi_i \in \mathcal{O}_l\left(\frac{\partial g_1}{\partial y'_1}, \ldots, \frac{\partial g_1}{\partial y'_l}\right).$$

It then follows from equation (3.8) that

$$\begin{aligned} g_2(y) &= f_2(0, y) = f_1 \circ \varphi(0, y) = f_1(\xi_1(y), \ldots, \xi_k(y), y') \\ &= g_1(y') + \sum_{i=1}^{k} \xi_i(y)^2. \end{aligned}$$

Thus g_1 is right equivalent to $g_2 - \sum_{i=1}^{k} \xi_i^2$. The proposition now follows from the following lemma. □

Lemma 3.10. *Let $g : (\mathbb{C}^l, 0) \to (\mathbb{C}, 0)$ be a holomorphic function germ with $g \in \mathfrak{m}^3$. Let*

$$J_g := \mathcal{O}_l \left(\frac{\partial g}{\partial z_1}, \dots, \frac{\partial g}{\partial z_l} \right).$$

If $h \in (J_g)^2$, then $g + h$ is right equivalent to g.

Proof. The proof proceeds similarly to that of Theorem 3.1.

We consider, as there,

$$G(z,t) = g(z) + th(z), \quad t \in [0,1].$$

Let $t_0 \in [0,1]$, let $\mathcal{E}$ be the ring of holomorphic function germs

$$(\mathbb{R} \times \mathbb{C}^l, (t_0, 0)) \to (\mathbb{C}, 0)$$

and let $\mathfrak{n}$ be the corresponding maximal ideal. Let

$$J_G = \mathcal{E} \left(\frac{\partial G}{\partial z_1}, \dots, \frac{\partial G}{\partial z_l} \right).$$

We need to show that

$$h \in \mathfrak{m} J_G.$$

Now

$$\begin{aligned} \mathfrak{m} J_g &\subset \mathfrak{m} J_G + t\mathfrak{m}\mathcal{E} \left(\frac{\partial h}{\partial z_1}, \dots, \frac{\partial h}{\partial z_l} \right) \\ &\subset \mathfrak{m} J_G + \mathfrak{n}\mathfrak{m} J_g. \end{aligned}$$

From Nakayama's lemma (Corollary 2.3) it follows that

$$\mathfrak{m} J_g \subset \mathfrak{m} J_G.$$

Since

$$h \in (J_g)^2 \subset \mathfrak{m} J_g,$$

it follows that $h \in \mathfrak{m} J_G$, as was to be shown. □

By Corollary 3.7 the simple function germs $f : (\mathbb{C}^2, 0) \to (\mathbb{C}, 0)$ play a special role. For this reason we shall classify them first of all in the next section. The general classification will then follow with the help of Proposition 3.23.

Proposition 3.24. *Let $f(z_1, z_2)$ be a homogeneous polynomial of degree 4 having an isolated singularity at 0. Then the corresponding function germ $f : (\mathbb{C}^2, 0) \to (\mathbb{C}, 0)$ is not simple.*

Proof. The proof proceeds as that of Proposition 3.22, where now we apply the fact that the vector space of homogeneous polynomials of degree 4 in the two variable z_1, z_2 has complex dimension

$$\binom{2+4-1}{4} = 5$$

while

$$\dim \mathrm{GL}(2, \mathbb{C}) = 4.$$

This proves Proposition 3.24. □

Immediately from this:

Corollary 3.8. *Let $f : (\mathbb{C}^2, 0) \to (\mathbb{C}, 0)$ be a simple holomorphic function germ with an isolated singularity at 0. Then $j^3 f \neq 0$.*

3.10. Classification of simple singularities

Our aim is now, first of all, to classify the simple singularities in $\mathbb{C}^2$. We wish to prove the following proposition:

Proposition 3.25. *Let $f : (\mathbb{C}^2, 0) \to (\mathbb{C}, 0)$ be a simple holomorphic function germ with an isolated singularity at 0 and* $\operatorname{grad} f(0) = 0$. *Then f is right equivalent to one of the following simple function germs:*

(a) $x^{k+1} + y^2$ *with* $k \geq 1$ *(A_k),*
(b) $x^2 y + y^{k-1}$ *with* $k \geq 4$ *(D_k),*
(c) $x^3 + y^4$ *(E_6),*
(d) $x^3 + xy^3$ *(E_7),*
(e) $x^3 + y^5$ *(E_8).*

Our proof of this proposition follows the representation in [**BK91**] which in turn rests on the original work [**Arn73**]. We need three lemmas for the proof:

Lemma 3.11. *Let $f(x, y)$ be a homogeneous polynomial of degree 3. Then f can be brought by a $\mathbb{C}$-linear transformation into one of the forms*

$$0, \quad x^2 y + y^3, \quad x^2 y, \quad x^3.$$

Proof. We have

$$f(x, y) = (a_1 x + b_1 y)(a_2 x + b_2 y)(a_3 x + b_3 y)$$

for suitable complex numbers a_i, b_i, $i = 1, 2, 3$. Now the equation

$$a_i x + b_i y = 0$$

is the equation of a line in $\mathbb{C}^2$, which can be regarded as a point of the complex projective line $\mathbb{P}_1\mathbb{C}$. Then Lemma 3.11 follows from the fact that three distinct points in $\mathbb{P}_1\mathbb{C}$ can be mapped to three arbitrary other points by a fractional linear transformation. □

Lemma 3.12. *The holomorphic function germs*

$$\begin{aligned} f_1 &: (\mathbb{C}^2,0) \to (\mathbb{C},0), \quad (x,y) \mapsto x^2, \\ f_2 &: (\mathbb{C}^2,0) \to (\mathbb{C},0), \quad (x,y) \mapsto x^2 y, \end{aligned}$$

are not simple.

Proof. The unfoldings $F_k(x,y,t) := x^2 + ty^k$ of f_1 are not right equivalent to one another for fixed t and $k = 1, 2, \ldots$. This follows from the fact that y^k is not right equivalent to y^l for $k \neq l$ (Exercise). Hence f_1 is not simple.

For f_2 one considers correspondingly the unfoldings $F_k(x,y,t) := x^2 y + ty^k$. □

Lemma 3.13. *Let $f_t : (\mathbb{C}^2,0) \to (\mathbb{C},0)$, $(x,y) \mapsto x^3 + xy^4 + ty^6$ with $t \in \mathbb{C}$. Then in general f_t and $f_{t'}$ are not right equivalent for $t \neq t'$.*

Proof. We write f_t in the form

$$f_t(x,y) = (x - \lambda_1 y^2)(x - \lambda_2 y^2)(x - \lambda_3 y^2),$$

where $\lambda_1, \lambda_2, \lambda_3$ are the roots of the equation $\lambda^3 + \lambda + t = 0$. The zero set

$$V_t := \{(x,y) \in \mathbb{C}^2 \mid f_t(x,y) = 0\}$$

of f_t consists of the three parabolas $x = \lambda_i y^2$, $i = 1,2,3$, with the same tangents at 0. We show that two systems V_t and $V_{t'}$ of three such parabolas for $t \neq t'$ cannot in general be mapped into one another by a biholomorphic map.

Not all three roots of the equation $\lambda^3 + \lambda + t = 0$ can be equal to 0. Without loss of generality let $\lambda_2 \neq 0$. We assume in the sequel that all three roots are pairwise distinct. By using the coordinate transformation

$$\begin{aligned} x' &:= x - \lambda_1 y^2, \\ y' &:= \sqrt{\lambda_2 - \lambda_1}\, y, \end{aligned}$$

we can ensure that for V_t,

$$V_t = \{(x',y') \mid x' = 0,\ x' = y'^2,\ x' = \theta y'^2\}.$$

Correspondingly we can find coordinates (x'',y'') for $V_{t'}$ such that

$$V_{t'} = \{(x'',y'') \mid x'' = 0,\ x'' = y''^2,\ x'' = \theta' y''^2\}.$$

We assume that $\theta \neq 0,1$ and $\theta' \neq 0,1$. Let $t \neq t'$. Then $\theta \neq \theta'$ too. Suppose V_t can be transported to $V_{t'}$ by a biholomorphic map. Then, since the common tangent $x' = 0$ of the parabolas V_t must be transported to the common tangent $x'' = 0$ of the parabolas $V_{t'}$, this map must have the form

$$\begin{aligned} x'' &= x'(a_{11} + u(x',y')), \quad u \in \mathfrak{m}, \\ y'' &= a_{21}x' + a_{22}y' + v(x',y'), \quad v \in \mathfrak{m}^2. \end{aligned}$$

Since the parabola $x' = y'^2$ must be mapped to the parabola $x'' = y''^2$, it follows that $a_{11} = a_{22}^2$. But the image of the parabola $x' = \theta y'^2$ is the curve $x'' = \theta y''^2 + w(y'')$ with $w \in \mathfrak{m}^3$. But this is not a parabola of the form $x'' = \theta' y''^2$ for any $\theta' \neq \theta$. □

Proof of Proposition 3.25. We denote the coordinates of $\mathbb{C}^2$ by (x, y). Let now $f : (\mathbb{C}^2, 0) \to (\mathbb{C}, 0)$, $(x, y) \mapsto f(x, y)$, be simple, and let $\operatorname{grad} f(0) = 0$. Then $j^1 f = 0$. Consider the rank of the Hessian matrix

$$r := \operatorname{rank}\left(\frac{\partial^2 f}{\partial z_i \partial z_j}(0)\right).$$

Then $r = 0$, 1 or 2.

(1) If $r = 2$, then, by the Morse lemma, f is right equivalent to $x^2 + y^2$ and we are in case (a) ($k = 1$, A_1).

(2) If $r = 1$, then, by the generalized Morse lemma (Proposition 3.20), f is right equivalent to

$$x^2 + g(y) \quad \text{with } g \in \mathfrak{m}^3.$$

We have two cases to distinguish:

(2.1) $g = 0$: By Lemma 3.12, f would not be simple.

(2.2) $g \neq 0$: Then $g(y) = ay^{k+1} + h(y)y^{k+1}$ with $a \neq 0$, $h \in \mathfrak{m}$ and $k \geq 2$. The transformation $y \mapsto y \sqrt[k+1]{a + h(y)}$ shows that g is right equivalent to y^{k+1}. Hence f is right equivalent to $x^2 + y^{k+1}$, $k \geq 2$, and we are in case (a) (A_k, $k \geq 2$).

(3) The case $r = 0$ remains. In this case we examine the 3-jet $j^3 f$. By Lemma 3.11 this is right equivalent to one of the following polynomials:

$$0, \quad x^2y + y^3, \quad x^2y, \quad x^3.$$

(3.1) $j^3 f = 0$: By Corollary 3.8, f is not simple.

(3.2) $j^3 f = x^2y + y^3$: Then $x^2y + y^3$ is 3-determined, by Theorem 3.1 (Exercise). Hence f is right equivalent to $x^2y + y^3$ and we are in case (b) ($k = 4$, D_4).

(3.3) $j^3 f = x^2y$: Then there are two possibilities:

(3.3.1) $f \sim x^2y$: By Lemma 3.12, f would not be simple.

(3.3.2) $f \not\sim x^2y$: Then

$$j^s f = x^2y + ay^s + 2bxy^{s-1} + x^2 g(x, y) \quad \text{with } g \in \mathfrak{m}^{s-2}, \quad s \geq 4.$$

We now put

$$\begin{aligned} x_1 &:= x + by^{s-2}, \\ y_1 &:= y + g(x, y). \end{aligned}$$

Then

$$j^s f = x_1^2 y_1 + a y_1^s.$$

We again distinguish cases.

(3.3.2.1) $a = 0$ for all $s \geq 4$: Then f would again be right equivalent to $x^2 y$ and so not simple.

(3.3.2.2) $a \neq 0$: By Theorem 3.1, $x^2 y + a y^s$ is s-determined (Exercise) and thus f is right equivalent to $x^2 y + y^{k-1}$, $k \geq 5$. We have then case (b) (D_k, $k \geq 5$).

(3.4) $j^3 f = x^3$: Next we consider the 4-jet $j^4 f$ of f. It has the form

$$j^4 f = x^3 + a y^4 + b x y^3 + 3x^2 g(x,y) \quad \text{with } g \in \mathfrak{m}^2.$$

After the substitution $x_1 := x + g(x,y)$ one obtains

$$j^4 f = x_1^3 + a y^4 + b x_1 y^3.$$

We again distinguish cases.

(3.4.1) $a \neq 0$: After the substitution

$$y_1 := \sqrt[4]{a} y + \frac{b}{4(\sqrt[4]{a})^3} x_1$$

$j^4 f$ becomes

$$j^4 f = x_1^3 + y_1^4 + 3x_1^2 h(x,y).$$

If we now put $x_2 := x_1 + h(x,y)$, we obtain

$$j^4 f = x_2^3 + y_1^4.$$

But now $x^3 + y^4$ is, by Theorem 3.1, 4-determined (Exercise). Hence f is right equivalent to $x^3 + y^4$ and we find ourselves in case (c) (E_6).

(3.4.2) $a = 0$, $b \neq 0$: By the substitution $y_1 := \sqrt[3]{b} y$ we obtain

$$j^4 f = x_1^3 + x_1 y_1^3.$$

By Theorem 3.1 the function germ $x^3 + x y^3$ is only 5-determined. Therefore we consider the 5-jet $j^5 f$ of f:

$$j^5 f = x_1^3 + x_1 y_1^3 + \alpha y_1^5 + \beta x_1 y_1^4 + 3x_1^2 h_1(x_1, y_1) \text{ with } h_1 \in \mathfrak{m}^3.$$

On substituting $x_2 := x_1 + \alpha y_1^2$, we obtain

$$j^5 f = x_2^3 + x_2 y_1^3 - 3\alpha x_2^2 y_1^2 + \beta' x_2 y_1^4 + 3x_2^2 h_2(x_2, y_1), \; h_2 \in \mathfrak{m}^3.$$

The substitution $y_2 := y_1 - \alpha x_2$ yields

$$j^5 f = x_2^3(1 - 2\alpha^3 x_2 - 3\alpha^2 y_2) + x_2 y_2^3 + \beta'' x_2 y_2^4 + 3x_2^2 h_3(x_2, y_2),$$

where $h_3 \in \mathfrak{m}^3$. Finally we put

$$\begin{aligned} x_3 &:= x_2 \sqrt[3]{1 - 2\alpha^3 x_2 - 3\alpha^2 y_2}, \\ y_3 &:= y_2 \frac{1}{\sqrt[9]{1 - 2\alpha^3 x_2 - 3\alpha^2 y_2}}. \end{aligned}$$

We then obtain

$$j^5 f = x_3^3 + x_3 y_3^3 + \beta''' x_3 y_3^4 + 3x_3^2 h_4(x_3, y_3) \text{ with } h_4 \in \mathfrak{m}^3.$$

Next we replace x_3 by $x_4 := x_3 + h_4(x_3, y_3)$ and then y_3 by $y_4 := y_3 \sqrt[3]{1 + \beta''' y_3}$ to obtain

$$j^5 f = x_4^3 + x_4 y_4^3.$$

Since $x^3 + xy^3$ is 5-determined, f is right equivalent to $x^3 + xy^3$ and we find ourselves in case (d) (E_7).

(3.4.3) $a = b = 0$: We consider again the 5-jet $j^5 f$ of f:

$$j^5 f = x^3 + \alpha y^5 + \beta x y^4 + 3x^2 h(x, y) \quad \text{with } h \in \mathfrak{m}^3.$$

The substitution $x_1 := x + h(x, y)$ yields

$$j^5 f = x_1^3 + \alpha y^5 + \beta x_1 y^4.$$

We again distinguish cases.

(3.4.3.1) $\alpha \neq 0$: Putting $y_1 := \sqrt[5]{\alpha} y$, one obtains

$$j^5 f = x_1^3 + y_1^5 + \beta' x_1 y_1^4.$$

If one now substitutes $y_2 := y_1 + (1/5)\beta' x_1$, one obtains

$$j^5 f = x_1^3 + y_2^5 + 3x_1^2 h_1(x_1, y_2) \quad \text{with } h_1 \in \mathfrak{m}^3.$$

The substitution $x_2 := x_1 + h_1(x_1, y_2)$ finally yields $j^5 f = x_2^3 + y_2^5$. It again follows from Theorem 3.1 that $x^3 + y^5$ is 5-determined. Hence f is right equivalent to $x^3 + y^5$ and we are in case (e) (E_8).

(3.4.3.2) $\alpha = 0$: Next we consider the 6-jet $j^6 f$ of f. With the same coordinate transformations as in case (3.4.2) we can achieve

$$j^6 f = x^3 + xy^4 + \lambda y^6, \quad \text{with } \lambda \in \mathbb{C}.$$

It follows then from Lemma 3.13 that f is not simple.

It remains to show that the function germs (a) – (e) are simple. To do this, we consider the universal unfoldings of these function germs, as listed in the following proposition. Let $F : (\mathbb{C}^2 \times \mathbb{C}^k, (0,0)) \to (\mathbb{C}, 0)$, $(x, y, u) \mapsto F(x, y, u)$, be such an unfolding and $p : \mathcal{X} \to S$ a suitable representative. If one works with the function germ $F(\cdot, \cdot, u) : (\mathbb{C}^2, (x, y)) \to (\mathbb{C}, F(x, y, u))$ in the above proof, one can show that this function germ is right equivalent to one of the finitely many function germs of type A_l, D_l or E_l with $l \leq k$. We leave the details to the reader as an exercise. □

Proposition 3.26. *The following functions represent universal unfoldings of the holomorphic function germs of type A_k, D_k, E_6, E_7 and E_8:*

$$\begin{aligned}
A_k &: F(x,y,u) := x^{k+1} + y^2 - u_0 + u_1 x + u_2 x^2 + \ldots + u_{k-1}x^{k-1},\\
D_k &: F(x,y,u) := x^2 y + y^{k-1} - u_0 + u_1 y + \ldots + u_{k-2}y^{k-2} + u_{k-1}x,\\
E_6 &: F(x,y,u) := x^3 + y^4 - u_0 + u_1 x + u_2 y + u_3 y^2 + u_4 xy + u_5 xy^2,\\
E_7 &: F(x,y,u) := x^3 + xy^3 - u_0 + u_1 x + u_2 y + u_3 y^2\\
&\qquad +u_4 y^3 + u_5 y^4 + u_6 xy,\\
E_8 &: F(x,y,u) := x^3 + y^5 - u_0 + u_1 x + u_2 y + u_3 y^2 + u_4 y^3\\
&\qquad +u_5 xy + u_6 xy^2 + u_7 xy^3.
\end{aligned}$$

Proof. This follows from Proposition 3.17 and the fact that the monomials by which the corresponding equations are perturbed form a basis of the vector space

$$\mathcal{O}_2/\mathcal{O}_2\left(\frac{\partial f}{\partial x}, \frac{\partial f}{\partial y}\right).$$

□

From the proof of Proposition 3.25 one can develop an algorithm to decide whether a given holomorphic function germ is simple, and, if so, to which of the function germs of Proposition 3.25 (a) – (e) it is right equivalent. Such an algorithm is presented in [**BK91**, p. 17].

We finally obtain the general classification of simple holomorphic function germs that goes back to V. I. Arnold [**Arn73**].

Theorem 3.2 (Arnold)**.** *Let $f : (\mathbb{C}^{n+1}, 0) \to (\mathbb{C}, 0)$ be a simple holomorphic function germ with* $\operatorname{grad} f(0) = 0$. *Then f is right equivalent to one of the following simple function germs:*

$$\begin{aligned}
A_k &: z_1^{k+1} + z_2^2 + \ldots + z_{n+1}^2 \quad \textit{for } n \geq 0,\ k \geq 1,\\
D_k &: z_1^2 z_2 + z_2^{k-1} + z_3^2 + \ldots + z_{n+1}^2 \quad \textit{for } n \geq 1,\ k \geq 4,\\
E_6 &: z_1^3 + z_2^4 + z_3^2 + \ldots + z_{n+1}^2,\\
E_7 &: z_1^3 + z_1 z_2^3 + z_3^2 + \ldots + z_{n+1}^2,\\
E_8 &: z_1^3 + z_2^5 + z_3^2 + \ldots + z_{n+1}^2.
\end{aligned}$$

Proof. Let $n \geq 1$. By Corollary 3.7

$$k := \operatorname{rank}\left(\frac{\partial^2 f}{\partial z_i \partial z_j}(0)\right) \geq n - 1.$$

By the generalized Morse lemma f is then right equivalent to

$$\tilde{f}(z) := z_1^2 + \ldots + z_{n-1}^2 + g(z_n, z_{n+1}) \quad \text{with } g \in \mathfrak{m}_2^2.$$

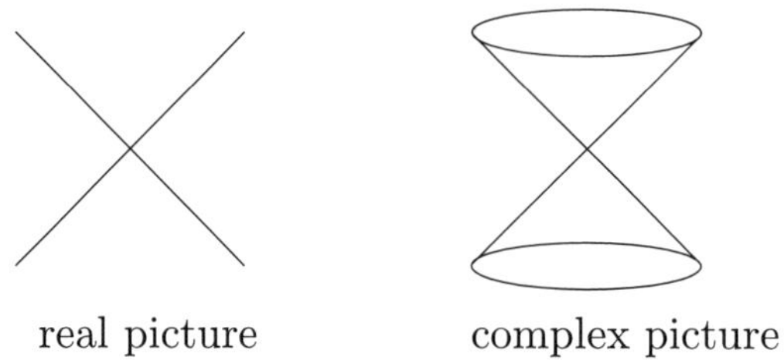

Figure 3.10. X_0

From the generalized Morse lemma it follows, besides, that each unfolding of $\tilde{f}$ is right equivalent to an unfolding of the form

$$z_1^2 + \ldots + z_{n-1}^2 + G(z_n, z_{n+1}, u)$$

where $G(z_n, z_{n+1}, u)$ represents an unfolding of g. If, now, g is simple, then $\tilde{f}$ is simple too. If on the other hand, g is not simple, it follows from Proposition 3.23 that $\tilde{f}$ is also not simple. Thus Theorem 3.2 follows from Proposition 3.25. □

3.11. Real morsifications of the simple curve singularities

We shall now motivate the notation of the simple singularities in the previous section. For this purpose we present a short introduction to the topology of singularities in the context of a singularity of type A_1.

We consider the function

$$\begin{array}{rccc} f: & \mathbb{C}^2 & \longrightarrow & \mathbb{C} \\ & (z_1, z_2) & \longmapsto & z_1^2 + z_2^2. \end{array}$$

The unique critical point of this function is the origin $(z_1, z_2) = (0, 0)$. This is also the unique singularity on the hypersurface

$$X_0 = \{(z_1, z_2) \in \mathbb{C}^2 \mid z_1^2 + z_2^2 = 0\}$$

that consists of two complex lines that intersect at the origin (see Figure 3.10). The space germ $(X_0, 0)$ thus has an isolated singularity at 0. We also called more briefly $(X_0, 0)$ an isolated singularity.

We now want to study the topology of the map f.

For $\lambda \in \mathbb{C}$, $\lambda \neq 0$ we put

$$X_\lambda = \{(z_1, z_2) \in \mathbb{C}^2 \mid z_1^2 + z_2^2 = \lambda\},$$

the fiber of f over the value λ. Since X_λ contains no singularities, X_λ is a Riemann surface. It coincides with the Riemann surface of the algebraic function

$$z_2 = \sqrt{\lambda - z_1^2}.$$

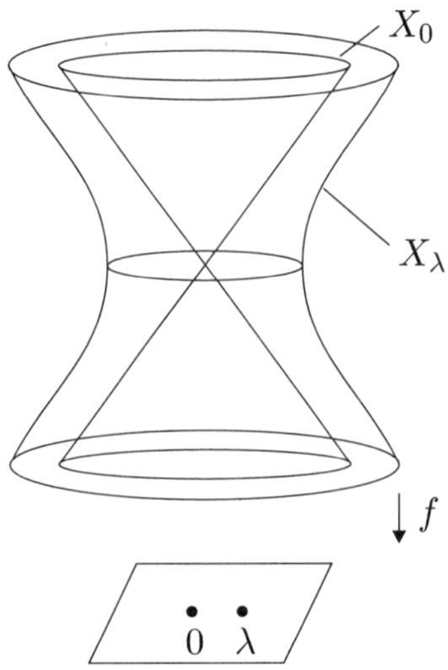

Figure 3.11. Fibers of the map f

We recall how one obtains this surface topologically: We take two copies of the complex z_1-plane and paste the edges of the cuts $(-\sqrt{\lambda}, \sqrt{\lambda})$ crosswise (see Figure 1.8). In this way one sees that X_λ is homeomorphic to a cylinder $S^1 \times \mathbb{R}$. The real 4-dimensional preimage space $\mathbb{C}^2$ thus decomposes as the singular fiber X_0 over 0 and the nonsingular fibers X_λ over $\lambda \neq 0$ which are homeomorphic to cylinders (see Figure 3.11).

We wish to describe the level surface X_λ in terms of coordinates. To do this, we put

$$\begin{aligned} z_1 &= x_1 + iy_1, \\ z_2 &= x_2 + iy_2. \end{aligned}$$

Let $\lambda \in \mathbb{R}$. Then X_λ is given by

$$z_1^2 + z_2^2 = \lambda \iff \begin{cases} x_1^2 + x_2^2 - y_1^2 - y_2^2 = \lambda, \\ x_1 y_1 + x_2 y_2 = 0. \end{cases}$$

With the help of the coordinate transformation

$$u = x_1, \quad v = x_2, \quad w = \pm\sqrt{y_1^2 + y_2^2}$$

(the sign preceding w is determined by the x_1, x_2, y_1, y_2), we can combine both these equations together in the following equation:

$$u^2 + v^2 - w^2 = \lambda.$$

This equation describes a hyperboloid in $\mathbb{R}^3$ (see Figure 3.12). As $\lambda \to 0$, its waist shrinks to a point.

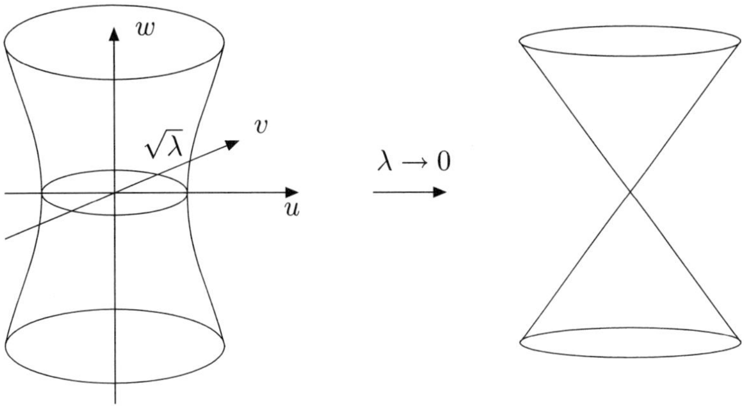

Figure 3.12. The level surface X_λ

We next want to examine how the fiber X_λ varies as λ goes once around the critical value 0. This is the idea of *monodromy* that will play a central role in Chapter 5. We therefore consider the path

$$\lambda(t) = \eta \exp(2\pi i t), \quad 0 \leq t \leq 1, \quad \eta > 0\,,$$

in the image plane. It goes once around the origin in the positive direction (counterclockwise) on the boundary of a circle of radius $\eta > 0$ in the λ-plane (see Figure 3.13). We look at how the fiber $X_{\lambda(t)}$ varies as we let t go from 0 to 1. For this we again use the description of the fiber $X_{\lambda(t)}$ as the Riemann surface of the function

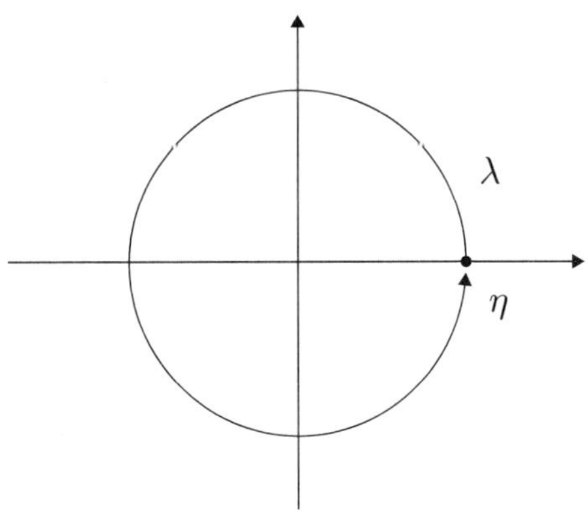

Figure 3.13. The path λ

$$z_2 = \sqrt{\lambda(t) - z_1^2}.$$

As t increases, the branch points $\pm\sqrt{\lambda(t)} = \pm\sqrt{\eta}\exp(\pi i t)$ wind round the point 0 in the positive direction. We thus obtain the sequence of Riemann surfaces represented in Figure 3.14. At $t = 1$ the cut has turned itself

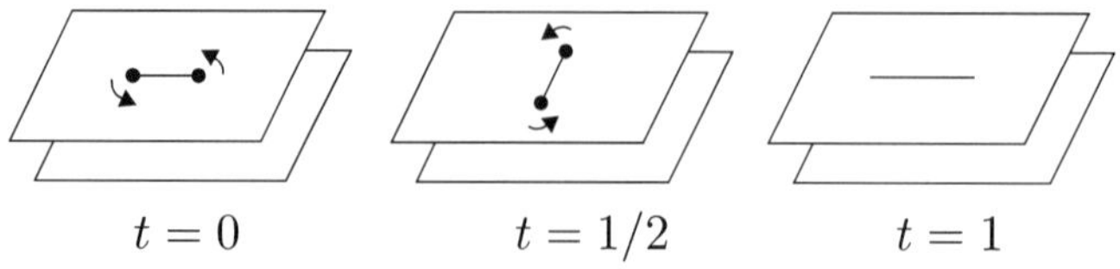

Figure 3.14. $X_{\lambda(t)}$ for $t = 0, 1/2, 1$

through 180° and we come back again to the surface $X_\eta = X_{\lambda(0)}$.

We can now construct a family of diffeomorphisms

$$h_t : X_\eta \longrightarrow X_{\lambda(t)}$$

with $h_0 = \mathrm{id}$ that is continuous in t. We first consider a differentiable bell function

$$\chi : \mathbb{R} \longrightarrow \mathbb{R}$$

with

$$\begin{aligned} \chi(\tau) &= 1 \text{ for } 0 \le |\tau| \le 2\sqrt{\eta}, \\ \chi(\tau) &= 0 \text{ for } 3\sqrt{\eta} \le |\tau| \end{aligned}$$

(see Figure 3.15). Then we put

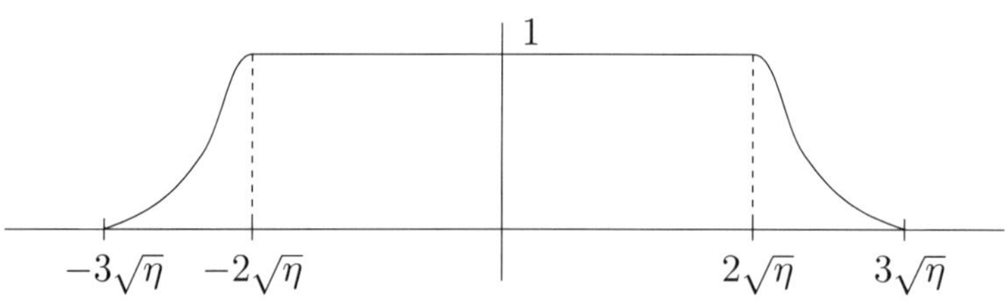

Figure 3.15. Graph of the bell function χ

$$\begin{aligned} g_t : \quad \mathbb{C} &\longrightarrow \mathbb{C} \\ z_1 &\longmapsto z_1 \exp[\pi i t \cdot \chi(|z_1|)]. \end{aligned}$$

Let

$$h_t : X_\eta \longrightarrow X_{\lambda(t)}$$

be a lifting of g_t to the covering

$$\begin{array}{ccc} X_\eta & \xrightarrow{h_t} & X_{\lambda(t)} \\ \downarrow & & \downarrow \\ \mathbb{C} & \xrightarrow{g_t} & \mathbb{C} \end{array}$$

The map

$$h := h_1 : X_\eta \longrightarrow X_\eta$$

is called the *geometric monodromy* of f. The diffeomorphisms h_t, and h_1 with them, are of course uniquely determined only up to homotopy.

We now want to study the diffeomorphism h more closely. We consider the circle

$$\delta := \{(u, v, w) \in \mathbb{R}^3 \mid u^2 + v^2 = \eta,\ w = 0\}$$

in the coordinates (u, v, w) (cf. Figure 3.16). The cycle δ shrinks to the point 0 as $\eta \to 0$. Following Picard-Lefschetz, δ is called a *vanishing cycle* (cf. §5.2).

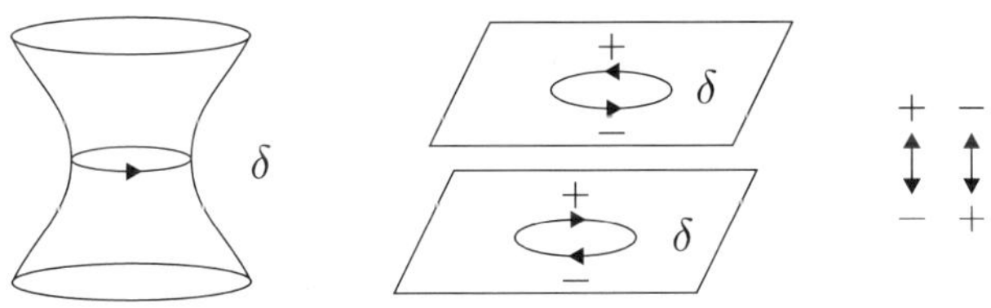

Figure 3.16. Vanishing cycle δ

On the other hand we consider the curve

$$\delta^* := \{(u, v, w) \in \mathbb{R}^3 \mid u = 0,\ v^2 - w^2 = \eta,\ v < 0\}$$

(cf. Figure 3.17). If instead of X_η we consider the corresponding compact Riemann surface $\bar{X}_\eta$ of $\sqrt{\eta - z_1^2}$ over $\mathbb{P}_1(\mathbb{C})$, then the boundary of the corresponding compactified δ^* consists of the two points over ∞. For this reason we speak of a relative cycle. The cycle δ^* will be called a "*covanishing cycle*". It intersects the cycle δ transversally in exactly one point. We orient δ^* so that the tangent vectors to δ^* and δ at the point of intersection, taken in this order, define the orientation of the complex z_1-plane. This means that the intersection number $\langle \delta^*, \delta \rangle$ between the cycles δ^* and δ is equal to $+1$.

We next look at what the diffeomorphisms h_t do to the cycles δ and δ^* (see Figure 3.18). Outside the compact subset of X_η determined by $|z_1| \leq 3\sqrt{\eta}$ the map $h : X_\eta \to X_\eta$ is the identity. Inside this subset it is as in Figure 3.18. Transported to the cylinder the action h looks as in Figure 3.19.

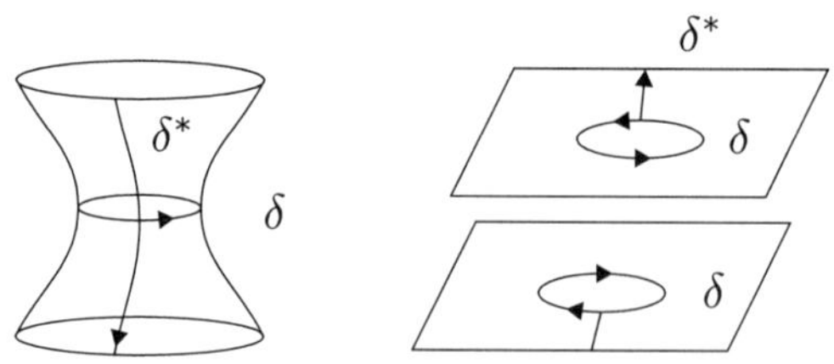

Figure 3.17. Covanishing cycle δ^*

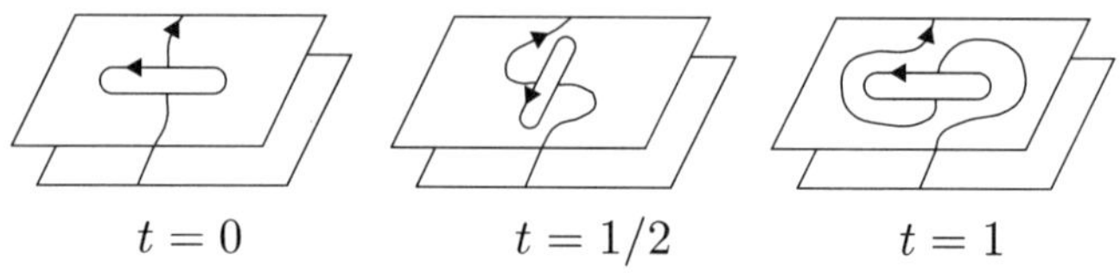

Figure 3.18. Image of δ and δ^* under h_t

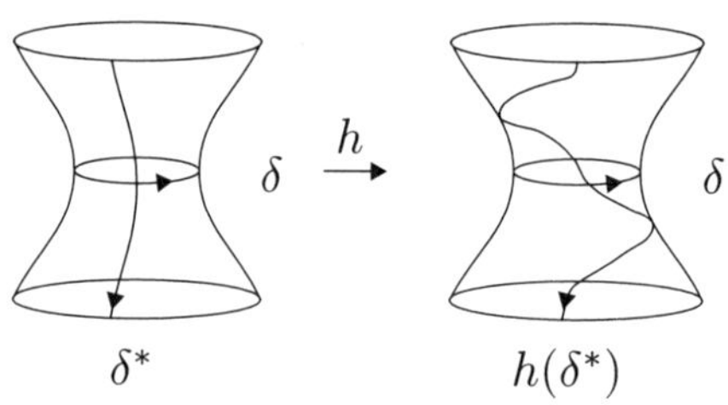

Figure 3.19. Effect of the monodromy h

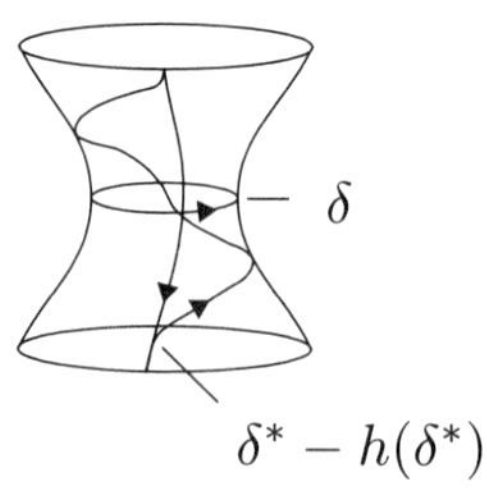

Figure 3.20. The cycle $\delta^* - h(\delta^*)$

The monodromy h is the identity outside a certain ring, while inside this ring the individual circles are twisted, through 0 at one end to 2π at the other end.

We see in this figure that this cycle is homologous to δ, i.e.,

$$\delta^* - h(\delta^*) \sim \delta.$$

The diffeomorphism h is the identity outside a compact set. The cycle $\delta^* - h(\delta^*)$ is represented in Figure 3.20. This is the simplest form of the *Picard-Lefschetz formula* (cf. §5.3).

Next we consider particular morsifications of the simple holomorphic function germs $f : (\mathbb{C}^2, 0) \to (\mathbb{C}, 0)$. The representatives described in Proposition 3.25 have the property that they are real, i.e., the restriction to $\mathbb{R}^2 \subset \mathbb{C}^2$ maps $\mathbb{R}^2$ to $\mathbb{R}$.

Definition. Let $f : (\mathbb{C}^2, 0) \to (\mathbb{C}, 0)$ be a holomorphic function germ with $f(M \cap \mathbb{R}^2) \subset \mathbb{R}$ for a suitable representative $f : M \to \mathbb{C}$. A morsification f_λ of f is called *real* if $f_\lambda(M \cap \mathbb{R}^2) \subset \mathbb{R}$.

Now we consider first of all a simple holomorphic function germ $f : (\mathbb{C}^2, 0) \to (\mathbb{C}, 0)$ of type A_k given by $f(x, y) = x^{k+1} - y^2$. One obtains a real morsification f_λ of f as follows: In the case when k is even we put

$$f_\lambda(x, y) = \lambda^{2k+2} T_{k+1}\left(\frac{x}{\lambda^2}\right) - y^2 + \frac{1}{2^k}\lambda^{2k+2}.$$

Here $T_n(x) = (1/2^{n-1})\cos(n \cdot \arccos x)$ is the n-th Chebyshev polynomial. When k is odd, we consider

$$f_\lambda(x, y) = (\lambda^{k+1} T_{(k+1)/2}\left(\frac{x}{\lambda^2}\right) - y)(\lambda^{k+1} T_{(k+1)/2}\left(\frac{x}{\lambda^2}\right) + y).$$

The real curve

$$X_{\mathbb{R},0} := \{(x, y) \in \mathbb{R}^2 \mid f_\lambda(x, y) = 0\}$$

appears as in Figure 3.21 (k even), resp. Figure 3.22 (k odd).

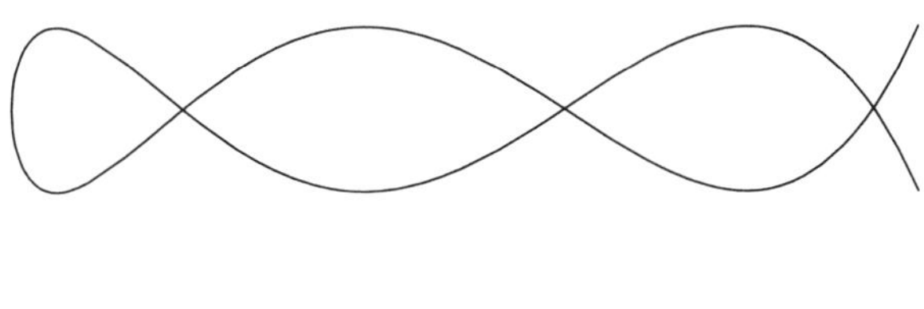

Figure 3.21. The curve $X_{\mathbb{R},0}$ for $k = 6$

Now let $\eta \in \mathbb{R}$, $\eta < 0$, with $|\eta|$ sufficiently small. Then the real level set

$$X_{\mathbb{R},\eta} := \{(x, y) \in \mathbb{R}^2 \mid f_\lambda(x, y) = \eta\}$$

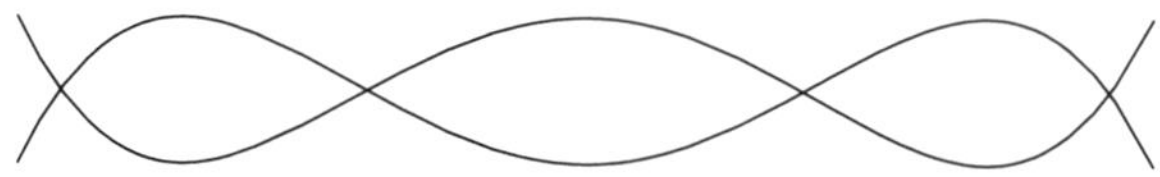

Figure 3.22. The curve $X_{\mathbb{R},0}$ for $k = 7$

consists of cycles. As we let η tend to the function value of the minima of f_λ, these cycles vanish. We can treat these cycles as vanishing cycles corresponding to the minima.

Next we consider the situation of a double point of $X_{\mathbb{R},\eta}$ that lies between two minima. In a neighborhood of such a double point we can introduce real coordinates $(\tilde{x}, \tilde{y})$ so that f_λ has the form

$$f_\lambda(\tilde{x}, \tilde{y}) = \tilde{x}^2 - \tilde{y}^2 .$$

In order to be able to apply our results on the local topology of the function $f(z_1, z_2) = z_1^2 + z_2^2$, we consider the complex coordinates

$$\begin{aligned} z_1 &= x_1 + iy_1, \\ z_2 &= x_2 + iy_2 \end{aligned}$$

with $x_1 = \tilde{x}$ and $y_2 = \tilde{y}$. The vanishing cycles corresponding to the two neighboring minima are the two branches of the hyperbola

$$\tilde{x}^2 - \tilde{y}^2 = \eta \quad \Leftrightarrow \quad x_1^2 - y_2^2 = \eta.$$

In the space $y_1 = 0$, with the coordinates (u, v, w), these are the curves

$$\begin{aligned} \delta_1^* &:= \{(u, v, w) \in \mathbb{R}^3 \mid v = 0,\ u^2 - w^2 = \eta,\ u < 0\}, \\ \delta_2^* &:= \{(u, v, w) \in \mathbb{R}^3 \mid v = 0,\ u^2 - w^2 = \eta,\ u > 0\}. \end{aligned}$$

These curves correspond exactly to the covanishing cycles. We orient the vanishing cycle δ of the double points and the covanishing cycles so that

$$\langle \delta_1^*, \delta \rangle = \langle \delta_2^*, \delta \rangle = 1.$$

In this way we obtain a vanishing cycle for each minimum and each double point. We enumerate the cycles beginning with the cycles for the minima. Let $\delta_1, \ldots, \delta_p$ be the cycles of the minima and $\delta_{p+1}, \ldots, \delta_k$ the cycles of the double points. Then we assign a graph to the system $(\delta_1, \ldots, \delta_k)$ of vanishing cycles as follows: The vertices of this graph correspond to the vanishing cycles. We connect the vertices δ_i and δ_j for $i < j$ by an edge if and only if $\langle \delta_i, \delta_j \rangle = 1$. This graph is called a *Coxeter-Dynkin diagram* of f (cf. §5.5).

In our case we obtain the graph represented in Figure 3.23. This is a

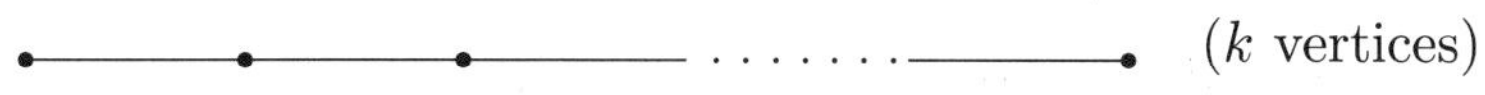

Figure 3.23. The Coxeter-Dynkin diagram of type A_k

Coxeter-Dynkin diagram of type A_k, as known from the theory of Lie groups.

We can also carry out an analogous construction for the simple holomorphic function germs $f : (\mathbb{C}^2, 0) \to (\mathbb{C}, 0)$ of type D_k. Let f be given by

$$f(x, y) = x^2 y - y^{k-1} = y(x^2 - y^{k-2}).$$

Then we consider the real morsification

$$f_\lambda(x, y) = (y - 2\lambda^2)\tilde{f}_\lambda(x, y),$$

where $\tilde{f}_\lambda$ is the above real morsification determined through the given singularity of type A_{k-3}. For the remaining three types E_6, E_7 and E_8 it is not so simple to describe real morsifications whose only critical points are minima and saddle points. But it is possible to do so in these cases too. In

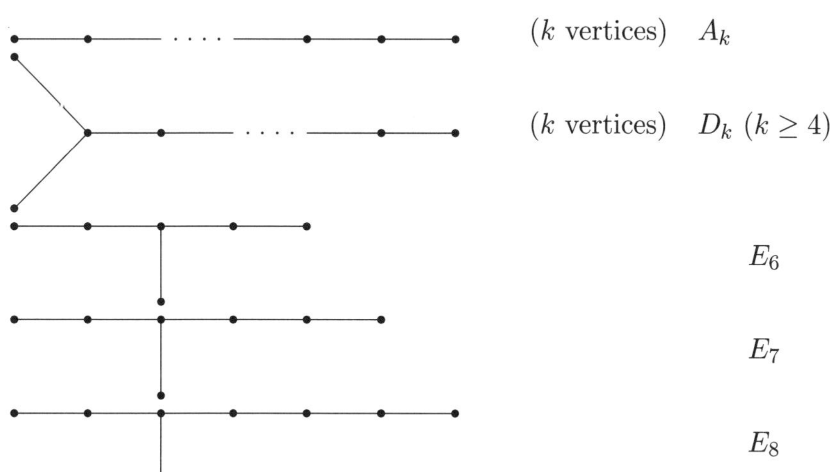

Figure 3.24. Coxeter-Dynkin diagrams of simple curve singularities

fact the simple holomorphic function germs $f : (\mathbb{C}^2, 0) \to (\mathbb{C}, 0)$ are the only holomorphic function germs for which there are real morsifications with only one or two critical values (cf. [**Dur79**, Characterization B6]).

In all one obtains the list of Coxeter-Dynkin diagrams presented in Figure 3.24. These diagrams are also known in the theory of Lie groups and take their name from there.

A definition and a thorough discussion of the concepts of monodromy, vanishing cycles and Coxeter-Dynkin diagrams for general isolated singularities of holomorphic functions form the subject matter of Chapter 5.

Chapter 4

Fundamentals of differential topology

4.1. Differentiable manifolds with boundary

Differentiable manifolds were introduced in §3.1. We now wish to consider a more general concept: manifolds with boundary. We define them now. To this purpose we consider the closed Euclidean half space (see Figure 4.1)

$$\mathbb{R}^n_+ := \{x \in \mathbb{R}^n \mid x_n \geq 0\}.$$

It is clear what the C^∞-differentiable maps are on open subsets of $\mathbb{R}^n_+$. (The open subsets of $\mathbb{R}^n_+$ are precisely the intersections of open subsets of $\mathbb{R}^n$ with $\mathbb{R}^n_+$.) The following definition is therefore meaningful.

Definition. An *n-dimensional topological manifold with boundary* is a Hausdorff space M with a countable basis for its topology such that for every point $a \in M$ there are an open neighborhood U and a homeomorphism

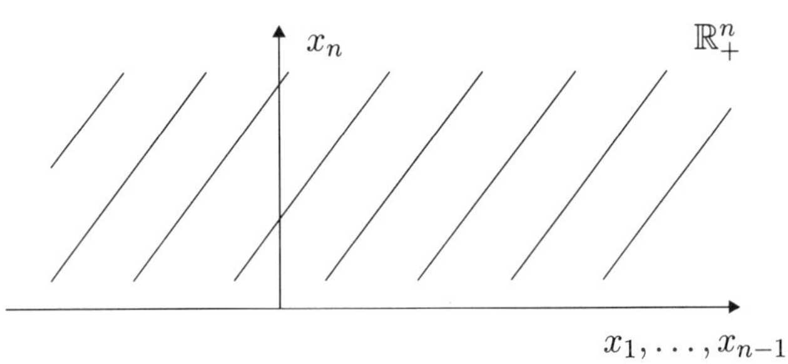

Figure 4.1. $\mathbb{R}^n_+$

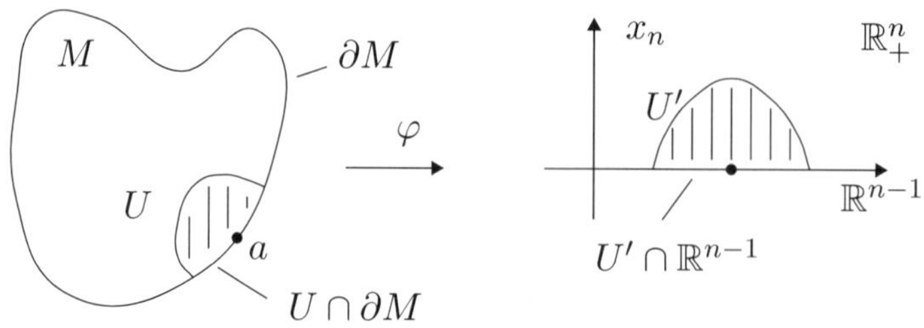

Figure 4.2. Chart of a manifold with boundary

$\varphi : U \to U'$ onto an open subset U' of $\mathbb{R}^n_+$. The map $\varphi : U \to U'$ is called a *chart around* a.

The concepts of atlas, differentiable atlas, equivalence of atlases, differentiable structure are defined word for word as for topological manifolds (without boundary). A *differentiable manifold with boundary* is a topological manifold with boundary together with a differentiable structure.

Let M be an n-dimensional differentiable manifold with boundary. A point $a \in M$ is called a *boundary point* of M if there is a chart $\varphi : U \to U'$, $a \in U$, with $x_n = \varphi_n(a) = 0$, that is, if it is mapped onto a boundary point of $\mathbb{R}^n_+$ by a chart φ.

Remark 4.1. This condition is independent of the chart: If $\psi : V \to V'$, $a \in V$, is another chart around a, then the transition function $\psi \circ \varphi^{-1} : \varphi(U \cap V) \to \psi(U \cap V)$ is a diffeomorphism that maps the points on the boundary of $\mathbb{R}^n_+$ again onto the boundary. This follows from the local invertibility theorem (interior points are mapped to interior points).

Definition. The set of boundary points of M is called the *boundary* of M and will be denoted by ∂M.

Remark 4.2. The boundary of M is in a natural way an $(n-1)$-dimensional (ordinary) differentiable manifold: From each chart $\varphi : U \to U'$ around a boundary point $a \in \partial M$ it is easy to obtain a chart of the boundary ∂M around a: for indeed $U \cap \partial M$ is open in ∂M, and

$$\varphi|_{U \cap \partial M} : U \cap \partial M \longrightarrow \{x \in U' \mid x_n = 0\} = U' \cap \mathbb{R}^{n-1}$$

is a chart of ∂M around a (see Figure 4.2).

Remark 4.3. The complement of the boundary $M - \partial M$ is in a natural way an n-dimensional differentiable manifold (without boundary) and is called the *interior* of M. In particular a manifold with boundary $\partial M = \emptyset$ is a manifold in the usual sense. A *closed* manifold is a compact manifold without boundary.

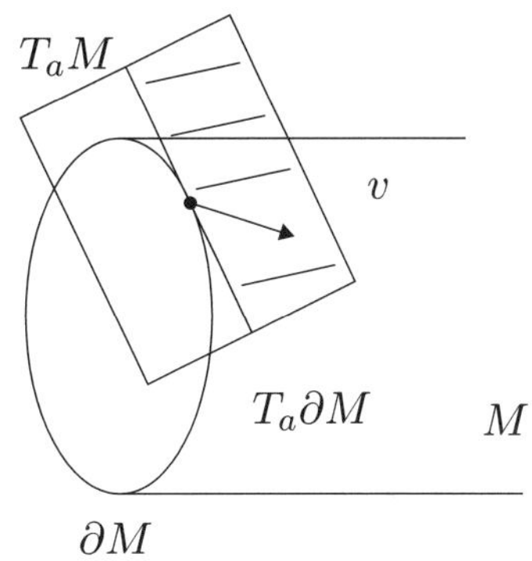

Figure 4.3. Tangent space at a boundary point

Example 4.1. The closed ball

$$B^n = \{x \in \mathbb{R}^n \mid |x|^2 \leq 1\}$$

(Euclidean norm) is a differentiable manifold with boundary $\partial B^n = S^{n-1}$.

Let M be an n-dimensional differentiable manifold with boundary. The tangent space at a point $a \in M$ is defined, as for manifolds without boundary, by

$$T_aM := \operatorname{Der} \mathcal{E}_{M,a}.$$

(The geometric definition can also be adapted, though not word for word.) The tangent space T_aM is also an n-dimensional vector space for $a \in \partial M$. The tangent space $T_a\partial M$ to the $(n-1)$-dimensional differentiable manifold ∂M is an $(n-1)$-dimensional subspace of T_aM that divides T_aM into two half spaces one of which lies on the side of M with respect to one, and therefore every, chart around a (see Figure 4.3). Depending on which half space a vector $v \in T_aM \setminus T_a\partial M$ belongs, it is said to *point inwards* or *outwards*.

4.2. Riemannian metric and orientation

We now come to the concept of a Riemannian manifold.

Definition. A *Riemannian metric* on a differentiable manifold M is a choice of a scalar product $\langle\ ,\ \rangle_a$ on each tangent space T_aM such that for every two vector fields X, Y on M the function $\langle X, Y\rangle : M \to \mathbb{R}$, $a \mapsto \langle X_a, Y_a\rangle_a$, is differentiable. A *Riemannian manifold* is a differentiable manifold together with a Riemannian metric.

Example 4.2. Let M be a Riemannian manifold and let

$$DM := \{X \in TM \mid \langle X_a, X_a\rangle_a \leq 1 \text{ for all } a \in M\}.$$

Then DM is a differentiable manifold with boundary and is called the (unit) *disc bundle* of the tangent bundle of M.

We assumed in the definition of a manifold M that M is a Hausdorff space *with a countable basis* for its topology. This means that there is a sequence $\mathfrak{A} = \{U_i\}_{i\in\mathbb{N}}$ of open sets of M such that every nonempty open set of M is a union of sets from $\mathfrak{A}$.

To show why this hypothesis is needed, we make a small excursion into set theoretic topology.

A Hausdorff space X is called *locally compact* if each point of X has a compact neighborhood. A topological manifold is locally compact, since each point has a neighborhood homeomorphic to an open subset of $\mathbb{R}^n$, and so in particular it contains a compact neighborhood.

Let X be a topological space. A covering of X is called *locally finite* if each $x \in X$ has a neighborhood that intersects only finitely many sets of the covering. Let $\mathfrak{A}$ and $\mathfrak{A}'$ be two coverings of X. The covering $\mathfrak{A}'$ is called a *refinement* of $\mathfrak{A}$ if each set from $\mathfrak{A}'$ lies in a set from $\mathfrak{A}$. A Hausdorff space X is called *paracompact* if every open covering of X has a locally finite open refinement.

Proposition 4.1. *A locally compact Hausdorff space with countable basis is paracompact. In particular a topological manifold is paracompact.*

Proof. See e.g. [**War83**]. □

Definition. Let X be a topological space and $\{U_\beta\}_{\beta\in B}$ a covering of X. A family $\{\rho_\alpha\}_{\alpha\in A}$ of continuous functions $\rho_\alpha : X \to [0,1]$ is called a *partition of unity subordinate to the covering* $\{U_\beta\}_{\beta\in B}$ if

(i) each point $x \in X$ has a neighborhood on which only finitely many of the ρ_α are different from zero,

(ii) for each $\alpha \in A$ there is a $\beta \in B$ such that

$$\operatorname{Supp} \rho_\alpha := \overline{\{x \in X \mid \rho_\alpha(x) \neq 0\}} \subset U_\beta,$$

(iii) for all $x \in X$

$$\sum_{\alpha\in A} \rho_\alpha(x) = 1.$$

In a paracompact Hausdorff space every open covering has a partition of unity subordinate to it. More holds for differentiable manifolds:

Proposition 4.2. *On a differentiable manifold every open covering has a* differentiable *partition of unity subordinate to it, i.e., the ρ_α can even be chosen to be differentiable.*

Proof. See e.g. [**BJ82**, (7.3) Satz]. □

Such a differentiable partition of unity is an important technical tool, which one needs, for example, in the proof of the following remark:

Remark 4.4. On every differentiable manifold there exists a Riemannian metric (Proof: Exercise, exploiting a differentiable partition of unity).

In the remainder of this section we shall discuss the concept of the orientation of a manifold.

Recall that an *orientation* of an n-dimensional real vector space V is an equivalence class of ordered bases with respect to the following equivalence relation: two ordered bases $(e_1, \ldots, e_n)$ and $(e'_1, \ldots, e'_n)$ are called equivalent if the change of basis has positive determinant.

Definition. Let (E, π, B, F) be an n-dimensional differentiable vector bundle. A family $\{o_b\}_{b \in B}$ of orientations of the fibers E_b is called an *orientation* of E if around each point b of B there is a bundle chart $\psi : \pi^{-1}(U) \to U \times \mathbb{R}^n$ such that the orientation o_x for each $x \in U$ is taken to the same orientation of $\mathbb{R}^n$ by $\psi_x : E_x \xrightarrow{\cong} \mathbb{R}^n$.

Definition. An *orientation* of a differentiable manifold M is an orientation of the tangent bundle TM. A differentiable manifold M is called *orientable* if the tangent bundle TM has an orientation.

Proposition 4.3. *Complex manifolds are orientable.*

Proof. Let X be a complex manifold of dimension n, let $a \in X$ and let $\{z_1, \ldots, z_n\}$ be a local coordinate system around a. Again we write $z_j = x_j + iy_j$ with $x_j, y_j \in \mathbb{R}$. Then

$$(x_1, y_1, \ldots, x_n, y_n)$$

is a local coordinate system of the differentiable manifold M underlying X around a and defines an orientation of T_aM. Since the transition functions of X are holomorphic, the corresponding real transition functions of M are real differentiable and satisfy the Cauchy-Riemann differential equations. Hence the functional determinants of the transition functions are all positive. Hence the above choice of an orientation of T_aM for all $a \in M$ furnishes an orientation of the tangent bundle TM. □

Definition. The orientation of the complex manifold X in the proof of Proposition 4.3 is called the *preferred orientation* of X.

Now let M be a differentiable manifold with boundary. An orientation of M determines an orientation of ∂M as follows: For $a \in \partial M$ let $(e_1, e_2, \ldots, e_n)$ be a positively oriented basis of T_aM such that $e_2, \ldots, e_n \in$

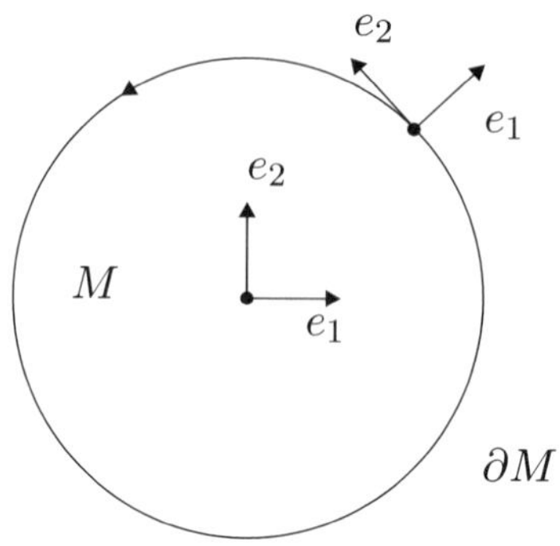

Figure 4.4. Preferred orientation of the boundary

$T_a\partial M$ and e_1 points outwards. Then $(e_2, \ldots, e_n)$ determines an orientation of ∂M at a (see Figure 4.4). This is the *preferred orientation of the boundary* of an oriented differentiable manifold with boundary.

4.3. The Ehresmann fibration theorem

We now want to a prove a fundamental theorem that plays a decisive role in the topological study of hypersurface singularities.

Let M, N be differentiable manifolds and $f : M \to N$ a differentiable map.

The following theorem is usually ascribed to Ehresmann.

Theorem 4.1 (Ehresmann fibration theorem). *Let M, B be differentiable manifolds and $f : M \to B$ a proper surjective submersion. Then $f : M \to B$ is the projection of a differentiable fiber bundle.*

Proof (following [**Kod86**, §2.3]). Let $t \in B$, $\dim B = m$, and let $\varphi : V \to V' \subset \mathbb{R}^m$, $t \in V$, be a chart of B around t. We assume that $\varphi(t) = 0$. Then V' contains a compact cube $\bar{U}'$, where

$$U' = \{x \in \mathbb{R}^m \mid |x_j| < r, \ j = 1, \ldots, m\}.$$

Let $U := \varphi^{-1}(U')$. We shall establish the

Claim. There is a diffeomorphism

$$\psi : U \times M_t \longrightarrow f^{-1}(U)$$

($M_t = f^{-1}(t)$) such that the following diagram commutes:

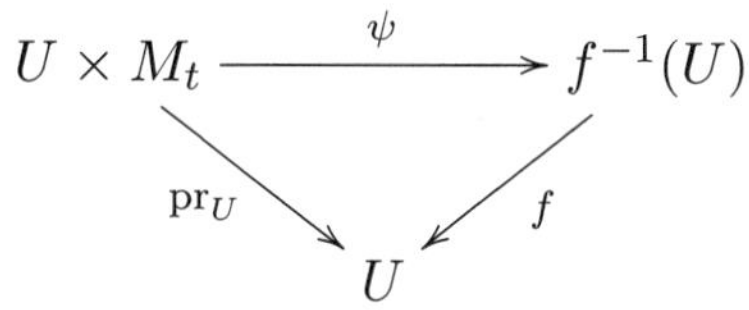

For the purpose of the proof we can, after identification by means of φ, assume that $t = 0$, $U = U' \subset \mathbb{R}^m$. Let $n = \dim M$.

By the rank theorem, for each point $a \in f^{-1}(\bar{U})$ there are charts $\varphi_a : U_a \to U'_a \subset \mathbb{R}^n$ around a and $\psi_a : V_a \to V'_a \subset \mathbb{R}^m$ around $f(a)$ such that $\psi_a \circ f \circ \varphi_a^{-1} : U'_a \to V'_a$ is given by $(x_1, \ldots, x_n) \mapsto (x_1, \ldots, x_m)$. Since f is proper, $f^{-1}(\bar{U})$ is compact. Hence there is a finite subcover $\{U_\alpha\}_{1 \le \alpha \le p}$ of $\{U_a\}_{a \in f^{-1}(\bar{U})}$.

We now proceed by induction on m.

a) The case $m = 1$: To accomplish the proof, we first construct a vector field and then a one parameter group of diffeomorphisms.

In the case $m = 1$ the sets $U = (-r, r)$ and V are open intervals. We denote the local coordinate system belonging to U_α by $(x_1^{(\alpha)}, \ldots, x_n^{(\alpha)})$. We now construct a vector field X on the neighborhood $\bigcup_\alpha U_\alpha$ of $f^{-1}(\bar{U})$ such that on each U_α

$$T_a f(X_a) = \left.\frac{\partial}{\partial x_1^{(\alpha)}}\right|_{f(a)} \quad \text{for all } a \in U_\alpha.$$

To do this, we consider the vector field $\partial/\partial x_1^{(\alpha)}$ on U_α (see Figure 4.5). Let $\{\rho_\alpha\}$ be a differentiable partition of unity subordinate to the covering $\{U_\alpha\}$, as guaranteed by Proposition 4.2. Then we put

$$X = \sum_{\alpha=1}^{p} \rho_\alpha \frac{\partial}{\partial x_1^{(\alpha)}}.$$

This vector field satisfies the desired condition. The vector field X is a vector field whose vectors are never tangential to a fiber (vertical).

By Proposition 3.3 for each $a \in M_0$ there are an open neighborhood W_a of a, an $\varepsilon(a) > 0$ and a local one parameter group $g^{(0,a)} : (-\varepsilon(a), \varepsilon(a)) \times W_a \to M$ that induces X on W_a. Since M_0 is compact, there are $a_1, \ldots, a_q \in M_0$ such that

$$M_0 \subset W_0 := \bigcup_{i=1}^{q} W_{a_i}.$$

Let $\varepsilon_0 = \min \varepsilon(a_i)$. If $W_{a_i} \cap W_{a_j} \neq \emptyset$, then $g_t^{(0,a_i)}$ and $g_t^{(0,a_j)}$ induce the same vector field on $W_{a_i} \cap W_{a_j}$, and then, by the uniqueness assertion of

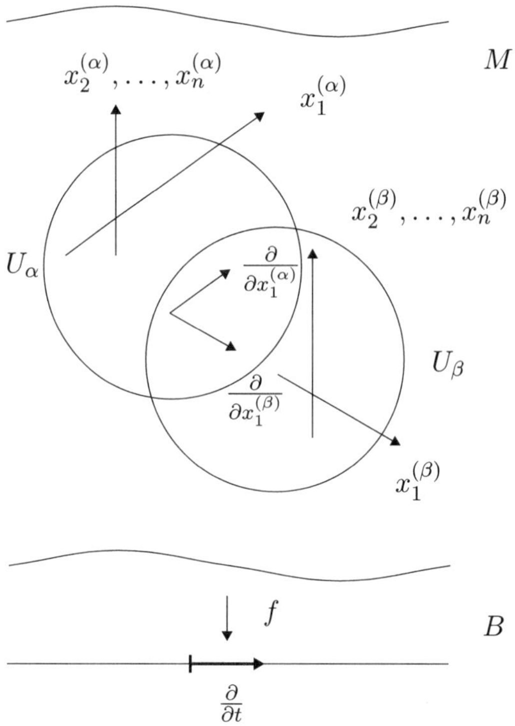

Figure 4.5. Construction of the vector field X

Proposition 3.3, they agree also for $|t| < \varepsilon_0$. We can thus define $g_t^{(0)}$ on W_0 as follows:

$$g_t^{(0)}(x) = g_t^{(0,a_i)}(x) \text{ if } x \in W_{a_i}.$$

This argument can now be repeated for each $s \in [-r, r]$ instead of 0. In this way we obtain a local one parameter group

$$g^{(s)} : (s - \varepsilon_s, s + \varepsilon_s) \times W_s \longrightarrow M.$$

For $t \in (s - \varepsilon_s, s + \varepsilon_s) \cap (s' - \varepsilon_{s'}, s' + \varepsilon_{s'})$ we have $g_t^{(s)} = g_t^{(s')}$ by the uniqueness assertion of Proposition 3.3. Hence we can extend $g^{(0)}$ to a map

$$g : (-r, r) \times M_0 \longrightarrow f^{-1}((-r, r))$$

with $f \circ g = \mathrm{pr}_U$. Since $g_t : M_0 \to M_t$ is a diffeomorphism for each $t \in U$, it follows that g is a diffeomorphism. The map $\psi := g$ is thus a diffeomorphism with the desired property.

b) The general case. We put $U = U_1 \times U_2$, where

$$\begin{aligned} U_1 &= \{t_1 \in \mathbb{R} \mid |t_1| < r\}, \\ U_2 &= \{(t_2, \dots, t_m) \in \mathbb{R}^{m-1} \mid |t_i| < r,\ i = 2, \dots, m\}, \\ f &= (f_1, \dots, f_m). \end{aligned}$$

As in part a) we construct a diffeomorphism $\psi_1 : U_1 \times f^{-1}(\bar{U}_2) \to f^{-1}(U_1 \times \bar{U}_2)$ with $f_1 \circ \psi_1 = \mathrm{pr}_{U_1}$ ($\mathrm{pr}_{U_1} : U_1 \times f^{-1}(\bar{U}_2) \to U_1$). By the induction hypothesis there exists a diffeomorphism

$$\psi_2 : U_2 \times M_0 \longrightarrow f^{-1}(U_2)$$

with $f \circ \psi_2 = \mathrm{pr}_{U_2}$ ($\mathrm{pr}_{U_2} : U_2 \times M_0 \to U_2$). The composition

$$\psi : U \times M_0 = U_1 \times U_2 \times M_0 \xrightarrow{(\mathrm{id} \times \psi_2)} U_1 \times f^{-1}(U_2) \xrightarrow{\psi_1} f^{-1}(U_1 \times U_2) = f^{-1}(U)$$

is then a diffeomorphism $\psi : U \times M_0 \to f^{-1}(U)$ as desired, such that the diagram

$$\begin{array}{ccc} U \times M_0 & \xrightarrow{\quad\psi\quad} & f^{-1}(U) \\ & {\scriptstyle \mathrm{pr}_U} \searrow \quad \swarrow {\scriptstyle f} & \\ & U & \end{array}$$

commutes.

This completes the proof of the theorem. □

Theorem 4.1 can be generalized as follows.

Theorem 4.2. *Let M be a differentiable manifold with boundary ∂M, and let B be a differentiable manifold. If both $f : M \to B$ and $f|_{\partial M} : \partial M \to B$ are proper surjective submersions, then f and $f|_{\partial M}$ are projections of differentiable fiber bundles.*

Proof. The proof of Theorem 4.1 can be transferred to the case that M is a differentiable manifold with boundary. From this it follows that $f : M \to B$ is the projection of a differentiable fiber bundle. The corresponding conclusion for $f|_{\partial M} : \partial M \to B$ follows directly from Theorem 4.1. □

4.4. The holonomy group of a differentiable fiber bundle

Let (E, π, B, F) be a differentiable fiber bundle and $\gamma : I = [0,1] \to B$ a differentiable path. We now present a construction that for each $x \in E_{\gamma(0)}$ allows us to lift the path γ in a unique way to a "horizontal" path $\hat{\gamma}_x : I \to E$ with $\hat{\gamma}_x(0) = x$. We follow the treatment of [**Wol64**].

In order to be able to lift γ uniquely, we need a decomposition of the tangent bundle of E into a vertical and a horizontal sub-bundle. We first

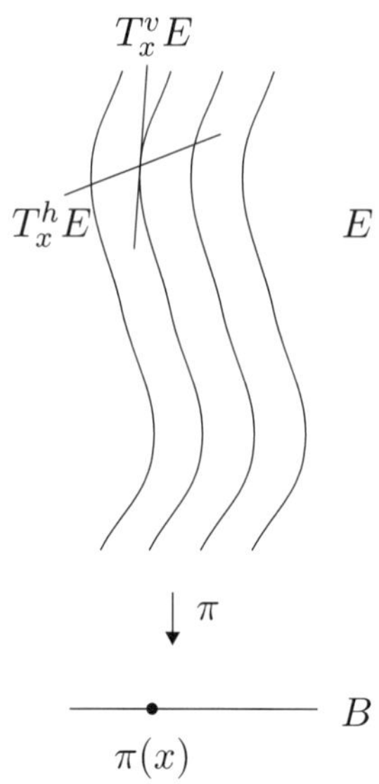

Figure 4.6. Vertical and horizontal tangent space

consider a decomposition of the tangent space $T_x E$ to E at a point $x \in E$. Let

$$T_x^v E = \{v \in T_x E \mid T_x\pi(v) = 0\}.$$

This subspace of $T_x E$ is called the *vertical tangent space* to E at x. It is isomorphic to the tangent space $T_x E_{\pi(x)}$ to the fiber $E_{\pi(x)}$ at the point x. A complement $T_x^h E$ of this subspace in $T_x E$ is called a *horizontal tangent space* to E in x (see Figure 4.6). Thus

$$T_x E = T_x^v E \oplus T_x^h E.$$

The linear map $T_x\pi$ maps $T_x^h E$ isomorphically onto the tangent space $T_{\pi(x)}B$ at the basis B in $\pi(x)$. But note that $T_x^h E$, being a complementary subspace, is in contrast to $T_x^v E$ not uniquely determined.

We now want to decompose the tangent bundle in the corresponding manner. We need several more definitions.

Definition. Let $(E, \pi, B, \mathbb{R}^n)$ be a vector bundle and $E' \subset E$ a subset of E such that to each $b \in B$ there are a neighborhood U of b and a bundle chart $\psi : \pi^{-1}(U) \to U \times \mathbb{R}^n$ with

$$\psi(\pi^{-1}(U) \cap E') = U \times \mathbb{R}^k \subset U \times \mathbb{R}^n .$$

Then $(E', \pi|_{E'}, B, \mathbb{R}^k)$ is a k-dimensional vector bundle and is called a *subbundle* of E.

Example 4.3. If (E, π, B, F) is a differentiable fiber bundle, then

$$T^vE := \operatorname{Ker} T\pi := \bigcup_{x \in E} \operatorname{Ker} T_x\pi$$

with the natural projection is a sub-bundle of the tangent bundle TE of E.

Definition. If $(E, \pi_1, B, \mathbb{R}^n)$, $(F, \pi_2, B, \mathbb{R}^m)$ are vector bundles over B, then $(E \oplus F, \pi_1 \oplus \pi_2, B, \mathbb{R}^n \times \mathbb{R}^m)$ with

$$E \oplus F := \bigcup_{b \in B} E_b \oplus F_b,$$

$$\begin{array}{rccc} \pi_1 \oplus \pi_2 : & E \oplus F & \longrightarrow & B \\ & (x, y) & \longmapsto & \pi_1(x) = \pi_2(y), \end{array}$$

and with the obvious bundle charts, is a vector bundle. This vector bundle is called the *Whitney sum* of E and F and for brevity will be denoted $E \oplus F$.

Definition. An *Ehresmann connection* for a differentiable fiber bundle (E, π, B, F) is a sub-bundle T^hE of the tangent bundle TE such that TE is the Whitney sum $TE = T^vE \oplus T^hE$ of T^vE and T^hE.

Proposition 4.4. *Every differentiable fiber bundle (E, π, B, F) has an Ehresmann connection.*

Proof. By Remark 4.4 there exists a Riemannian metric on E. For $x \in E$ define the subspace T_x^hE as the orthogonal complement $(T_x^vE)^\perp$ of T_x^vE in T_xE with respect to this metric. By definition of the Riemannian metric this defines a sub-bundle of TE. □

Proposition 4.5. *Let (E, π, B, F) be a differentiable fiber bundle with compact fiber F and an Ehresmann connection. If $\gamma : I \to B$ is a differentiable path and $x \in E_{\gamma(0)}$, then there is exactly one lifting $\hat{\gamma} : I \to E$ with $\hat{\gamma}(0) = x$ that is horizontal, i.e., $\hat{\gamma}'(t) \in T^h_{\hat{\gamma}(t)}E$ for all $t \in I$.*

Proof. Since the map $T_x\pi|_{T_x^hE} : T_x^hE \to T_{\pi(x)}B$ is a linear isomorphism, for each $x \in E$ there is exactly one vector field X on $\pi^{-1}(\gamma(I))$ with $T_a\pi(X_a) = \gamma'(t)$ for all $t \in I$ and $a \in \pi^{-1}(\gamma(t))$. Since $\pi^{-1}(\gamma(I))$ is compact, by hypothesis, it follows, as in part a) of the proof of the Ehresmann fibration theorem, that the vector field X is induced by a uniquely determined one parameter group $g_t : E_{\gamma(0)} \to E_{\gamma(t)}$ $(t \in [0,1])$ of diffeomorphisms. The lifting sought is then the phase curve of g_t through x, i.e.,

$$\hat{\gamma}(t) := g_t(x) \text{ for } t \in [0,1].$$

This proves Proposition 4.5. □

Let (E, π, B, F) be a differentiable fiber bundle with compact fiber F and an Ehresmann connection. Furthermore, let $\gamma : I \to B$ be a differentiable

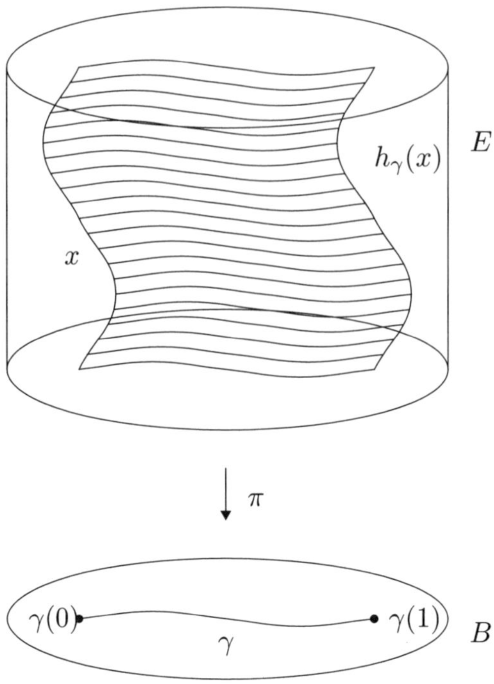

Figure 4.7. Parallel transport along the path γ

path and $g_t : E_{\gamma(0)} \to E_{\gamma(t)}$ ($t \in [0,1]$) the corresponding one parameter group as in the proof of Proposition 4.5. The diffeomorphism

$$h_\gamma := g_1 : E_{\gamma(0)} \longrightarrow E_{\gamma(1)}$$

is called the *parallel transport* of $E_{\gamma(0)}$ *along the path* γ *with respect to the connection* (see Figure 4.7).

We next examine the extent to which the diffeomorphism h_γ depends on the path γ.

Definition. Let M, N be differentiable manifolds and $f, g : M \to N$ diffeomorphisms. The diffeomorphisms f and g are called *isotopic* if there exists a differentiable map

$$H : M \times I \longrightarrow N$$

with the following properties:

(i) Each of the maps $H_t : M \to N$, $x \mapsto H(x,t)$, is a diffeomorphism.

(ii) We have $H_0 = f$, $H_1 = g$.

Such a map $H : M \times I \to N$ is called an *isotopy* between f and g.

The set of diffeomorphisms $f : M \to M$ of a differentiable manifold M onto itself forms a group, which we denote by $\mathrm{Diff}(M)$. The diffeomorphisms

$f : M \to M$ which are isotopic to the identity form a subgroup $\mathrm{Is}(M)$. We put

$$\mathrm{Diff}_0(M) := \mathrm{Diff}(M)/\,\mathrm{Is}(M).$$

$\mathrm{Diff}_0(M)$ is the group of isotopy classes of diffeomorphisms of M onto itself.

Proposition 4.6 (Homotopy lifting theorem). *Let (E, π, B, F) be a differentiable fiber bundle with compact fiber F and an Ehresmann connection. If $\gamma_1, \gamma_2 : I \to B$ are two homotopic differentiable paths with the same start point $a = \gamma_1(0) = \gamma_2(0)$ and end point $b = \gamma_1(1) = \gamma_2(1)$, then the diffeomorphisms h_{γ_1} and h_{γ_2} are isotopic.*

Proof. Let $H : I \times I \to B$ be a homotopy between γ_1 and γ_2. We can assume that H is differentiable. By Proposition 4.5 each path $H_t : I \to B$, $s \mapsto H(s,t)$, determines a parallel transport $\hat{H}_t : E_a \to E_b$. Then $\hat{H} : E_a \times I \to E_b$, $(x,t) \mapsto \hat{H}_t(x)$, is an isotopy between h_{γ_1} and h_{γ_2}. □

Now let an Ehresmann connection be chosen and let $b \in B$ be a fixed point in B. A closed differentiable path $\gamma : I \to B$ with start and end point b determines a diffeomorphism $h_\gamma : E_b \to E_b$.

Definition. The set of such diffeomorphisms h_γ forms a subgroup of $\mathrm{Diff}(E_b)$ and is called the *holonomy group* of the connection.

From Proposition 4.6 it follows that the assignment $\gamma \mapsto h_\gamma$ determines a homomorphism

$$\rho_0 : \pi_1(B, b) \longrightarrow \mathrm{Diff}_0(E_b)\,.$$

This homomorphism is independent of the Ehresmann connection chosen.

Definition. The image of $\rho_0 : \pi_1(B,b) \to \mathrm{Diff}_0(E)$ is called the *geometric monodromy group* of the differentiable fiber bundle (E, π, B, F).

Definition. Let X, Y be topological spaces. Two continuous maps $f, g : X \to Y$ are called *homotopic* when there is a continuous map $H : X \times I \to Y$ with $H(x, 0) = f(x)$ and $H(x,1) = g(x)$ for all $x \in X$. A topological space X is called *contractible* when the identity $\mathrm{id} : X \to X$ is homotopic to a constant map $g : X \to \{p\}$ with $p \in X$.

Proposition 4.7. *Every differentiable fiber bundle (E, π, B, F) with compact fiber F over a contractible base space B is trivial.*

Proof. By Proposition 4.4, (E, π, B, F) has an Ehresmann connection. Let $p \in B$ and let $H : B \times I \to B$ be a homotopy between $\mathrm{id} : B \to B$ and $g : B \to B$ with $g(b) = p$ for all $b \in B$. We can again assume that H is differentiable. Then for each $b \in B$ the map $H_b : I \to B$, $t \mapsto H(b,t)$, is

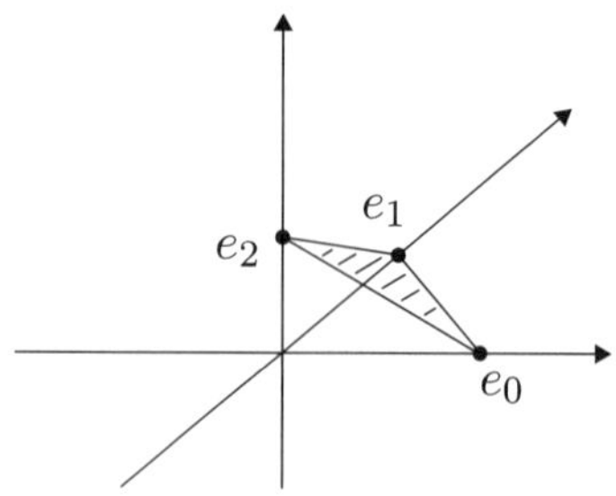

Figure 4.8. Standard 2-simplex

a differentiable path from b to p. Let $\hat{H}_b : E_b \to E_p$ be the corresponding parallel transport. We define a map

$$\psi : E \longrightarrow B \times E_p$$

by $\psi(x) = (\pi(x), \hat{H}_{\pi(x)}(x))$. This map is a diffeomorphism with $\mathrm{pr}_1 \circ \psi = \pi$ and it is the desired trivialization. □

Remark 4.5. Note that the trivialization in Proposition 4.7 depends on the Ehresmann connection chosen.

4.5. Singular homology groups

In this section we present a short introduction to singular homology theory.

Let X be a topological space. We denote the *standard q-simplex* in $\mathbb{R}^{q+1}$ by Δ^q, i.e.,

$$\Delta^q = \left\{ \sum_{i=0}^{q} \lambda_i e_i \,\middle|\, \sum_{i=0}^{q} \lambda_i = 1, \quad 0 \le \lambda_i \le 1 \right\},$$

where $e_0, e_1, \ldots, e_q$ is the standard basis of $\mathbb{R}^{q+1}$ (see Figure 4.8).

Definition. A *singular q-simplex* of X is a continuous map $\sigma : \Delta^q \to X$. A *singular q-chain* of X is a formal linear combination

$$c = \sum_{i=1}^{m} n_i \sigma_i,$$

where $n_i \in \mathbb{Z}$ and the σ_i are singular q-simplexes. Let $C_q(X)$ be the set of all singular q-chains.

We introduce further notation. The *support* $|\sigma|$ *of a singular q-simplex* is the set $|\sigma| := \sigma(\Delta^q)$. The *support of a q-chain* $c = \sum n_i \sigma_i$ is the set $|c| := \bigcup |\sigma_i|$.

We now define the boundary operator $\partial_q : C_q(X) \to C_{q-1}(X)$. For $q > 0$ we define $F_i^q : \Delta^{q-1} \to \Delta^q$ by $\sum_{j=0}^{q-1} \lambda_j e_j \mapsto \sum_{j=0}^{i-1} \lambda_j e_j + \sum_{j=i}^{q-1} \lambda_j e_{j+1}$.

This means that F_i^q maps the standard $(q-1)$-simplex Δ^{q-1} onto the face $[e_0, \ldots, \hat{e}_i, \ldots, e_q]$ of Δ^q. If $\sigma : \Delta^q \to X$ is a singular q-simplex, the *i-th face* $\sigma^{(i)}$ of σ is the singular $(q-1)$-simplex $\sigma \circ F_i^q : \Delta^{q-1} \to X$. The *boundary* of σ is

$$\partial_q \sigma = \sum_{i=0}^{q} (-1)^i \sigma^{(i)}.$$

If $c = \sum n_i \sigma_i$ is a q-chain, we put

$$\partial_q c = \partial_q \left(\sum n_i \sigma_i \right) = \sum n_i \partial_q \sigma_i.$$

In this way we have defined a homomorphism

$$\partial_q : C_q(X) \longrightarrow C_{q-1}(X).$$

Next we put

$$C_q(X) = 0 \text{ for } q < 0, \ \partial_q = 0 \text{ for } q \leq 0.$$

We have an important result:

Proposition 4.8. *$\partial_q \partial_{q+1} = 0$ for all q.* □

Proof. It follows immediately from the definition that

$$F_i^{q+1} \, F_j^q = F_j^{q+1} \, F_{i-1}^q \qquad \text{for } j < i.$$

It is enough to show that $\partial_q \circ \partial_{q+1}(\sigma) = 0$ for a $(q+1)$-simplex σ. Now

$$\begin{aligned}
\partial_q(\partial_{q+1}\sigma) &= \sum_{i=0}^{q+1} (-1)^i \, \partial_q \sigma^{(i)} \\
&- \sum_{i=0}^{q+1} (-1)^i \sum_{j=0}^{q} (-1)^j (\sigma \circ F_i^{q+1}) \circ F_j^q \\
&= \sum_{0 \leq j < i}^{q+1} (-1)^{i+j} \sigma \circ \left(F_j^{q+1} F_{i-1}^q \right) + \sum_{0 \leq i \leq j}^{q} (-1)^{i+j} \sigma \circ (F_i^{q+1} F_j^q) \\
&= 0.
\end{aligned}$$

The last equality follows on putting $i' = j$ and $j' = i - 1$ in the first summand. □

A singular q-chain c with $\partial_q(c) = 0$ is called a *q-cycle.* If $c = \partial_{q+1}(c')$ for some singular $(q+1)$-chain c', then c is called a *q-boundary.* We put

$$\begin{aligned}
Z_q(X) &:= \{q\text{-cycles}\} = \operatorname{Ker} \partial_q, \\
B_q(X) &:= \{q\text{-boundaries}\} = \operatorname{Im} \partial_{q+1}.
\end{aligned}$$

By Proposition 4.8

$$B_q(X) \subset Z_q(X).$$

Definition. The factor group

$$H_q(X) = Z_q(X)/B_q(X)$$

is called the q-th (*singular*) *homology group* of X. Two q-chains $c_1, c_2 \in C_q(X)$ are called *homologous*, symbolically, $c_1 \sim c_2$, if $c_1 - c_2$ is a q-boundary. The homology class of a q-cycle c is denoted by $[c] \in H_q(X)$.

We now want to define the reduced homology groups. Towards this end, we define a homomorphism $\varepsilon : C_0(X) \to \mathbb{Z}$ as follows. If σ is a singular 0-simplex, we put $\varepsilon(\sigma) = 1$. We extend this map to a homomorphism $\varepsilon : C_0(X) \to \mathbb{Z}$ (*augmentation*) by

$$\varepsilon\left(\sum n_i\sigma_i\right) = \sum n_i$$

for a 0-chain $\sum n_i\sigma_i$. So

$$(*) \qquad \varepsilon \circ \partial_1 = 0.$$

This formula follows from the fact that $\varepsilon(\partial_1\sigma) = \varepsilon(\sigma^{(0)} - \sigma^{(1)}) = 0$ for a singular 1-simplex σ. Put

$$\tilde{Z}_0(X) := \operatorname{Ker} \varepsilon.$$

Since $\varepsilon \circ \partial_1 = 0$, we have $B_0(X) \subset \tilde{Z}_0(X)$.

Definition. Let $X \neq \emptyset$. The factor group

$$\tilde{H}_0(X) = \tilde{Z}_0(X)/B_0(X)$$

is called the *reduced* 0*-th homology group* of X. We put

$$\tilde{H}_q(X) := H_q(X) \text{ for } q > 0.$$

What is the connection between $H_0(X)$ and $\tilde{H}_0(X)$? Since $\tilde{Z}_0(X) \subset Z_0(X)$ is a subgroup, we can interpret $\tilde{H}_0(X)$ as a subgroup of $H_0(X)$. Let $\iota_* : \tilde{H}_0(X) \to H_0(X)$ be the inclusion. Since $\varepsilon \circ \partial_1 = 0$, the map $\varepsilon : C_0(X) \to \mathbb{Z}$ induces a homomorphism

$$\varepsilon_* : H_0(X) \longrightarrow \mathbb{Z}.$$

Proposition 4.9. *The sequence (for $X \neq \emptyset$)*

$$0 \longrightarrow \tilde{H}_0(X) \xrightarrow{\iota_*} H_0(X) \xrightarrow{\varepsilon_*} \mathbb{Z} \longrightarrow 0$$

of groups and homomorphisms is exact. In particular $\tilde{H}_0(X) \cong \operatorname{Ker} \varepsilon_*$, and $H_0(X) \cong \tilde{H}_0(X) \oplus \mathbb{Z}$ (under a noncanonical isomorphism).

Proof. Easy exercise. □

Remark 4.6. If $X \neq \emptyset$ is path connected, then $\varepsilon_* : H_0(X) \to \mathbb{Z}$ is an isomorphism (Proof: Exercise); hence $H_0(X) \cong \mathbb{Z}$ and thus, by Proposition 4.9, $\tilde{H}_0(X) = 0$. More generally: if r is the number of path connected components of X, then $H_0(X) \cong \mathbb{Z}^r$ and $\tilde{H}_0(X) \cong \mathbb{Z}^{r-1}$.

Example 4.4. Let S^n be the n-dimensional unit sphere. Then (cf. [**GH81**, Part II, 15])

$$\tilde{H}_q(S^n) = \begin{cases} 0 & \text{for } q \neq n, \\ \mathbb{Z} & \text{for } q = n. \end{cases}$$

Definition. Let X be a topological space such that all its homology groups are finitely generated and only finitely many of them are nonnull. Let $b_q(X)$ be the rank of $H_q(X)$. This number is called the q-th *Betti number* of X. The number

$$\chi(X) := \sum_{q=0}^{\infty} (-1)^q b_q(X)$$

is called the *Euler characteristic* of X.

Example 4.5. The Euler characteristic of the n-dimensional unit sphere is

$$\chi(S^n) = 1 + (-1)^n.$$

Now let $f : X \to Y$ be a continuous map between topological spaces. If $\sigma : \Delta^q \to X$ is a singular q-simplex in X, then $f \circ \sigma : \Delta^q \to Y$ is a singular q-simplex in Y. We obtain a homomorphism $C_q(f) : C_q(X) \to C_q(Y)$ by

$$C_q(f)\left(\sum n_i \sigma_i\right) = \sum n_i (f \circ \sigma_i).$$

Proposition 4.10. *We have* $\partial_q C_q(f) = C_{q-1}(f)\partial_q$.

Proof. Exercise. □

It follows from Proposition 4.10 that $C_q(f)$ maps q-cycles to q-cycles and q-boundaries to q-boundaries. Hence $C_q(f)$ induces a homomorphism

$$f_* : H_q(X) \longrightarrow H_q(Y)$$

by $f_*([z]) = [C_q(f)(z)]$. This homomorphism is called the *homomorphism of the q-th homology group induced by* $f : X \to Y$.

If $\varepsilon : C_0(X) \to \mathbb{Z}$ is the augmentation, it follows immediately from the definition of $C_q(f)$ that the following diagram commutes:

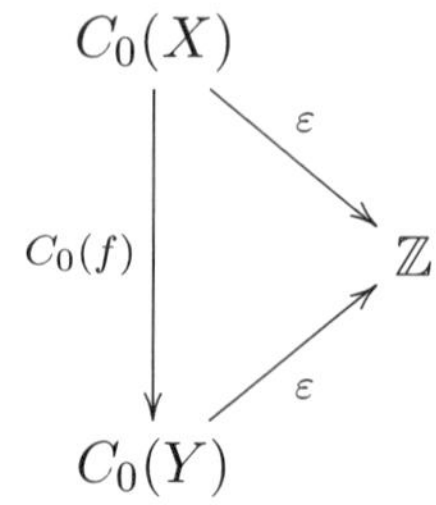

Hence $C_0(f)$ maps the subgroup $\tilde{Z}_0(X)$ of $Z_0(X)$ to $\tilde{Z}_0(Y)$ and so induces a homomorphism $\tilde{H}_0(X) \to \tilde{H}_0(Y)$ which will also be denoted by

$$f_* : \tilde{H}_0(X) \longrightarrow \tilde{H}_0(Y)\,.$$

We now introduce the relative homology groups. Let A be a subspace of X, that is, a subset $A \subset X$ endowed with the relative topology. One also calls (X, A) a *pair*. We next study the connection between the homology groups of X and those of A.

First one can consider the group $C_q(A)$. This is a subgroup of $C_q(X)$.

Definition. A q-chain $c \in C_q(X)$ is called a *relative q-cycle of X with respect to A* if $\partial_q c \in C_{q-1}(A)$. Let

$$Z_q(X, A) := \{\text{relative } q\text{-cycles}\}.$$

A q-chain $c \in C_q(X)$ is called a *relative q-boundary of X with respect to A* if there is a q-chain $c' \in C_q(A)$ in A with $c \sim c'$ (in X). Let

$$B_q(X, A) := \{\text{relative } q\text{-boundaries}\}.$$

Remark 4.7. $B_q(X, A) \subset Z_q(X, A)$ (Proof: Exercise).

Definition. The factor group

$$H_q(X, A) := Z_q(X, A)/B_q(X, A)$$

is called the q-th *relative homology group of X with respect to A*. We write $c \sim c' \bmod A$ for $c, c' \in Z_q(X, A)$ if $c - c' \in B_q(X, A)$, and $[z]_{\mathrm{mod}\,A}$ for the class of $z \in Z_q(X, A)$ in $H_q(X, A)$.

Example 4.6. $X = I \times S^1$ a cylinder and $A = \{0\} \times S^1 \cup \{1\} \times S^1$. See Figure 4.9.

Remark 4.8. If $A = \emptyset$, then $C_q(A) = 0$ for all q. It follows that $H_q(X, \emptyset) = H_q(X)$.

Now let $i : A \to X$ be the inclusion. Then i induces a homomorphism $i_* : H_q(A) \to H_q(X)$ for each q.

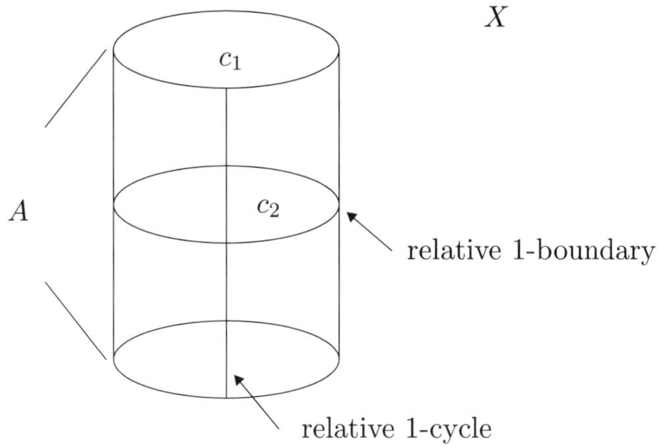

Figure 4.9. Example of a relative 1-cycle and of a relative 1-boundary

Definition. We define a map $j_* : H_q(X) \to H_q(X, A)$ by $[z] \mapsto [z]_{\operatorname{mod} A}$.

We define a map $\partial_* : H_q(X, A) \to H_{q-1}(A)$ (the *connecting homomorphism*) by $[z]_{\operatorname{mod} A} \mapsto [\partial_q z]$.

Remark 4.9. ∂_* is well defined: Let $z, z' \in Z_q(X, A)$ with $z \sim z' \operatorname{mod} A$. Then there are a $c \in C_q(A)$ and $z'' \in C_{q+1}(X)$ with $z - z' - c = \partial_{q+1} z''$. Then $\partial_q z - \partial_q z' = \partial_q c$, and so $\partial_q z \sim \partial_q z'$ in A.

Definition. The infinite sequence of homology groups and homomorphisms

$$\dots \xrightarrow{\partial_*} H_q(A) \xrightarrow{i_*} H_q(X) \xrightarrow{j_*} H_q(X, A) \xrightarrow{\partial_*} H_{q-1}(A) \xrightarrow{i_*} \dots$$

is called the *homology sequence of the pair* (X, A).

Proposition 4.11. *The homology sequence of the pair (X, A) is exact.*

Proof. We show exactness at the place $H_q(X, A)$, i.e., we show that $\operatorname{Im} j_* = \operatorname{Ker} \partial_*$. We leave the proof of exactness at the other places as an exercise.

a) To show $\operatorname{Im} j_* \subset \operatorname{Ker} \partial_*$: Let $z \in Z_q(X)$, i.e., $\partial_q z = 0$. Then $\partial_* j_*[z] = [\partial_q z] = 0$, and so $j_*[z] \in \operatorname{Ker} \partial_*$.

b) To show $\operatorname{Ker} \partial_* \subset \operatorname{Im} j_*$: Let $z \in Z_q(X, A)$ with $\partial_*[z]_{\operatorname{mod} A} = 0$. This means that there is a $c \in C_q(A)$ with $\partial_q z = \partial_q c$, whence $\partial_q(z - c) = 0$. Thus $z - c \in Z_q(X)$ and $z - c \sim z \mod A$. Hence $j_*[z - c] = [z - c]_{\operatorname{mod} A} = [z]_{\operatorname{mod} A}$ and thus $[z]_{\operatorname{mod} A} \in \operatorname{Im} j_*$. □

Proposition 4.12. *Let (X, A) be a pair with $A \neq \emptyset$. Then the image of the connecting homomorphism $\partial_* : H_1(X, A) \to H_0(A)$ lies in $\tilde{H}_0(A)$ and the following sequence is exact:*

$$\dots \xrightarrow{j_*} H_1(X, A) \xrightarrow{\partial_*} \tilde{H}_0(A) \xrightarrow{i_*} \tilde{H}_0(X) \xrightarrow{j_*} H_0(X, A) \longrightarrow 0.$$

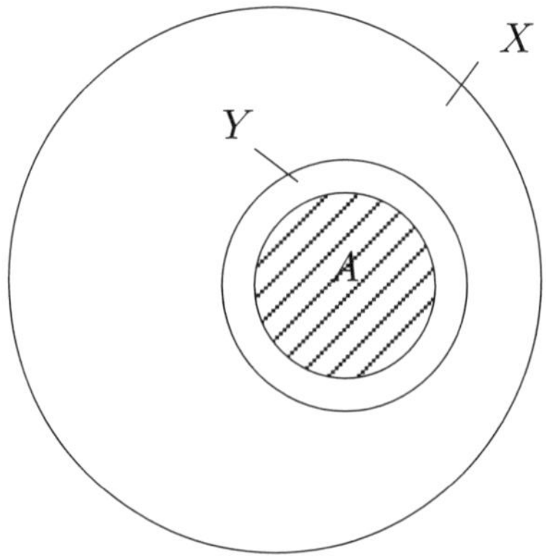

Figure 4.10. The excision theorem

Proof. Exercise. □

We introduce yet more notation:

$$H_*(X) := \bigoplus_q H_q(X),$$
$$H_*(X, A) := \bigoplus_q H_q(X, A).$$

We now cite without proof two standard results from algebraic topology that we shall need later. For proofs see [**GH81**].

Theorem 4.3 (Homotopy invariance of homology). *If $f, g : X \to Y$ are homotopic continuous maps of topological spaces, then $f_* = g_*$.*

Theorem 4.4 (Excision theorem). *Let $A \subset Y \subset X$ be subspaces such that the closure of A is included in the interior of Y. Then A can be excised, i.e., there exists an isomorphism*

$$H_q(X, Y) \cong H_q(X - A, Y - A)$$

for all q.

4.6. Intersection numbers

We now want to define the intersection numbers between cycles. A thorough presentation of these intersection numbers demands a high conceptual investment. Here we give a geometric definition for differentiable manifolds, following [**GH78**, Chap. 0, §4]. We aim at an understandable presentation without proving all the technical details.

In the sequel M will be a differentiable manifold. A singular q-simplex $\sigma : \Delta^q \to M$ is called *differentiable* if there are an open neighborhood U

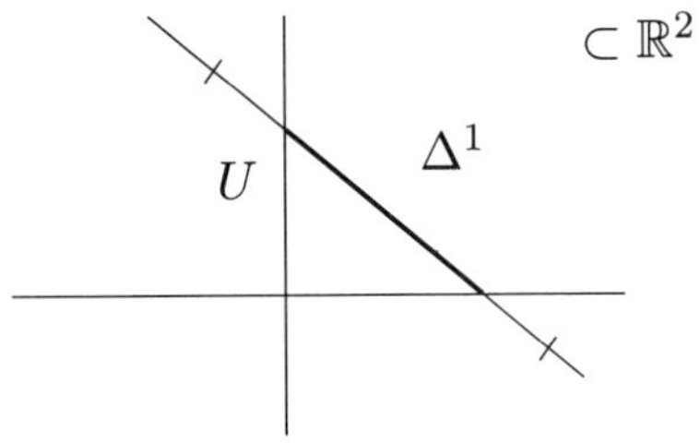

Figure 4.11. Neighborhood U of Δ^1

of Δ^q in the affine q-dimensional hyperplane of $\mathbb{R}^{q+1}$ in which Δ^q lies and which is isomorphic to $\mathbb{R}^q$, and a continuation $\tilde{\sigma} : U \to M$ of σ such that $\tilde{\sigma}$ is differentiable (see Figure 4.11). A singular q-chain $c = \sum n_i\sigma_i$ of M is called *(piecewise) differentiable* if the individual singular q-simplexes σ_i are differentiable. We put

$$C_q^{\mathrm{pd}}(M) := \{\text{piecewise differentiable } q\text{-chains of } M\}.$$

If c is a piecewise differentiable q-chain of M, then the boundary $\partial_q c$ is also piecewise differentiable. Hence we can put

$$\begin{aligned} Z_q^{\mathrm{pd}}(M) &:= \{c \in C_q^{\mathrm{pd}}(M) \mid \partial_q c = 0\}, \\ B_q^{\mathrm{pd}}(M) &:= \partial_{q+1}(C_{q+1}^{\mathrm{pd}}(M)), \\ H_q^{\mathrm{pd}}(M) &:= Z_q^{\mathrm{pd}}(M)/B_q^{\mathrm{pd}}(M). \end{aligned}$$

Since every continuous map $\Delta^q \to M$ can be approximated by a differentiable map, each homology class in $H_q(M)$ can be represented by a piecewise differentiable q-cycle, and each piecewise differentiable null homologous q-cycle is the boundary of a piecewise differentiable $(q+1)$-chain. It follows that one has a canonical isomorphism

$$H_q^{\mathrm{pd}}(M) \cong H_q(M)\,.$$

In an analogous manner one can also introduce the relative homology groups $H_q^{\mathrm{pd}}(M, A)$, for which $H_q^{\mathrm{pd}}(M, A) \cong H_q(M, A)$.

Given a piecewise differentiable q-chain C of M, we denote by $|C|_0$ the set of *smooth* points of the support $|C|$ of C, i.e., the points of $|C|$ at which $|C|$ is locally a submanifold of M.

Let A be a piecewise differentiable k-cycle, B a piecewise differentiable $(n-k)$-cycle of M. We say that A and B *intersect transversally at the point* $x \in |A|_0 \cap |B|_0$ if the corresponding simplexes, on whose support x lies, are transversal to one another. We say that A and B *intersect transversally* if

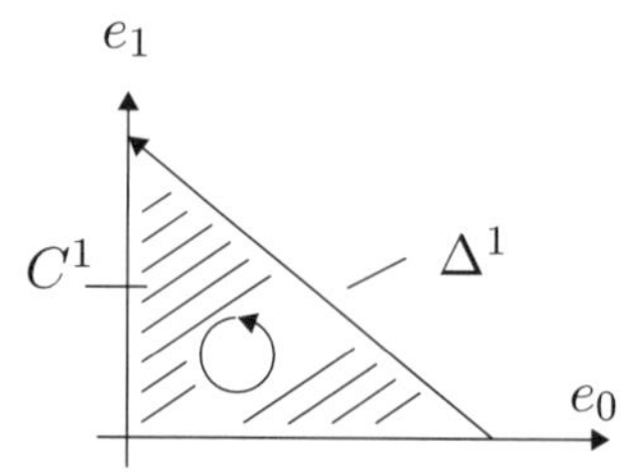

Figure 4.12. Orientation of Δ^1

$|A| \cap |B| = |A|_0 \cap |B|_0$ and A and B intersect transversally in all points $x \in |A|_0 \cap |B|_0$.

We now want to establish an orientation for a differentiable singular q-simplex. To do this, we first choose an orientation for the standard q-simplex Δ^q. We choose the standard orientation of $\mathbb{R}^{q+1}$ and orient Δ^q as a subset of the boundary of

$$C^q = \left\{ \sum_{i=0}^{q} \lambda_i e_i \,\middle|\, \sum_{i=0}^{q} \lambda_i \leq 1, \quad 0 \leq \lambda_i \leq 1 \right\},$$

where we treat C^q as a manifold with boundary. (We neglect the points where C^q is not smooth.) This means that for an a from the interior of Δ^q a basis $(v_1, \ldots, v_q)$ of $T_a\Delta^q$ is positively oriented if with

$$v_0 = a + \sum_{i=0}^{q} \frac{1}{q+1} e_i$$

the basis $(v_0, v_1, \ldots, v_q)$ determines the standard orientation of $\mathbb{R}^{q+1}$ (see Figure 4.12).

If $\sigma : \Delta^q \to M$ is a differentiable singular q-simplex, then $|\sigma|$ inherits the orientation of Δ^q through $T\sigma$.

Now let M be an oriented n-dimensional differentiable manifold, let $A = \sum n_i \sigma_i$ be a piecewise differentiable k-cycle, and let $B = \sum m_j \tau_j$ be a piecewise differentiable $(n-k)$-cycle. Let $p \in |A_0| \cap |B|_0$ be a point at which A and B intersect transversally. We shall assume that $p \in |\sigma_i| \cap |\tau_j|$. Let $v_1, \ldots, v_k$ be a positively oriented basis of $T_p|\sigma_i| \subset T_pM$ and $w_1, \ldots, w_{n-k}$ a positively oriented basis of $T_p|\tau_j| \subset T_pM$. We define the *intersection index* $i_p(\sigma_i, \tau_j)$ of σ_i and τ_j at p as follows:

$$i_p(\sigma_i, \tau_j) := \begin{cases} +1 & \text{if } v_1, \ldots, v_k, w_1, \ldots, w_{n-k} \text{ is a positively} \\ & \text{oriented basis of } T_pM, \\ -1 & \text{otherwise.} \end{cases}$$

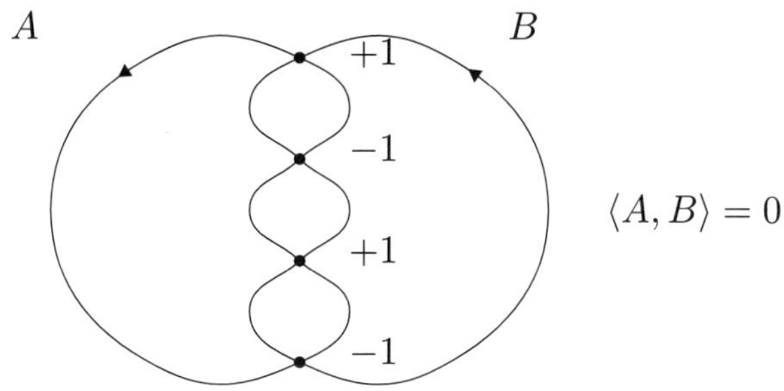

Figure 4.13. Example of $\langle A, B\rangle = 0$

The *intersection index* $i_p(A,B)$ of the cycles A and B at p is then defined by

$$i_p(A,B) := \sum_{p\in|\sigma_i|\cap|\tau_j|} n_i m_j i_p(\sigma_i,\tau_j).$$

Suppose now that A and B intersect transversally. Then, by Proposition 3.5, $|A|\cap|B| = |A|_0 \cap |B|_0$ is a 0-dimensional submanifold, so, in particular, it is discrete. Since A and B have compact support, $|A|\cap|B|$ is a finite set of points. We can therefore define the *intersection number* $\langle A, B\rangle$ of A and B as follows:

$$\langle A, B\rangle := \sum_{p\in|A|\cap|B|} i_p(A,B)$$

(see Figure 4.13).

We next show that the intersection number $\langle A, B\rangle$ depends only on the homology classes of A and B.

Proposition 4.13. *If A is null homologous, then $\langle A, B\rangle = 0$.*

Proof. One can show that it is enough to establish the claim for the case $A = \partial C$ with $C = \sum C_i$, where each C_i is a $(k+1)$-dimensional piecewise differentiable submanifold of M with boundary. We assume that each C_i is oriented so that for all smooth points $p \in |A|$ we have that $(v_1, \ldots, v_k, w)$ gives the orientation of C_i, where $(v_1, \ldots, v_k)$ is a positively oriented basis of $T_p(|A|)$ and w is an inward pointing normal vector to $|A|$ at p.

One can deform C homotopically so that C intersects the cycle B transversally only in smooth points. This follows from a general transversality theorem that we shall not prove (cf. e.g. [**BJ82**, §14]). The intersection $C \cap |B|$ then, by Proposition 3.5, consists of a disjoint union of piecewise differentiable paths $\{\gamma_\alpha\}$ such that the end points of γ_α lie in $|A|\cap|B|$ (see Figure 4.14). Let $\gamma : I \to M$ be such a path with $\gamma(0) = p_0 \in |A|\cap|B|$, $\gamma(1) = p_1 \in |A|\cap|B|$.

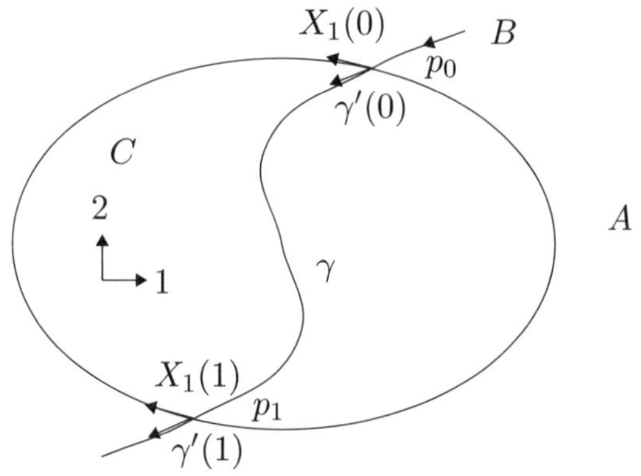

Figure 4.14. Proof of the Claim

Claim. $i_{p_0}(A, B) = -i_{p_1}(A, B)$.

Proof of the Claim: Let $X_1(t), \ldots, X_k(t) \in T_{\gamma(t)}C$ be vector fields on C along γ and $X_{k+2}(t), \ldots, X_n(t) \in T_{\gamma(t)}|B|$ vector fields on B along γ such that

a) $X_1(t), \ldots, X_k(t), \gamma'(t)$ is a positively oriented basis of $T_{\gamma(t)}C$ for all t,

b) $\gamma'(t), X_{k+2}(t), \ldots, X_n(t)$ is a positively oriented basis of $T_{\gamma(t)}|B|$ for all t,

c) $X_1(t), \ldots, X_k(t), \gamma'(t), X_{k+2}(t), \ldots, X_n(t)$ is a positively oriented basis of $T_{\gamma(t)}M$ for all t,

d) $X_1(0), \ldots, X_k(0)$ is a positively oriented basis of $T_{p_0}|A|$, and $X_1(1), \ldots, X_k(1)$ is a basis of $T_{p_1}|A|$.

Such vector fields can be found by using the parameter transformation $t \mapsto 1 - t$, if necessary.

It follows from these conditions that

$$i_{p_0}(A, B) = +1, \quad i_{p_1}(A, B) = -1,$$

as was to be shown.

The conclusion of Proposition 4.13 follows from the validity of the Claim for each path γ_α. □

Generally, we can now define the *intersection number* $\langle \alpha, \beta \rangle$ for two homology classes $\alpha \in H_k(M)$ and $\beta \in H_{n-k}(M)$ as follows: We represent α and β by piecewise differentiable cycles A and B that intersect transversally and put

$$\langle \alpha, \beta \rangle := \langle A, B \rangle.$$

By Proposition 4.13 the intersection number $\langle \alpha, \beta \rangle$ is well defined by this.

From the definition it follows that the map

$$\begin{array}{rccc} \langle\ ,\ \rangle: & H_k(M) \times H_{n-k}(M) & \longrightarrow & \mathbb{Z} \\ & (\alpha, \beta) & \longmapsto & \langle \alpha, \beta \rangle \end{array}$$

is bilinear. This map is called the *intersection form*. From the definition of the intersection index it follows further that

Proposition 4.14. *For all $\alpha \in H_k(M)$, $\beta \in H_{n-k}(M)$ we have*

$$\langle \beta, \alpha \rangle = (-1)^{k(n-k)} \langle \alpha, \beta \rangle.$$

For n even, $k = n/2$, it follows from this that

$$\langle \beta, \alpha \rangle = (-1)^{\frac{n}{2}} \langle \alpha, \beta \rangle,$$

i.e., the intersection form is symmetric for $n \equiv 0 \pmod 4$ and skew symmetric for $n \equiv 2 \pmod 4$.

We can at last explain what we should understand by the intersection number of two arbitrary cycles A, B, of dimension k, $n-k$, that do not intersect transversally: We define

$$\langle A, B \rangle := \langle [A], [B] \rangle,$$

where $[A]$, $[B]$ are the corresponding homology classes of A, B. In particular the intersection number $\langle X, Y \rangle$ is defined for two oriented closed submanifolds X and Y of M of complementary dimension.

Remark 4.10. If M is a complex manifold, if X, Y are complex submanifolds of M of complementary dimension, and if M, X, Y carry their preferred orientation, then the intersection number $\langle X, Y \rangle$ is always nonnegative (Proof: Exercise).

We note finally that if M, N are n-dimensional oriented differentiable manifolds and $f : M \to N$ is an orientation preserving diffeomorphism, then

$$\langle f_*\alpha, f_*\beta \rangle = \langle \alpha, \beta \rangle$$

for all $\alpha \in H_k(M)$, $\beta \in H_{n-k}(M)$.

If M is an n-dimensional oriented differentiable manifold with boundary ∂M, we can in the same way define an intersection number $\langle \alpha, \beta \rangle$ for $\alpha \in H_k(M)$, $\beta \in H_{n-k}(M, \partial M)$ or for $\alpha \in H_k(M, \partial M)$ and $\beta \in H_{n-k}(M)$.

We now want as an example to determine the self-intersection number of the zero section S^n in the tangent bundle TS^n. One obtains this intersection number with the help of a general result which we now present.

Let N be an n-dimensional compact differentiable manifold and $X : N \to TN$ a vector field on N that has only isolated zeros. Let $p \in N$ be such a zero. We define the index of the vector field at p as follows. Let

$\varphi : U \to U' \subset \mathbb{R}^n$ be a chart around p with $\varphi(p) = 0$, where U is chosen so small that U contains no other zeros of X. We then consider the map

$$g = T\varphi \circ X \circ \varphi^{-1} : U' \longrightarrow \mathbb{R}^n.$$

Let S^{n-1}_ε be a small $(n-1)$-sphere around 0 in $\mathbb{R}^n$ with $S^{n-1}_\varepsilon \subset U'$. Since p is the only zero of X on U, we have $g(S^{n-1}_\varepsilon) \subset \mathbb{R}^n \setminus \{0\}$. Let $\tilde{g}$ be the map defined by

$$\begin{array}{rccc} \tilde{g}: & S^{n-1}_\varepsilon & \longrightarrow & S^{n-1} \\ & x & \longmapsto & \frac{g(x)}{\|g(x)\|}\,. \end{array}$$

Let $x \in S^{n-1}_\varepsilon$ be a regular point of $\tilde{g}$. Then $T_x\tilde{g} : T_x S^{n-1}_\varepsilon \to T_{\tilde{g}(x)} S^{n-1}$ is a linear isomorphism. We define sign $T_x\tilde{g}$ as $+1$ or -1 as the determinant of the map $T_x\tilde{g}$ is positive or negative. Let $y \in S^{n-1}$ be a regular value of $\tilde{g}$. We define

$$\operatorname{ind}_p(X) := \sum_{x \in \tilde{g}^{-1}(y)} \operatorname{sign} T_x\tilde{g}.$$

One can show that the number $\operatorname{ind}_p(X)$ is independent of the regular value $y \in S^{n-1}$ chosen and of the chart chosen (cf. [**Mil65**, §5 and §6]). The number $\operatorname{ind}_p(X)$ is called the *index* of the vector field X at p.

Proposition 4.15. *Let N be an n-dimensional compact oriented differentiable manifold and $X : N \to TN$ a vector field on N transversal to the zero section $N \subset TN$. Then, for the self-intersection number $\langle N, N\rangle$ of the zero section in the tangent bundle TN,*

$$\langle N, N\rangle = \sum_p \operatorname{ind}_p(X),$$

where the sum extends over the zeros p of the vector field X.

Proof. Since the vector field X is transversal to the zero section $N \subset TN$, the zeros of X are isolated. Since N is compact, the sum on the right side is finite, so well defined.

Let $\tilde{N}$ be the image of the vector field X. Since the vector field is transversal to the zero section N, the submanifolds N and $\tilde{N}$ of TN intersect transversally in the zeros of the vector field X. (One can think of $\tilde{N}$ as the zero section displaced in the direction of the vector field; see Figure 4.15.) Now let p be a zero of the vector field X. We must show that

$$i_p(N, \tilde{N}) = \operatorname{ind}_p(X)\,.$$

The tangent space to TN at p is however

$$T_pTN \cong T_pN \oplus T_pN.$$

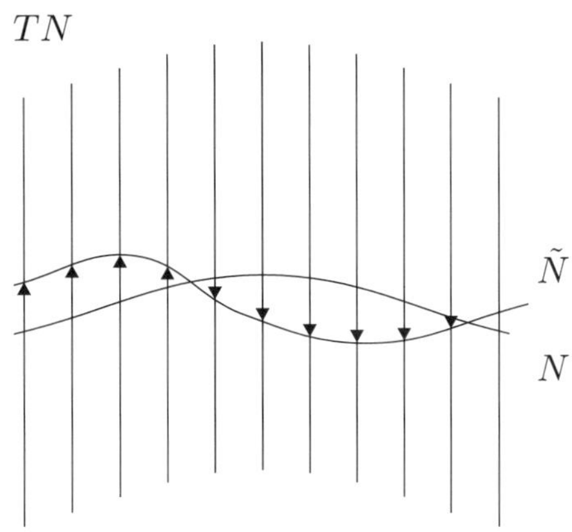

Figure 4.15. Displacement of the zero section

If $v_1, \ldots, v_n$ is a positively oriented basis of T_pN, then $v_1, \ldots, v_n, v_1, \ldots, v_n$ is a positively oriented basis of T_pTN. Since X is transversal to N at p,

$$T_pN + T_pX(T_pN) = T_pTN.$$

It follows that the linear map

$$T_pX : T_pN \longrightarrow T_pX(T_pN) \subset T_pTN$$

is an isomorphism onto its image in T_pTN. By definition, $\mathrm{ind}_p(X) = +1$ or $\mathrm{ind}_p(X) = -1$ as the determinant of this linear map is positive or negative. As this determinant is positive or negative, so correspondingly is the basis $v_1, \ldots, v_n, T_p(X)(v_1), \ldots, T_p(X)(v_n)$ a positively or negatively oriented basis of T_pTN. It follows from the definition of the intersection index that

$$i_p(N, \tilde{N}) = \mathrm{ind}_p(X).$$

This proves the proposition. □

Now, with the help of Proposition 4.15, we shall compute the self-intersection number of S^n in the tangent bundle TS^n. The total space of the tangent bundle TS^n can be described as follows (Exercise):

$$TS^n = \{u + iv \in \mathbb{C}^{n+1} \mid |u| = 1,\ \langle u, v \rangle = 0\}.$$

Let $p = (0, \ldots, 0, 1) + i0$ be the north pole of S^n in TS^n. Then we consider the vector field $X : S^n \to TS^n$ defined by

$$X(u) = u + i(p - \langle p, u \rangle u)$$

where we identify S^n with the set $\{u \in \mathbb{R}^{n+1} \mid |u| = 1\}$. This vector field is directed northwards at every point and has zeros at both poles (see Figure 4.16).

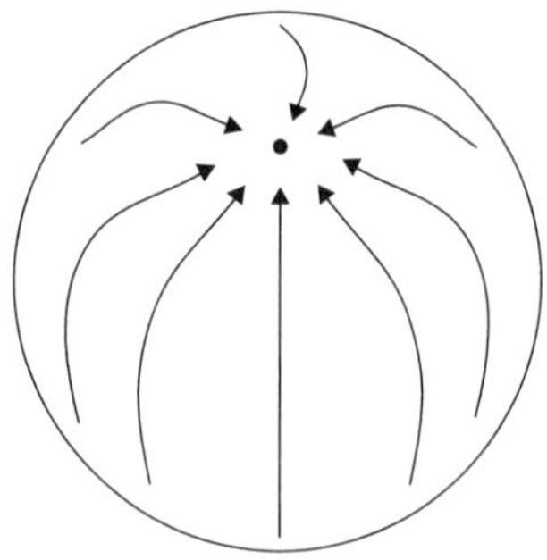

Figure 4.16. The vector field X

At the south pole all the vectors are directed outwards. Hence the index at the south pole is equal to $+1$. At the north pole all the vectors are directed inwards. The map $g|_{S_\varepsilon^{n-1}}$ on a small circle around the north pole in a suitable local coordinate system is the antipodal map. Hence

$$\mathrm{ind}_p(X) = (-1)^n.$$

From Proposition 4.15 follows

Corollary 4.1. $\langle S^n, S^n \rangle = 1 + (-1)^n$.

We note that the following more general result holds, but we shall not use it and therefore we cite it without proof.

Proposition 4.16 (Poincaré-Hopf theorem). *Let N be a compact differentiable manifold and X a vector field on N with isolated zeros. Furthermore, let $\chi(N)$ be the Euler characteristic of N. Then*

$$\sum_p \mathrm{ind}_p(X) = \chi(N),$$

where the sum extends over the zeros of X.

We record another theorem that we shall use in Chapter 5 but do not want to prove. For the proof we refer to [**GH78**, p. 53].

Notation. If L is a $\mathbb{Z}$-module, we denote the dual module by $L^\# = \mathrm{Hom}(L, \mathbb{Z})$.

Theorem 4.5 (Lefschetz-Poincaré duality). *Let M be an m-dimensional compact oriented differentiable manifold with boundary ∂M, $1 \le k \le m$, and let $H_{k-1}(M)$ be torsion free. Then the map*

$$\begin{array}{rccll} \sigma : & H_{m-k}(M, \partial M) & \longrightarrow & H_k(M)^\# & \\ & \alpha & \longmapsto & l_\alpha & \textit{with } l_\alpha(\beta) := \langle \alpha, \beta \rangle \end{array}$$

is an isomorphism.

4.7. Linking numbers

We now introduce the linking number of two n-cycles in S^{2n+1}.

Let a and b be two piecewise differentiable n-cycles in the $(2n+1)$-dimensional unit sphere S^{2n+1} that do not intersect. For $n = 0$ we assume that the cycles a, b are null homologous. For $n > 0$ this is automatically the case. Let A be a piecewise differentiable $(n+1)$-chain in S^{2n+1} with $\partial A = a$. Then the intersection number $\langle A, b\rangle$ is well defined, for the boundary $a = \partial A$ of A does not intersect the cycle b (see Figure 4.17).

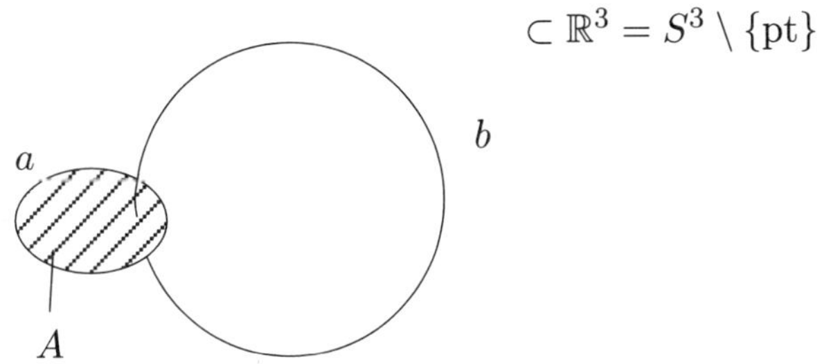

Figure 4.17. Definition of the linking number

Lemma 4.1. *The intersection number $\langle A, b\rangle$ is independent of the choice of the piecewise differentiable $(n+1)$-chain A with $\partial A = a$.*

Proof. Let A' be another piecewise differentiable $(n+1)$-chain with $\partial A' = a$. Then $A - A'$ is a piecewise differentiable $(n+1)$-cycle in S^{2n+1}. Since b is null homologous, we have $\langle A - A', b\rangle = 0$ and thus $\langle A, b\rangle = \langle A', b\rangle$. □

Definition. The *linking number* of the cycles $a, b \in B_n^{Sd}(S^{2n+1})$ is the number

$$l(a, b) = \langle A, b\rangle.$$

For the proof of the following proposition we require yet another concept from topology.

Definition. Let Y be a topological space and A a subspace. The topological space Y/A is the quotient space of Y under the equivalence relation

$$x \sim y \text{ for all } x, y \in A, \quad x \sim x \text{ for all } x \in Y$$

with the quotient topology. It is called the space which arises from Y *by identification of A with a point* (or *through shrinking A to a point*).

Proposition 4.17. *For $a, b \in B_n^{Sd}(S^{2n+1})$*

$$l(a, b) = (-1)^{n+1} l(b, a).$$

Proof. For the proof we consider another way of defining the linking number: Let a, b be null homologous piecewise differentiable n-cycles in S^{2n+1}. We consider S^{2n+1} as the boundary of the $(2n+2)$-dimensional unit ball D^{2n+2}. Let $\tilde{A}$ and $\tilde{B}$ be piecewise differentiable $(n+1)$-chains in D^{2n+2} with $\partial\tilde{A} = a$, $\partial\tilde{B} = b$ such that $|\tilde{A}| \setminus |\partial\tilde{A}|$ and $|\tilde{B}| \setminus |\partial\tilde{B}|$ are contained in the interior of D^{2n+2}. Then the intersection number $\langle \tilde{A}, \tilde{B}\rangle_D$ in D^{2n+2} is well defined, i.e., it is independent of the choice of the chains $\tilde{A}, \tilde{B}$ with $\partial\tilde{A} = a$, $\partial\tilde{B} = b$. For if $\tilde{A}'$ is another piecewise differentiable $(n+1)$-chain in D^{2n+2} with $\partial\tilde{A}' = a$, then $\tilde{A} - \tilde{A}'$ is a null homologous piecewise differentiable $(n+1)$-cycle in D^{2n+2}. The corresponding result follows for an $(n+1)$-chain $\tilde{B}'$ with $\partial\tilde{B}' = b$. Hence we can take for $\tilde{A}$ and $\tilde{B}$ the following $(n+1)$-chains

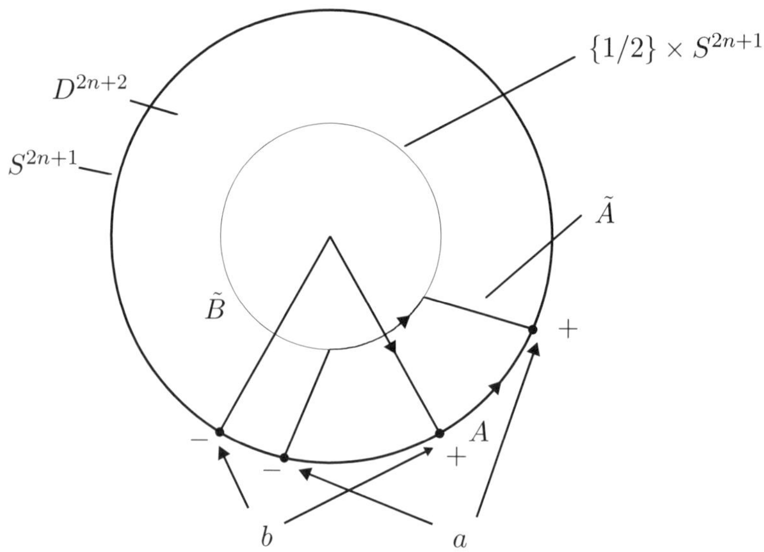

Figure 4.18. Another definition of the linking number

in D^{2n+2}. We use the representation of D^{2n+2} as the quotient space

$$D^{2n+2} = [0,1] \times S^{2n+1}/\{0\} \times S^{2n+1}.$$

Here the subsets $\{t\} \times S^{2n+1}$ for $t \in [0,1]$ correspond to concentric circles around 0 of radius t. Let A be a piecewise differentiable $(n+1)$-chain in S^{2n+1} with $\partial A = a$. Then we put

$$\begin{aligned} \tilde{A} &= \left[\frac{1}{2}, 1\right] \times a \cup \left\{\frac{1}{2}\right\} \times A, \\ \tilde{B} &= [0,1] \times b/\{0\} \times b \end{aligned}$$

(see Figure 4.18 for $n = 0$). Then $\tilde{A}$ and $\tilde{B}$ are piecewise differentiable $(n+1)$-chains in D^{2n+2} with $\partial\tilde{A} = a$ and $\partial\tilde{B} = b$. These chains intersect in points of the form $\tilde{p} = (1/2, p)$, where p is a point of intersection of the chain A with the cycle b. Let $(e_1, \ldots, e_{n+1})$ be a positively oriented basis of T_pA, and let $(e'_1, \ldots, e'_n)$ be a positively oriented basis of T_pb. We assume that $(e_1, \ldots, e_{n+1}, e'_1, \ldots, e'_n)$ specifies the orientation of T_pS^{2n+1}. If e_0 is an outward pointing normal vector to T_pS^{2n+1} in T_pD^{2n+2}, then $(e_1, \ldots, e_{n+1})$ is a positively oriented basis of $T_{\tilde{p}}\tilde{A}$ and $(e_0, e'_1, \ldots, e'_n)$ is a positively oriented basis of $T_{\tilde{p}}\tilde{B}$. According to the orientation convention for the boundary of an oriented manifold,

$$(e_0, e_1, \ldots, e_{n+1}, e'_1, \ldots, e'_n)$$

is a positively oriented basis of $T_{\tilde{p}}D^{2n+2}$. The basis

$$(e_1, \ldots, e_{n+1}, e_0, e'_1, \ldots, e'_n)$$

is, however, obtained by a permutation with sign $(-1)^{n+1}$ from the above basis. Hence it follows that

$$l(a,b) = \langle A, b\rangle = (-1)^{n+1}\langle\tilde{A}, \tilde{B}\rangle_D.$$

From this it follows that

$$\begin{aligned} l(a,b) &= (-1)^{n+1}\langle\tilde{A}, \tilde{B}\rangle_D \\ &= \langle\tilde{B}, \tilde{A}\rangle_D \\ &= (-1)^{n+1} l(b,a), \end{aligned}$$

as was to be shown. □

4.8. The braid group

We introduce the braid group in this section.

Definition. Let B_μ be the group with the generators $\alpha_1, \ldots, \alpha_{\mu-1}$ and relations

$$\begin{aligned} \alpha_j\alpha_{j+1}\alpha_j &= \alpha_{j+1}\alpha_j\alpha_{j+1} \text{ for } j = 1, \ldots, \mu-2, \\ \alpha_i\alpha_j &= \alpha_j\alpha_i \text{ for } |i-j| > 1. \end{aligned}$$

This group is called the *(Artin) braid group with* μ *strands.*

The braid group B_μ was introduced by E. Artin in 1925, after it had appeared implicitly in work of A. Hurwitz. To explain the name "braid group", we give another definition of this group.

Let S_μ be the symmetric group of permutations of μ elements. We let S_μ act on $\mathbb{C}^\mu$ by permuting the coordinates. Consider the quotient space $\mathbb{C}^\mu/S_\mu$ under this group action. This is the μ-fold symmetric product of $\mathbb{C}$. In a natural way this space bears the structure of an algebraic variety, for the algebra $\mathbb{C}[z_1, \ldots, z_\mu]^{S_\mu}$ of S_μ-invariant polynomials is finitely generated, namely by the μ elementary symmetric functions. One sees also that $\mathbb{C}^\mu/S_\mu$ is biholomorphically equivalent to $\mathbb{C}^\mu$. Next for $i \neq j$ we consider the hypersurface

$$H_{ij} = \{(z_1, \ldots, z_\mu) \in \mathbb{C}^\mu \mid z_i = z_j\} \subset \mathbb{C}^\mu.$$

We put

$$Y_\mu := \left(\mathbb{C}^\mu \setminus \bigcup_{1 \leq i < j \leq \mu} H_{ij}\right) \Big/ S_\mu.$$

The subspace $Y_\mu \subset \mathbb{C}^\mu/S_\mu$ can be considered as the space of all sets of μ distinct complex numbers $\{z_1, \ldots, z_\mu\}$. It can also be identified with the space of all complex polynomials of degree μ with simple distinct zeros; one can associate to the set $\{z_1, \ldots, z_\mu\}$ the polynomial

$$\prod_{i=1}^{\mu}(x - z_i) = x^\mu + \sigma_1(z_1, \ldots, z_\mu)x^{\mu-1} + \ldots + \sigma_\mu(z_1, \ldots, z_\mu)$$

where $\sigma_1, \ldots, \sigma_\mu$ are the elementary symmetric functions. We distinguish the base point $\bar{y} = (1, \ldots, \mu)$ in Y_μ.

Definition. A *braid with μ strands* is the homotopy class of a closed path in Y_μ with start and end point $\bar{y}$, hence an element of $\pi_1(Y_\mu, \bar{y})$.

One can imagine an element $b \in \pi_1(Y_\mu, \bar{y})$ as follows. The element b is represented by a closed path $\beta : [0,1] \to Y_\mu$. Using β, a set $\beta(t) = \{\beta_1(t), \ldots, \beta_\mu(t)\}$ of μ distinct complex numbers is associated to every $t \in [0,1]$ so that the $\beta_i : [0,1] \to \mathbb{C}$ are continuous functions and $\beta(0) = \beta(1) = \{1, \ldots, \mu\}$. Thus the path β determines continuous functions $\beta_i : [0,1] \to \mathbb{C}$, $i = 1, \ldots, \mu$, with $\beta_i(t) \neq \beta_j(t)$ for all $t \in [0,1]$ and $i \neq j$ and $\beta_i(0) = i$, $\{\beta_1(1), \ldots, \beta_\mu(1)\} = \{1, \ldots, \mu\}$. The graphs of the functions β_i in $[0,1] \times \mathbb{C}$ look as follows: We have sketched an example for $\mu = 3$ in Figure 4.19.

This explains the name braid. To represent such a braid, one draws a flat projection of such a graph with the crossings made suitably recognizable; see Figure 4.20. Here we meet the convention that at a crossover the solid line has a smaller imaginary part.

One can now easily give a generating system for $\pi_1(Y_\mu, \bar{y})$. Let a_j be the braid described by Figure 4.21. Then $a_1, \ldots, a_{\mu-1}$ form a generating system for $\pi_1(Y_\mu, \bar{y})$.

Example 4.7. The "moroccan braid" from the diploma thesis of Eberhard Voigt [**Voi80**] has the representation (see Figure 4.22)

$$(\alpha_4\alpha_1^{-1}\alpha_2^{-1}\alpha_3)^5.$$

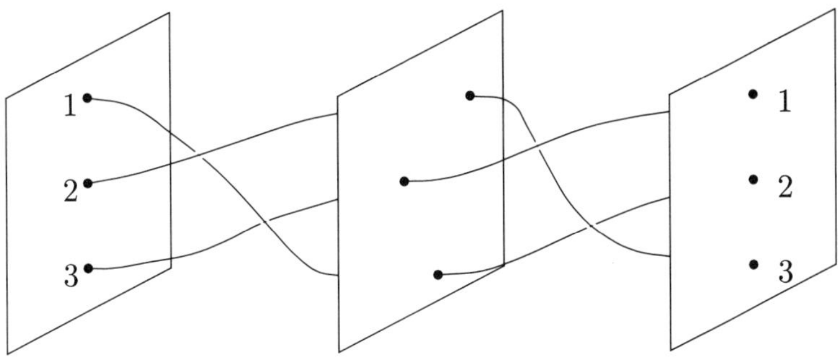

Figure 4.19. A braid with 3 strands

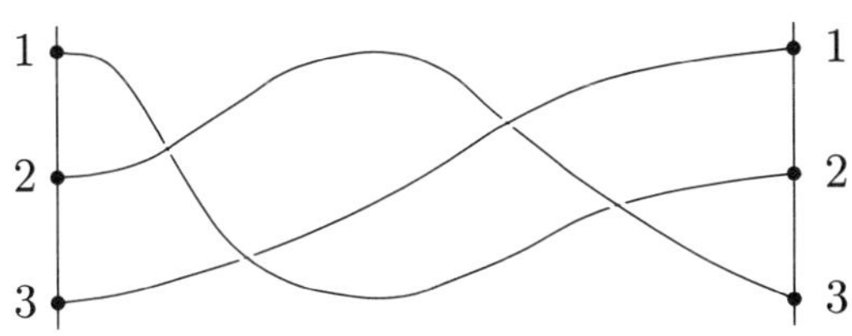

Figure 4.20. Plane projection of a braid

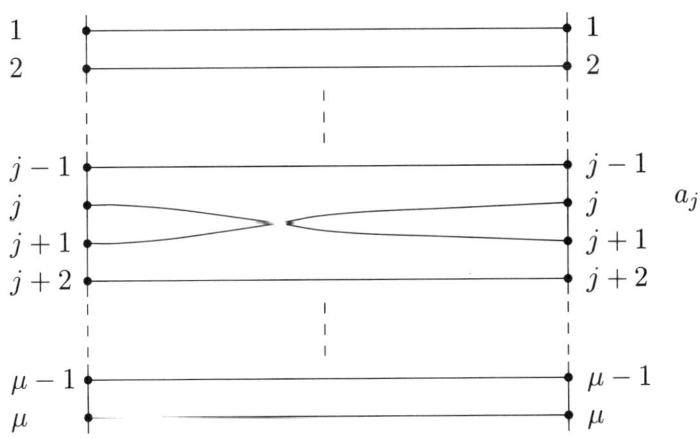

Figure 4.21. The braid a_j

This thesis contains the following assembly instructions for a leather bracelet with this braid pattern. One takes a long strip of leather (see Figure 4.23) and cuts 4 slits lengthwise (or, equivalently, one can take 5 strips bound together). To this one applies the following braid operations: f_i (resp. f_i^{-1}) denotes passing the left end L through the i-th slit from under (resp. from

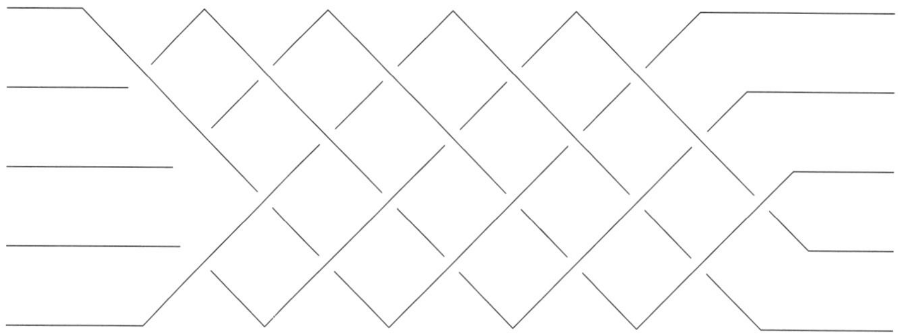

Figure 4.22. A moroccan braid

Figure 4.23. Leather strip with slits

over). The assembly instructions are now

$$f_3 f_1^{-1} f_4^{-1} f_1 f_2^{-1} f_4 f_3^{-1} f_2.$$

4.9. The homotopy sequence of a differentiable fiber bundle

Let X be a topological space, $x_0 \in X$. We wish to define the higher homotopy groups of X. Analogously to the fundamental group $\pi_1(X, x_0)$ one can define a group $\pi_q(X, x_0)$ for $q \geq 1$, where one replaces the unit interval by the q-dimensional unit ball I^q.

So let

$$\begin{aligned} I^q &= \underbrace{I \times \ldots \times I}_{q} \\ &= \{t = (t_1, \ldots, t_q) \in \mathbb{R}^q \mid 0 \leq t_i \leq 1, i = 1, \ldots, q\}. \end{aligned}$$

Let ∂I^q be the boundary of I^q.

Definition. A subspace A of a topological space X is called a *deformation retract* of X if there is a map $r : X \to A$ such that $\mathrm{id} : X \to X$ is homotopic to ir, where $i : A \to X$ is the inclusion.

Definition. Let X, Y be topological spaces, $A_1, \ldots, A_k$ subspaces of X, $B_1, \ldots, B_k$ subspaces of Y. A map

$$f : (X, A_1, \ldots, A_k) \longrightarrow (Y, B_1, \ldots, B_k)$$

is a continuous map $f : X \to Y$ with $f(A_j) \subset B_j$ for $j = 1, \ldots, k$. If one of the subspaces $A_1, \ldots, A_k, B_1, \ldots, B_k$ consists of a single point, we shall omit the set brackets around this point.

Two maps

$$f, g : (X, A_1, \ldots, A_k) \longrightarrow (Y, B_1, \ldots, B_k)$$

are called *homotopic relative to* $(A_1, B_1; \ldots; A_k, B_k)$, symbolically,

$$f \sim g \ \mathrm{rel}\, (A_1, B_1; \ldots; A_k, B_k),$$

if there is a homotopy $H : X \times I \to Y$ between f and g with $H_t(A_i) \subset B_i$ for all $t \in [0, 1]$ and $i = 1, \ldots, k$, where $H_t : X \to Y$ is defined by $H_t(x) = H(x, t)$ for all $x \in X$.

Definition. For $q \geq 1$ let $\pi_q(X, x_0)$ be the set of all homotopy classes (relative to $(\partial I^q, x_0)$) of maps

$$f : (I^q, \partial I^q) \longrightarrow (X, x_0).$$

We define an addition in $\pi_q(X, x_0)$ as follows: For $[f], [g] \in \pi_q(X, x_0)$ let $[f] + [g] := [f + g]$, where $f + g : (I^q, \partial I^q) \to (X, x_0)$ is defined by

$$(f+g)(t) = \begin{cases} f(2t_1, t_2, \ldots, t_q) & \text{for } 0 \leq t_1 \leq \frac{1}{2}, \\ g(2t_1 - 1, t_2, \ldots, t_q) & \text{for } \frac{1}{2} \leq t_1 \leq 1, \end{cases}$$

for $t = (t_1, \ldots, t_q) \in I^q$. With this addition $\pi_q(X, x_0)$ forms a group.

Definition. The group $\pi_q(X, x_0)$ $(q \geq 1)$ is called the *q-th homotopy group of X with base point x_0*. For $q = 0$ let $\pi_0(X, x_0)$ be the set of path connected components of X. The path connected component of X containing the point x_0 is called the *neutral element* of $\pi_0(X, x_0)$ and will be denoted by 0.

Remark 4.11. One can show that $\pi_q(X, x_0)$ is an abelian group for each $q > 1$. For this reason one writes the composition additively.

We next want to introduce the relative homotopy groups too. Let $A \subset X$ be a subspace with $x_0 \in A$. One calls (X, A, x_0) with $X \supset A \ni x_0$ a *triple*. Again let $q \geq 1$.

We consider the q-dimensional unit ball I^q. A $(q-1)$-*face* is a set

$$F_{i0}^{q-1} := \{t = (t_1, \ldots, t_q) \in I^q \mid t_i = 0\}$$

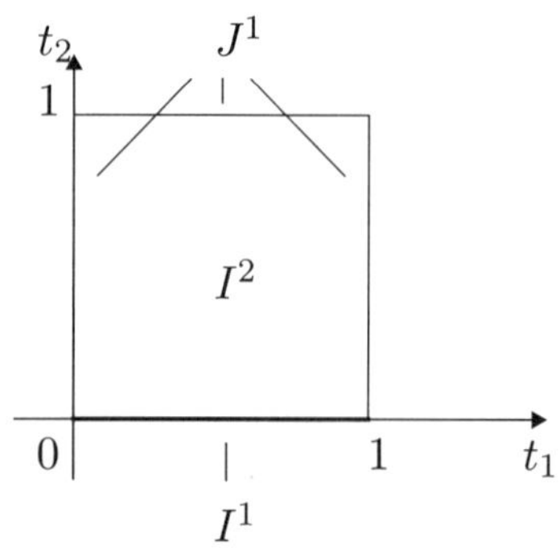

Figure 4.24. The unit ball I_2

or

$$F_{i1}^{q-1} := \{t = (t_1, \dots, t_q) \in I^q \mid t_i = 1\}$$

for $i = 1, \dots, q$. We put

$$\begin{aligned} I^{q-1} &:= F_{q0}^{q-1} = \{t \in I^q \mid t_q = 0\}, \\ J^{q-1} &:= \bigcup_{i=1}^{q-1} F_{i0}^{q-1} \cup \bigcup_{i=1}^{q-1} F_{i1}^{q-1} \cup F_{q1}^{q-1} \end{aligned}$$

(see Figure 4.24 for $q = 2$). We have

$$\begin{aligned} \partial I^q &= I^{q-1} \cup J^{q-1}, \\ \partial I^{q-1} &= I^{q-1} \cap J^{q-1}. \end{aligned}$$

Definition. The set of all homotopy classes (relative to $(I^{q-1}, A; J^{q-1}, x_0)$) of maps

$$f : (I^q, I^{q-1}, J^{q-1}) \longrightarrow (X, A, x_0)$$

will be denoted by $\pi_q(X, A, x_0)$ (see Figure 4.25 for $q = 2$). We define an addition on $\pi_q(X, A, x_0)$ as above. Then $\pi_q(X, A, x_0)$ is called the q-th *relative homotopy group of X with respect to A with base point x_0.*

Now let (Y, B, y_0) be another triple, and $f : (X, A, x_0) \to (Y, B, y_0)$ a map. If $[\varphi] \in \pi_q(X, A, x_0)$, then $[f \circ \varphi] \in \pi_q(Y, B, y_0)$. The assignment

$$[\varphi] \longrightarrow [f \circ \varphi]$$

thus defines a map

$$f_* : \pi_q(X, A, x_0) \longrightarrow \pi_q(Y, B, y_0).$$

For $q = 0$ the map f_* takes the neutral element of $\pi_0(X, x_0)$ to the neutral element of $\pi_0(Y, y_0)$. For $q \geq 1$ it is easy to see that f_* is a homomorphism. In this case the homomorphism f_* is called the *homomorphism induced* by $f : (X, A, x_0) \to (Y, B, y_0)$.

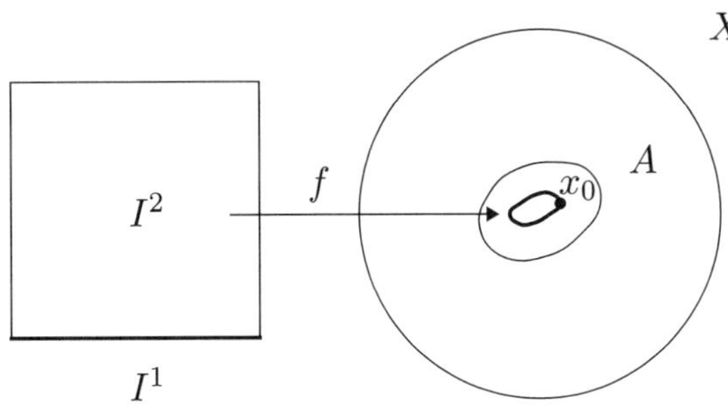

Figure 4.25. A map $f : (I^2, I^1, J^1) \to (X, A, x_0)$

In particular the inclusion $i : (A, x_0) \to (X, x_0)$ induces a map $i_* : \pi_q(A, x_0) \to \pi_q(X, x_0)$ for $q \geq 0$ and the inclusion $j : (X, x_0) \to (X, A, x_0)$ induces a homomorphism $j_* : \pi_q(X, x_0) \to \pi_q(X, A, x_0)$ for $q \geq 1$.

Definition. We define a map

$$\partial_* : \pi_q(X, A, x_0) \longrightarrow \pi_{q-1}(A, x_0)$$

(the *boundary operator* or *connecting homomorphism*) by

$$[f] \longmapsto [f|_{I^{n-1}}].$$

Remark 4.12. It is easy to see that ∂_* is well defined and is a homomorphism for $q \geq 2$. For $q = 1$ the map ∂_* takes the neutral elements one to the other.

Definition. The sequence

$$\dots \xrightarrow{\partial_*} \pi_q(A, x_0) \xrightarrow{i_*} \pi_q(X, x_0) \xrightarrow{j_*} \pi_q(X, A, x_0) \xrightarrow{\partial_*} \pi_{q-1}(A, x_0) \xrightarrow{i_*} \dots$$
$$\dots \xrightarrow{j_*} \pi_1(X, A, x_0) \xrightarrow{\partial_*} \pi_0(A, x_0) \xrightarrow{i_*} \pi_0(X, x_0)$$

is called the *homotopy sequence of the triple* (X, A, x_0).

Proposition 4.18. *The homotopy sequence of the triple* (X, A, x_0) *is exact.*

Proof. Again we shall show exactness only at one place and leave the demonstration of exactness at the other places as an exercise (cf. §4.5). We shall show exactness at the place $\pi_q(X, A, x_0)$.

To show: $\operatorname{Im} j_* = \operatorname{Ker} \partial_*$.

a) We show that $\operatorname{Im} j_* \subset \operatorname{Ker} \partial_*$. Let $\alpha \in \pi_q(X, x_0)$, $\alpha = [f]$, where $f : (I^q, \partial I^q) \to (X, x_0)$. Then $\partial_* j_* \alpha = [f|_{I^{q-1}}] = 0$, for $f(I^{q-1}) = x_0$.

b) We show that $\operatorname{Ker} \partial_* \subset \operatorname{Im} j_*$. Let $\alpha \in \pi_q(X, A, x_0)$, $\alpha = [f]$, where $f : (I^q, I^{q-1}, J^{q-1}) \to (X, A, x_0)$, and let $\partial_* \alpha = 0$.

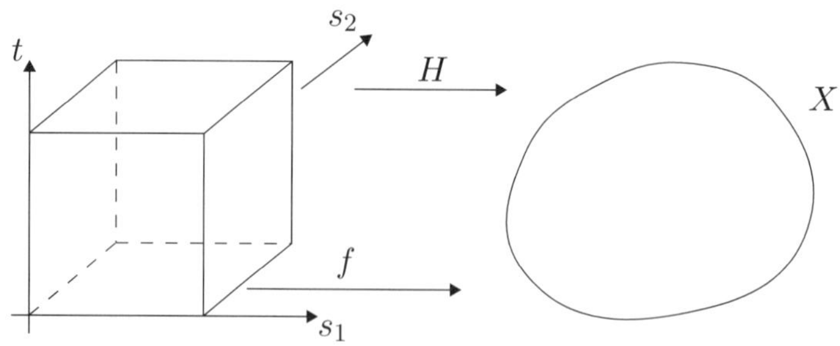

Figure 4.26. The homotopy H

b1) First let $q > 1$. From the condition $\partial_* \alpha = 0$ it follows that there is then a homotopy $G_t : I^{q-1} \to A$, $t \in [0,1]$, with $G_0 = f|_{I^{q-1}}$, $G_1(I^{q-1}) = x_0$ and $G_t(\partial I^{q-1}) = x_0$ for all $t \in [0,1]$. We define a homotopy

$$H : \partial I^q \times I \longrightarrow A$$

by

$$H(s,t) = \begin{cases} G_t(s) & \text{for } s \in I^{q-1}, t \in I, \\ x_0 & \text{for } s \in J^{q-1}, t \in I. \end{cases}$$

Since $H(s,0) = f(s)$ for $s \in \partial I^q$, the map

$$\begin{array}{rccl} \tilde{H}: & \partial I^q \times I \cup I^q \times \{0\} & \longrightarrow & X \\ & (s,t) & \longmapsto & \begin{cases} H(s,t) & \text{for } (s,t) \in \partial I^q \times I, \\ f(s) & \text{for } (s,0) \in I^q \times \{0\} \end{cases} \end{array}$$

is continuous (see Figure 4.26). Since $\partial I^q \times I \cup I^q \times \{0\}$ is a (deformation) retract of $I^q \times I$, there is a continuous extension

$$F : I^q \times I \longrightarrow X$$

of $\tilde{H}$. For the map $F_1 : I^q \to X$, $s \mapsto F(s,1)$, we have $F_1(s) = H(s,1) = x_0$ for all $s \in \partial I^q$. Thus F_1 is a map $F_1 : (I^q, \partial I^q) \to (X, x_0)$ and represents an element $\beta \in \pi_q(X, x_0)$. Since F is a homotopy between f and F_1 relative to $(I^{q-1}, A; J^{q-1}, x_0)$, we get $j_*\beta = \alpha$.

b2) In the case $q = 1$, $f : I \to X$ is a path with $f(0) \in A$ and $f(1) = x_0$. The condition $\partial_*\alpha = 0$ means that $f(0)$ lies in the very same path connected component of A as x_0. Then a path $\gamma : [0,1] \to A$ which joins $f(0)$ and $f(1) = x_0$ in A defines a homotopy $F : I \times I \to X$ with $F(s,0) = f(s)$ for all $s \in I$, $F(0,t) \in A$, $F(1,t) = x_0$ for all $t \in I$ and $F(0,1) = x_0$ (see Figure 4.27). Then $F_1 : I \to X$, $s \mapsto F(s,1)$, is a closed path in X with start and end point x_0, whence $\beta := [F_1] \in \pi_1(X, x_0)$ and F is a homotopy relative to $(I^0, A; J^0, x_0)$ between f and F_1. Thus $j_*\beta = \alpha$. □

Now let (E, π, B, F) be a differentiable fiber bundle with compact fiber F. We choose a base point $b_0 \in B$. Let $F = \pi^{-1}(b_0)$, $e_0 \in F$. Then

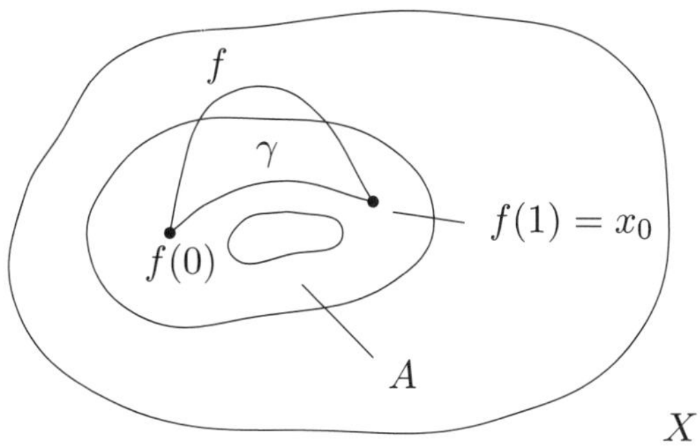

Figure 4.27. The paths f and γ

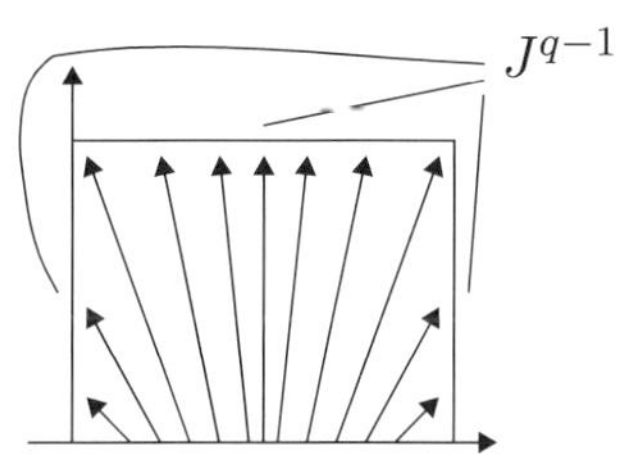

Figure 4.28. The retraction of I^q onto J^{q-1}

(E, F, e_0) is a triple and π defines a map

$$p : (E, F, e_0) \longrightarrow (B, b_0).$$

Furthermore, let us choose an Ehresmann connection of the given bundle.

Proposition 4.19. *The map*

$$p_* : \pi_q(E, F, e_0) \longrightarrow \pi_q(B, b_0)$$

is an isomorphism for each $q > 0$.

Proof. a) We show first that p_* is surjective.

Let $\alpha \in \pi_q(B, b_0)$, $\alpha = [f]$, $f : (I^q, \partial I^q) \to (B, b_0)$. Now J^{q-1} is a deformation retract of I^q (see Figure 4.28). This means that there are a retraction $r : I^q \to J^{q-1}$ and a homotopy $H : I^q \times I \to I^q$ between $\mathrm{id} : J^{q-1} \to I^q$ and ir, where $i : J^{q-1} \to I^q$ is the inclusion. We can assume that f, r and H are differentiable maps. For $s \in I^q$ let $\gamma_s : I \to B$ be defined by $\gamma_s(t) = f \circ H(s, 1-t)$ for $t \in I$. By Proposition 4.9 there is exactly one horizontal lifting $\hat{\gamma}_s : I \to E$ of γ_s with $\hat{\gamma}_s(0) = e_0$ for all $s \in J^{q-1}$. For $s \in I^{q-1}$ we have $\hat{\gamma}_s(1) \in F$ since

$$\gamma_s(1) = f \circ H(s, 0) = f(s) = b_0.$$

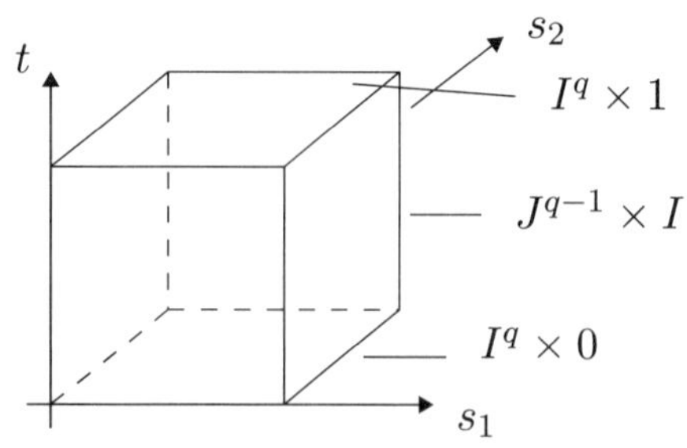

Figure 4.29. The cube $I^q \times I$

In this way we obtain a lifting

$$\hat{f} : (I^q, I^{q-1}, J^{q-1}) \longrightarrow (E, F, e_0)$$

of $f : (I^q, \partial I^q) \to (B, b_0)$ with respect to $\pi : E \to B$ by putting $s \mapsto \hat{\gamma}_s(1)$ for $s \in I^q$. This map represents an element $\beta \in \pi_q(E, F, e_0)$. Since $f = p \circ \hat{f}$, we have $p_* \beta = \alpha$. Thus p_* is surjective.

b) We show that p_* is injective.

To achieve this, let $\alpha \in \pi_q(E, F, e_0)$, $\alpha = [f]$, $f : (I^q, I^{q-1}, J^{q-1}) \to (E, F, e_0)$ and $\alpha \in \ker p_*$. Since $p_* \alpha = 0$, we see that $g := p \circ f : (I^q, \partial I^q) \to (B, b_0)$ is nullhomotopic, i.e., there is a homotopy $H : I^q \times I \to B$, where for $H_t : I^q \to B$, $s \mapsto H(s,t)$ $(t \in I)$, we have $H_0 = g$, $H_1(s) = b_0$ for all $s \in I^q$, and $H_t(\partial I^q) = b_0$ for all $t \in I$. Again we can assume that f and H are differentiable.

Consider

$$T = (I^q \times 0) \cup (J^{q-1} \times I) \cup (I^q \times 1) \subset I^q \times I$$

(see Figure 4.29). We define a map

$$G : T \to E$$

as follows:

$$G(s,t) = \begin{cases} f(s) & \text{for } s \in I^q, t = 0, \\ e_0 & \text{for } s \in J^{q-1}, t \in I \text{ and for } s \in I^q, t = 1. \end{cases}$$

Since T is a deformation retract of $I^q \times I$, we can, by the same argument as for f in a), lift H to a homotopy

$$\hat{H} : I^q \times I \longrightarrow E$$

with $\hat{H}|_T = G$. It follows from $\hat{H}|_T = G$ that for $\hat{H}_t : I^q \to E$, $s \mapsto H(s,t)$ we have $\hat{H}_0 = f$, $\hat{H}_1(s) = e_0$ for all $s \in I^q$, $\hat{H}_t(I^{q-1}) \subset F$, $\hat{H}_t(J^{q-1}) = e_0$ for all $t \in I$. Hence it follows that f is nullhomotopic relative to $(I^{q-1}, F; J^{q-1}, e_0)$. Hence $\alpha = 0$. It follows that p_* is injective. □

Next we consider the homotopy sequence of the triple (E, F, e_0). By Proposition 4.19 we can replace the term $\pi_q(E, F, e_0)$ by $\pi_q(B, b_0)$:

$$\cdots \longrightarrow \pi_q(E,e_0) \xrightarrow{j_*} \pi_q(E,F,e_0) \xrightarrow{\partial_*} \pi_{q-1}(F,e_0) \longrightarrow \cdots$$

$$\pi_q(E,F,e_0) \xrightarrow[\cong]{p_*} \pi_q(B,b_0), \qquad \text{(a)}: \pi_q(E,e_0) \dashrightarrow \pi_q(B,b_0), \qquad \text{(b)}: \pi_q(B,b_0) \dashrightarrow \pi_{q-1}(F,e_0)$$

The homomorphism (a) is induced by $\pi = p \circ j : (E, e_0) \to (B, b_0)$, and therefore it is the map π_*. The homomorphism (b) is $\partial_* \circ p_*^{-1}$ and will be denoted by d_*. It can be described as follows: One represents an $\alpha \in \pi_q(B, b_0)$ by a map $f : (I^q, \partial I^q) \to (B, b_0)$, lifts this map to a map $\hat{f} : (I^q, I^{q-1}, J^{q-1}) \to (E, F, e_0)$ and then considers the restriction $\hat{f}|_{I^{q-1}} : (I^{q-1}, \partial I^{q-1}) \to (F, e_0)$.

Definition. The exact sequence

$$\cdots \xrightarrow{d_*} \pi_q(F,e_0) \xrightarrow{i_*} \pi_q(E,e_0) \xrightarrow{\pi_*} \pi_q(B,b_0) \xrightarrow{d_*} \pi_{q-1}(F,e_0) \xrightarrow{i_*} \cdots$$
$$\cdots \xrightarrow{\pi_*} \pi_1(B,b_0) \xrightarrow{d_*} \pi_0(F,e_0) \xrightarrow{i_*} \pi_0(E,e_0)$$

is called the *homotopy sequence of the differentiable fiber bundle* (E, π, B, F).

Proposition 4.20. *If the differentiable fiber bundle (E, π, B, F) has a section $\sigma : B \to E$, then for each $q \geq 1$ there is a split short exact sequence*

$$0 \longrightarrow \pi_q(F,e_0) \xrightarrow{i_*} \pi_q(E,e_0) \underset{\sigma_*}{\overset{\pi_*}{\rightleftarrows}} \pi_q(B,b_0) \longrightarrow 0.$$

Hence $\pi_q(E, e_0) \cong \pi_q(F, e_0) \times \pi_q(B, b_0)$.

Proof. Since σ is a section, we have $\pi \circ \sigma = \mathrm{id}$. It follows that $\pi_* \circ \sigma_* = \mathrm{id}$, so π_* is surjective. From the exactness of the homotopy sequence of the differentiable fiber bundle (E, π, B, F) it follows that i_* is injective, for surjectivity of π_* implies $\operatorname{Ker} d_* = \operatorname{Im} \pi_* = \pi_q(B, b_0)$. But it follows that $\operatorname{Ker} i_* = \operatorname{Im} d_* = 0$. Hence we have the given short exact sequence for all $q \geq 1$.

From $\pi_* \circ \sigma_* = \mathrm{id}$ it follows that σ_* splits the sequence. From this one deduces that

$$\begin{array}{rccl} (i_*, \sigma_*): & \pi_q(F,e_0) \times \pi_q(B,b_0) & \longrightarrow & \pi_q(E,e_0) \\ & (\alpha, \beta) & \longmapsto & i_*\alpha + \sigma_*\beta \end{array}$$

is an isomorphism. $\square$

Chapter 5

Topology of singularities

5.1. Monodromy and variation

We now begin the investigation of the topology of isolated critical points of holomorphic functions. In §3.11 we encountered by way of a simple example the fundamental concepts of Picard-Lefschetz theory: vanishing cycles and monodromy. We next want to define these concepts in the general case of an isolated hypersurface singularity. First, some preparatory steps.

Let M be an $(n+1)$-dimensional complex manifold and $f : M \to \mathbb{C}$ a holomorphic function. Furthermore, let X be an open subset of M, so that its closure $\bar{X}$ is a $(2n+2)$-dimensional differentiable manifold with boundary $\partial\bar{X}$. Let $\bar{\Delta}$ be a closed disk in $\mathbb{C}$. Suppose that the following three conditions are satisfied:

(1) There is a neighborhood U of $\bar{\Delta}$ in $\mathbb{C}$ for which the map $f|_{\bar{X}\cap f^{-1}(U)} : \bar{X} \cap f^{-1}(U) \to U$ is proper.

(2) $f|_{\partial\bar{X}\cap f^{-1}(U)} : \partial\bar{X} \cap f^{-1}(U) \to U$ is a submersion.

(3) The function f has finitely many critical points p_i $(i = 1, \dots, m)$ on $\bar{X} \cap f^{-1}(U)$ with critical values $s_i = f(p_i) \in \Delta$.

Let us draw some consequences from these hypotheses.

a) From (1), (2) and the Ehresmann fibration theorem it follows that $f|_{\partial\bar{X}\cap f^{-1}(U)} : \partial\bar{X} \cap f^{-1}(U) \to \bar{\Delta}$ is the projection of a differentiable fiber bundle. Since $\bar{\Delta}$ is contractible, it follows from Proposition 4.11 that this fiber bundle is trivial. The trivialization depends on the Ehresmann connection and is determined uniquely up to isotopy. We choose an Ehresmann connection and so a fixed trivialization.

b) Let $\Delta' = \bar{\Delta} - \{s_1, \ldots, s_m\}$. From (1), (2), (3) and the Ehresmann fibration theorem (Theorem 4.3) it follows that the map

$$f|_{\bar{X}\cap f^{-1}(\Delta')} : \bar{X} \cap f^{-1}(\Delta') \longrightarrow \Delta'$$

is the projection of a differentiable fiber bundle. We write

$$\bar{X}_s := \bar{X} \cap f^{-1}(s)$$

for the fiber of this bundle above $s \in \Delta'$. This is a compact $2n$-dimensional differentiable manifold with boundary $\partial\bar{X}_s = \bar{X}_s \cap \partial\bar{X}$. We choose an Ehresmann connection on $\bar{X} \cap f^{-1}(\Delta')$ that agrees with the one chosen in a) on $\partial\bar{X} \cap f^{-1}(\Delta')$.

Now let $s \in \partial\Delta$, but let s not be a critical value of f, so that $s \in \Delta'$. Let $\gamma : [0,1] \to \Delta'$ be a closed path in Δ' starting and ending at s. Without loss of generality we can assume that all the paths considered are piecewise differentiable: we shall make this a standing general assumption. It is easy to convince oneself that the results of §4.4 hold for piecewise differentiable paths too. By §4.4 the path γ determines a parallel transport $h_\gamma : \bar{X}_s \to \bar{X}_s$ along γ with respect to the connection chosen. By a), h_γ is the identity on the boundary $\partial\bar{X}$ of $\bar{X}$. By Proposition 4.10 the diffeomorphism h_γ is determined uniquely up to isotopy by the class of γ in $\pi_1(\Delta', s)$ and is independent of the connection chosen, up to isotopy.

Definition. The diffeomorphism $h_\gamma : \bar{X}_s \to \bar{X}_s$ is called the *geometric monodromy* with respect to γ. The homomorphism $h_{\gamma*} : H_q(\bar{X}_s) \to H_q(\bar{X}_s)$ induced on the q-th homology group of $\bar{X}_s$ (for each $q \in \mathbb{Z}$) is called the *monodromy* (or the *monodromy operator*) with respect to γ.

The monodromy $h_{\gamma*}$ is, by Theorem 4.3, determined uniquely by the class of γ in the fundamental group $\pi_1(\Delta', s)$ so it too does not depend on the connection chosen.

If $\gamma = \gamma_1 * \gamma_2$ is the composition of two closed paths γ_1 and γ_2 with $[\gamma_1], [\gamma_2] \in \pi_1(\Delta', s)$, then

$$h_{\gamma*} = h_{\gamma_2*} \circ h_{\gamma_1*}.$$

We therefore have a homomorphism

$$\begin{array}{rccc} \rho: & \pi_1(\Delta', s) & \longrightarrow & \operatorname{Aut} H_q(\bar{X}_s) \\ & [\gamma] & \longmapsto & h_{\gamma*} \end{array}$$

from $\pi_1(\Delta', s)$ into the automorphism group of the group $H_q(\bar{X}_s)$.

The diffeomorphism h_γ also induces a homomorphism

$$h_{\gamma *} : H_q(\bar{X}_s, \partial \bar{X}_s) \to H_q(\bar{X}_s, \partial \bar{X}_s)$$

on the relative homology group of the pair $(\bar{X}_s, \partial \bar{X}_s)$. Let c be a relative q-cycle of $(\bar{X}_s, \partial \bar{X}_s)$. Now $h_\gamma(c) - c$ is, because $h_\gamma|_{\partial \bar{X}_s} = \mathrm{id}_{\partial \bar{X}_s}$, an absolute q-cycle of $\bar{X}_s$. It is easy to see that the map $H_q(\bar{X}_s, \partial \bar{X}_s) \to H_q(\bar{X}_s)$, $[c] \mapsto [h_\gamma(c) - c]$, is a well defined homomorphism.

Definition. The homomorphism

$$\begin{array}{rccc} \operatorname{var}_\gamma : & H_q(\bar{X}_s, \partial \bar{X}_s) & \longrightarrow & H_q(\bar{X}_s) \\ & [c] & \longmapsto & [h_\gamma(c) - c] \end{array}$$

is called the *variation* (or the *variation operator*) with respect to γ.

If $j_* : H_q(\bar{X}_s) \to H_q(\bar{X}_s, \partial \bar{X}_s)$ is the natural homomorphism defined in §4.5 it follows from this definition that the following diagram commutes:

$$\begin{array}{ccc} H_q(\bar{X}_s) & \xrightarrow{h_{\gamma *} - \mathrm{id}} & H_q(\bar{X}_s) \\ {\scriptstyle j_*}\downarrow & \nearrow{\scriptstyle \operatorname{var}_\gamma} & \downarrow{\scriptstyle j_*} \\ H_q(\bar{X}_s, \partial \bar{X}_s) & \xrightarrow{h_{\gamma *} - \mathrm{id}} & H_q(\bar{X}_s, \partial \bar{X}_s) \end{array}$$

We now prove another proposition on the connection between the variation and the intersection form.

Proposition 5.1. *Let*

$$[\gamma] \in \pi_1(\Delta', s), \quad a \in H_q(\bar{X}_s, \partial \bar{X}_s), \quad b \in H_{2n-q}(\bar{X}_s, \partial \bar{X}_s).$$

Then

(i) $\langle \operatorname{var}_\gamma a, \operatorname{var}_\gamma b \rangle + \langle a, \operatorname{var}_\gamma b \rangle + \langle \operatorname{var}_\gamma a, b \rangle = 0$,

(ii) $\langle h_{\gamma *} a, \operatorname{var}_\gamma b \rangle + \langle \operatorname{var}_\gamma a, b \rangle = 0$.

Proof. For (i), we choose representatives α, β of a, b, whose boundaries, which lie on $\partial \bar{X}_s$, do not intersect. This is possible since $\partial \alpha$ is a $(q-1)$-chain and $\partial \beta$ is a $(2n-q-1)$-chain on $\partial \bar{X}_s$, and

$$q - 1 + 2n - q - 1 = 2n - 2 < 2n - 1 = \dim \partial \bar{X}_s \, .$$

Then the intersection number $\langle \alpha, \beta \rangle$ of α and β is also defined. Of course it is not an invariant of the homology classes of α and β. Then

$$\begin{aligned} & \langle \operatorname{var}_\gamma a, \operatorname{var}_\gamma b \rangle + \langle a, \operatorname{var}_\gamma b \rangle + \langle \operatorname{var}_\gamma a, b \rangle \\ = \; & \langle h_\gamma \alpha, h_\gamma \beta \rangle - \langle \alpha, h_\gamma \beta \rangle - \langle h_\gamma \alpha, \beta \rangle + \langle \alpha, \beta \rangle \\ & + \langle \alpha, h_\gamma \beta \rangle - \langle \alpha, \beta \rangle + \langle h_\gamma \alpha, \beta \rangle - \langle \alpha, \beta \rangle \\ = \; & 0, \end{aligned}$$

since $\langle h_\gamma \alpha, h_\gamma \beta \rangle = \langle \alpha, \beta \rangle$ (cf. §4.6).

For (ii), since

$$h_{\gamma_*} = \mathrm{id} + j_* \operatorname{var}_\gamma, \quad \langle j_* \operatorname{var}_\gamma a, \operatorname{var}_\gamma b\rangle = \langle \operatorname{var}_\gamma a, \operatorname{var}_\gamma b\rangle,$$

we have

$$\begin{aligned}\langle h_{\gamma_*} a, \operatorname{var}_\gamma b\rangle + \langle \operatorname{var}_\gamma a, b\rangle &= \langle j_* \operatorname{var}_\gamma a, \operatorname{var}_\gamma b\rangle + \langle a, \operatorname{var}_\gamma b\rangle + \langle \operatorname{var}_\gamma a, b\rangle \\ &= 0\end{aligned}$$

by (i). □

5.2. Monodromy group and vanishing cycles

We shall now assume that $f : M \to \mathbb{C}$ is a Morse function:

(4) We assume that all the critical points $p_1, \dots, p_m$ of $f : M \to \mathbb{C}$ on $\bar{X} \cap f^{-1}(U)$ are nondegenerate and that all the critical values $s_1, \dots, s_n$ are distinct, i.e., $f|_{\bar{X} \cap f^{-1}(U)}$ is a Morse function.

Definition. The image Γ of the homomorphism

$$\begin{array}{rccc}\rho : & \pi_1(\Delta', s) & \longrightarrow & \operatorname{Aut} H_q(\bar{X}_s) \\ & [\gamma] & \longmapsto & h_{\gamma *}\end{array}$$

is called the *monodromy group* of the function f.

Now let $\gamma : [0,1] \to \Delta$ be a path joining the critical value s_i to s and not passing through any other critical value, i.e., $\gamma(0) = s_i$, $\gamma(1) = s$ and $\gamma((0,1]) \subset \Delta'$. By the Morse lemma (Proposition 3.15) there exist a neighborhood B_i of the nondegenerate critical point p_i over s_i in $\bar{X} \cap f^{-1}(\bar{\Delta})$ and a local coordinate system $(z_1, \dots, z_{n+1})$ such that in B_i we can write f in the form

$$f(z_1, \dots, z_{n+1}) = s_i + z_1^2 + \ldots + z_{n+1}^2$$

and such that, in these coordinates, B_i is a ball of radius ε around 0. For $t > 0$ sufficiently small the fiber $\bar{X}_{\gamma(t)}$ then contains an n-sphere

$$S(t) := \sqrt{\gamma(t) - s_i}\, S^n,$$

where

$$S^n = \{(x_1, \dots, x_{n+1}) \in \mathbb{R}^{n+1} \mid x_1^2 + \ldots + x_{n+1}^2 = 1\}$$

is the n-dimensional unit sphere. Then for

$$z_j = \sqrt{\gamma(t) - s_i} x_j, \ (x_1, \dots, x_{n+1}) \in S^n,$$

we have

$$\begin{aligned}|z_1|^2 + \ldots + |z_{n+1}|^2 &= |\gamma(t) - s_i|(x_1^2 + \ldots + x_{n+1}^2) \\ &= |\gamma(t) - s_i| < \varepsilon\end{aligned}$$

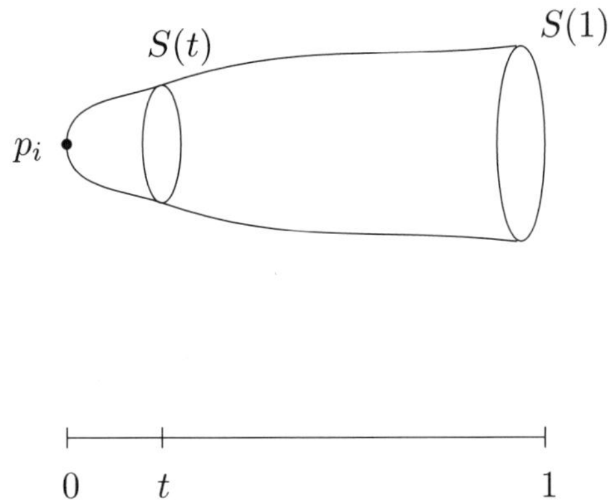

Figure 5.1. Vanishing cycle

for t suffficiently small, and

$$f(z_1, \ldots, z_{n+1}) = s_i + \gamma(t) - s_i = \gamma(t).$$

By parallel transport along γ one obtains an n-sphere $S(t) \subset \bar{X}_{\gamma(t)}$ for each $t \in (0,1]$. For $t = 0$ the sphere $S(t)$ shrinks to the critical point p_i (see Figure 5.1). We next choose an orientation of $S(1)$. Then $S(1)$ is an n-cycle, so it represents a homology class $\delta \in H_n(\bar{X}_s)$.

Definition. The homology class $\delta = [S(1)] \in H_n(\bar{X}_s)$ is called a *vanishing cycle* of f (along γ).

From the homotopy lifting theorem (Proposition 4.6) it follows that the homology class δ up to orientation depends only on the homotopy class of γ in the set of all paths $\psi : I \to \Delta$ with $\psi(0) = s_i$, $\psi(1) = s$ and $\psi((0,1]) \subset \Delta'$.

The path $\gamma : I \to \Delta$ now determines a closed path around s_i in the following way: Let Δ_i be a disc of sufficiently small radius η around s_i such that $\gamma(I)$ intersects the boundary $\partial\bar{\Delta}_i$ of $\bar{\Delta}_i$ exactly once, namely at the point $s_i + u_i$ at the time $t = \theta$. Let $\tau : I \to \bar{\Delta}_i$, $t \mapsto s_i + u_i e^{2\pi\sqrt{-1}t}$, be the path with start point $s_i + u_i$ that goes round s_i in the positive direction on the boundary of $\bar{\Delta}_i$. Furthermore, we put $\tilde{\gamma} = \gamma|_{[\theta,1]}$.

Definition. The closed path $\omega = \tilde{\gamma}^{-1}\tau\tilde{\gamma}$ with start and end point s is called a *simple loop associated to* γ (see Figure 5.2).

Definition. The monodromy

$$h_i = h_{\omega *} : H_q(\bar{X}_s) \longrightarrow H_q(\bar{X}_s)$$

associated to the simple loop ω, that belongs to γ, is called the *Picard-Lefschetz transformation* associated to the path γ (or to the vanishing cycle δ).

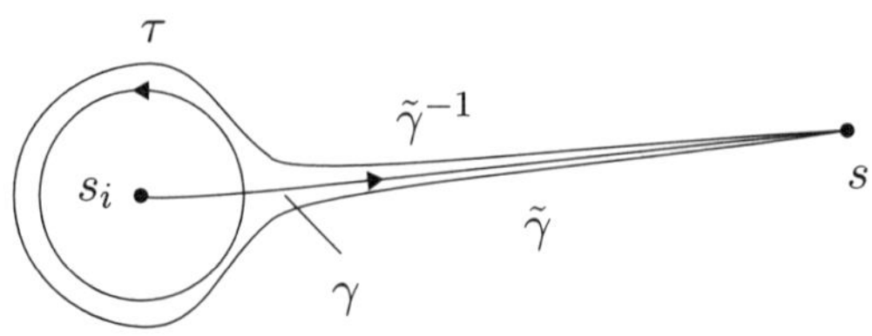

Figure 5.2. Simple loop associated to γ

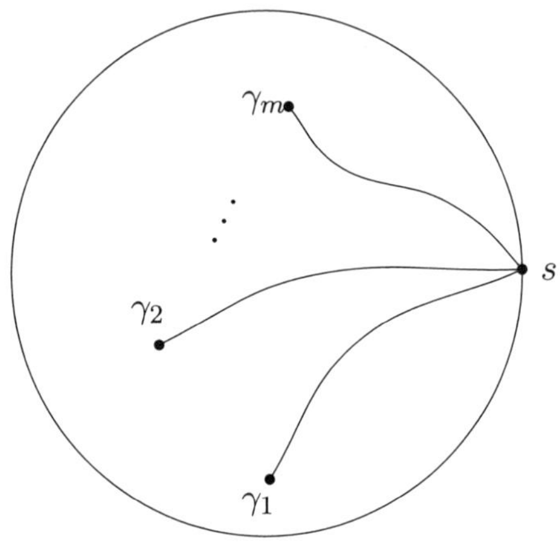

Figure 5.3. (Strongly) distinguished system of paths

We assume that ε and η are chosen so small that all the balls $\bar{B}_i$ and all the discs $\bar{\Delta}_i$ are disjoint. Next we consider an ordered system $(\gamma_1, \ldots, \gamma_m)$ of paths $\gamma_i : I \to \bar{\Delta}$ with $\gamma(0) = s_i$, $\gamma_i(1) = s$ and $\gamma_i((0,1]) \subset \Delta'$.

Definition. The system of paths $(\gamma_1, \ldots, \gamma_m)$ is called *(strongly) distinguished* if the following conditions are satisfied:

(i) The paths γ_i are not self intersecting.

(ii) The only common point of γ_i and γ_j for $i \neq j$ is s.

(iii) The paths are numbered in the order in which they meet s, where we count clockwise from the boundary of the disc (see Figure 5.3).

A system $(\delta_1, \ldots, \delta_m)$ of cycles $\delta_i \in H_q(\bar{X}_s)$ is called *(strongly) distinguished* if there is a (strongly) distinguished system $(\gamma_1, \ldots, \gamma_m)$ of paths such that δ_i is a vanishing cycle along γ_i.

Since Δ' is a disc from which m points have been deleted, its fundamental group $\pi_1(\Delta', s)$ is the free group on m generators (cf. Proposition 1.13). If $(\gamma_1, \ldots, \gamma_m)$ is a strongly distinguished system of paths, then $\pi_1(\Delta', s)$ is the free group on the generators $\omega_1, \ldots, \omega_m$, where ω_i is the simple loop associated to γ_i.

Definition. The system $(\gamma_1, \ldots, \gamma_m)$ of paths is called *weakly distinguished* if $\pi_1(\Delta', s)$ is the free group on the generators $(\omega_1, \ldots, \omega_m)$, where ω_i is the simple loop belonging to γ_i.

A system $(\delta_1, \ldots, \delta_m)$ of cycles $\delta_i \in H_n(\bar{X}_s)$ is called *weakly distinguished* if δ_i is a vanishing cycle along a path γ_i of a weakly distinguished system $(\gamma_1, \ldots, \gamma_m)$ of paths.

Note that the numbering is of no significance for a weakly distinguished system of paths.

Remark 5.1. A strongly distinguished system of paths is of course also weakly distinguished.

Proposition 5.2. *If $(\gamma_1, \ldots, \gamma_m)$ is a weakly distinguished system of paths, then the monodromy group of f is generated by the Picard-Lefschetz transformations h_i of the paths γ_i.*

Proof. This follows from the definition. □

Corollary 5.1. *The monodromy group of the function f is a group with m generators.*

5.3. The Picard-Lefschetz theorem

The assumptions are the same as in the preceding section. Let γ be a path from s_i to s with $\gamma((0,1]) \subset \Delta'$ and ω the corresponding associated simple loop. We now want to study the action of var_ω and $h_{\omega *}$ on the corresponding homology groups.

Without loss of generality let $s_i = 0$. Let the function f have the representation

$$f(z_1, \ldots, z_{n+1}) = z_1^2 + \ldots + z_{n+1}^2$$

on a neighborhood isomorphic to the ball

$$B_\varepsilon = \{(z_1, \ldots, z_{n+1}) \in \mathbb{C}^{n+1} \mid |z_1|^2 + \ldots + |z_{n+1}|^2 < \varepsilon^2\}$$

around $p_i = 0$. We assume further that s is the point $s = \eta \in \mathbb{R}$ and that $\eta < \varepsilon^2$. We assume yet further (as is possible modulo a homotopy) that ω is the loop

$$\begin{array}{rccc} \omega: & I & \longrightarrow & \mathbb{C} \\ & t & \longmapsto & \eta e^{2\pi\sqrt{-1}t}. \end{array}$$

Finally we assume further that no other critical values lie in the disc with radius η.

Let $r : \mathbb{C}^{n+1} \to \mathbb{R}$ with

$$\begin{aligned} r(z_1, \ldots, z_{n+1}) &= |z| = \sqrt{|z_1|^2 + \ldots + |z_{n+1}|^2} \\ &= \sqrt{z_1\bar{z}_1 + \ldots + z_{n+1}\bar{z}_{n+1}} \end{aligned}$$

be the norm on $\mathbb{C}^{n+1}$. We put

$$\bar{Y}_w = \bar{X}_w \cap \bar{B}_\varepsilon \text{ for } w \in \Delta.$$

Lemma 5.1. *If $w \in \Delta$ with $|w| \leq \eta$, then $\bar{X}_w$ intersects the boundary S_ε of $\bar{B}_\varepsilon$ transversally.*

Proof. Let $a \in \bar{X}_w \cap S_\varepsilon$. Suppose that $\bar{X}_w$ does not intersect the sphere S_ε transversally at a. We write the (real) gradient of a function $g : \mathbb{C}^{n+1} \to \mathbb{C}$ in the form

$$\operatorname{grad} g(z) = \left(\frac{\partial g}{\partial z_1}(z), \ldots, \frac{\partial g}{\partial z_{n+1}}(z), \frac{\partial g}{\partial \bar{z}_1}(z), \ldots, \frac{\partial g}{\partial \bar{z}_{n+1}}(z) \right).$$

Since the gradient is perpendicular to the level surface of a function, $\operatorname{grad} r^2(a)$ must be a (complex) linear combination of $\operatorname{grad} f(a)$ and $\operatorname{grad} \bar{f}(a)$, i.e.,

$$\operatorname{grad} r^2(a) = \alpha \operatorname{grad} f(a) + \beta \operatorname{grad} \bar{f}(a), \ \alpha, \beta \in \mathbb{C}.$$

Now

$$\begin{aligned} \operatorname{grad} r^2(z) &= (\bar{z}_1, \ldots, \bar{z}_{n+1}, z_1, \ldots, z_{n+1}), \\ \operatorname{grad} f(z) &= (2z_1, \ldots, 2z_{n+1}, 0, \ldots, 0), \\ \operatorname{grad} \bar{f}(z) &= (0, \ldots, 0, 2\bar{z}_1, \ldots, 2\bar{z}_{n+1}). \end{aligned}$$

It follows that

$$\bar{a}_j = 2\alpha a_j, \quad j = 1, \ldots, n+1.$$

Since $a \in S_\varepsilon$, not all coordinates a_j of a are equal to 0. For a j with $a_j \neq 0$ it follows that

$$|\bar{a}_j| = |2\alpha a_j| \Rightarrow |2\alpha| = 1.$$

From

$$r^2(a) = \sum_{j=1}^{n+1} a_j \bar{a}_j = 2\alpha \sum_{j=1}^{n+1} a_j^2 = 2\alpha f(a)$$

it then follows that

$$|w| = |f(a)| = r^2(a) = \varepsilon^2 > \eta.$$

This proves Lemma 5.1. □

From Lemma 5.1 and the Ehresmann fibration theorem for manifolds with boundary (Theorem 4.2) it follows that the manifolds with $\bar{Y}_w = \bar{X}_w \cap \bar{B}_\varepsilon$ for $w \in \Delta$ with $0 < |w| \leq \eta$ are diffeomorphic to one another.

Lemma 5.2. *The differentiable manifold with boundary $\bar{Y}_w$ is, for $w \in \Delta$ with $0 < |w| \le \eta$, diffeomorphic to the disc bundle DS^n of the tangent bundle to the n-dimensional unit sphere S^n.*

Proof. By the above remark it is enough to establish this for $w = \eta$. In what follows we write $\langle\ ,\ \rangle$ for the Euclidean scalar product on $\mathbb{R}^{n+1}$ and $|x|$ for the Euclidean norm of $x \in \mathbb{R}^{n+1}$. Furthermore, we write $z_j = x_j + iy_j$, where $x_j, y_j \in \mathbb{R}$, $j = 1, \dots, n+1$. The differentiable manifold with boundary $\bar{Y}_\eta$ will then be given by

$$\bar{Y}_\eta = \{x + iy \in \mathbb{C}^{n+1} \mid |x|^2 + |y|^2 \le \varepsilon^2,\ |x|^2 - |y|^2 = \eta,\ \langle x, y\rangle = 0\}.$$

The unit disc bundle DS^n of TS^n can be described as follows:

$$DS^n = \{u + iv \in \mathbb{C}^{n+1} \mid |u| = 1,\ |v| = 1,\ \langle u, v\rangle = 0\}.$$

Under the coordinate transformation

$$u = \frac{x}{|x|}, \qquad v = \sqrt{\frac{2}{\varepsilon^2 - \eta}}\, y$$

one can identify $\bar{Y}_\eta$ with DS^n. It is then easy to calculate that the desired diffeomorphism $\varphi : DS^n \to \bar{Y}_\eta$ is given by

$$\varphi(u + iv) = \sqrt{\frac{1}{2}(\varepsilon^2 - \eta)|v|^2 + \eta}\, u + i\sqrt{\frac{1}{2}(\varepsilon^2 - \eta)}\, v\,.$$

□

Since the sphere S^n is a deformation retract of DS^n, it follows from Lemma 5.2 that the n-sphere

$$S_\eta = \{x + iy \in \mathbb{C}^{n+1} \mid |x|^2 = \eta,\ y = 0\}$$

is a deformation retract of $\bar{Y}_\eta$. Hence $\bar{Y}_\eta$ has the same homology group as the n-dimensional unit sphere S^n. From Example 4.4, we have

$$\begin{aligned} H_q(\bar{Y}_\eta) &= 0 \text{ for } q \ne 0, n, \\ H_0(\bar{Y}_\eta) &= \mathbb{Z}, \\ H_n(\bar{Y}_\eta) &= \mathbb{Z}, \end{aligned}$$

and $H_n(\bar{Y}_\eta)$ is generated by the homology class δ of S_η, the vanishing cycle; that is,

$$H_n(\bar{Y}_\eta) = \mathbb{Z} \cdot \delta.$$

Proposition 5.3. *The vanishing cycle $\delta \in H_n(\bar{Y}_\eta)$ has the self-intersection number*

$$\langle \delta, \delta \rangle = (-1)^{\frac{n(n-1)}{2}}(1 + (-1)^n) = \begin{cases} 0 & \textit{for } n \equiv 1 \pmod 2, \\ +2 & \textit{for } n \equiv 0 \pmod 4, \\ -2 & \textit{for } n \equiv 2 \pmod 4. \end{cases}$$

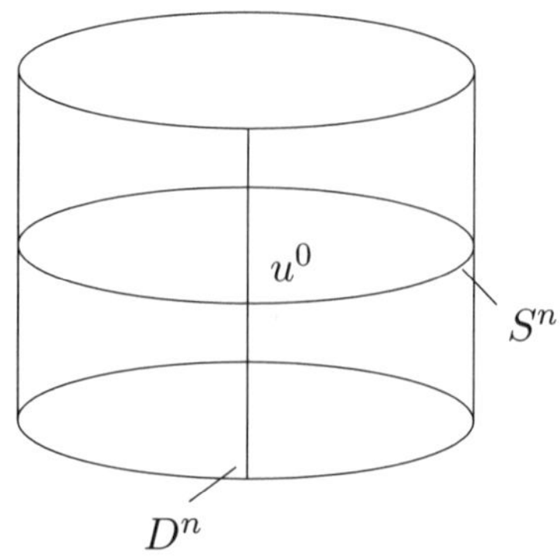

Figure 5.4. The disc bundle DS^n

Proof. The interior Y_η of $\bar{Y}_\eta$ is a complex manifold and the preferred orientation of Y_η at the point $p = (\sqrt{\eta}, 0, \ldots, 0)$ is given by $(e_2, ie_2, e_3, ie_3, \ldots, e_{n+1}, ie_{n+1})$, where $(e_1, \ldots, e_{n+1})$ is the canonical basis of $\mathbb{R}^{n+1}$.

The natural orientation of DS^n at the point $u^0 = (1, 0, \ldots, 0)$ is given by $(e_2, e_3, \ldots, e_{n+1}, ie_2, ie_3, \ldots, ie_{n+1})$. This orientation is carried over to that above by a permutation with sign $(-1)^{n(n-1)/2}$. Hence we must multiply the self-intersection number $\langle S^n, S^n \rangle = 1 + (-1)^n$ found in Corollary 4.1 by this sign. □

Next we consider the submanifold

$$T_\eta = \{x + iy \in \bar{Y}_\eta \mid x_1 > 0,\ x_2 = \ldots = x_{n+1} = 0\}$$

of $\bar{Y}_\eta$. After a choice of orientation T_η represents a relative cycle $\delta^* \in H_n(\bar{Y}_\eta, \partial \bar{Y}_\eta)$. Then

$$H_n(\bar{Y}_\eta, \partial \bar{Y}_\eta) = \mathbb{Z} \cdot \delta^*.$$

The submanifold T_η intersects the sphere S_η transversally at the point with coordinates $x_1 = \sqrt{\eta}$, $x_2 = \ldots = x_{n+1} = y_1 = \ldots = y_{n+1} = 0$. We orient T_η so that $\langle \delta^*, \delta \rangle = 1$. By the diffeomorphism φ of Lemma 5.2, T_η is mapped onto the fiber of the disc bundle DS^n at the point $u^0 = (1, 0, \ldots, 0)$, i.e., onto the disc D^n of radius 1 around 0 in $T_{u^0} S^n$ (see Figure 5.4).

Theorem 5.1 (Picard-Lefschetz theorem). *With the notation already introduced we have*

$$\operatorname{var}_\omega(\delta^*) = -(-1)^{\frac{n(n-1)}{2}} \delta.$$

Proof. We follow [**Loo84**].

Recall that ω is the loop given by $\omega(t) = \eta e^{2\pi i t}$. The map $g''_t : \mathbb{C}^{n+1} \to \mathbb{C}^{n+1}$, $z \mapsto e^{\pi i t} z$, for $t \in [0, 1]$, maps $\bar{Y}_\eta$ diffeomorphically onto $\bar{Y}_{\eta e^{2\pi i t}}$. For the map $g_1'' : \bar{Y}_\eta \to \bar{Y}_\eta$, $z \mapsto -z$, however, we have $g_1''|_{\partial\bar{Y}_\eta} \neq \mathrm{id}_{\partial\bar{Y}_\eta}$. Thus g_1'' does not represent the monodromy and we have more to do.

We define first a one parameter family of homeomorphisms $\psi_t : DS^n \to DS^n$ $(t \in \mathbb{R})$ by

$$\psi_t(u+iv) := \left[u\cos(\pi|v|\cdot t) + \frac{v}{|v|}\sin(\pi|v|\cdot t)\right] + i[-u|v|\sin(\pi|v|\cdot t) + v\cdot\cos(\pi|v|\cdot t)].$$

Since $\lim_{|v|\to 0}(1/|v|)\sin(\pi|v|\cdot t) = 1$, ψ_t is also defined for $v = 0$. We have

$$\psi_t(u+iv) = e^{-\pi i t}(u+iv) \text{ for } u+iv \in \partial DS^n.$$

Next we consider the homeomorphism

$$G_t' := g_t'' \circ \varphi \circ \psi_t : DS^n \longrightarrow \bar{Y}_{\eta e^{2\pi i t}}.$$

We have $G_0'|_{\partial DS^n} = G_1'|_{\partial DS^n}$. Let

$$\bar{Y} := \{z \in \mathbb{C}^{n+1} \mid |z| \leq \varepsilon,\ |f(z)| \leq \eta\}.$$

We can then define a trivialization of $\bar{Y}$ over $\partial\bar{\Delta}_\eta$ as follows:

$$\begin{array}{rccc} \tau_0 : & \partial\bar{\Delta}_\eta \times \partial DS^n & \longrightarrow & \partial\bar{Y}_{\partial\bar{\Delta}_\eta} \\ & (\eta e^{2\pi i t}, u+iv) & \longmapsto & G_t'(u+iv). \end{array}$$

This trivialization can be extended to all of $\bar{\Delta}_\eta$: Define

$$\begin{array}{rccc} \tau : & \bar{\Delta}_\eta \times \partial DS^n & \to & \partial\bar{Y} \\ & (\rho e^{2\pi i t}, u+iv) & \longmapsto & g_t'' \circ \varphi_\rho \circ \psi_t(u+iv), \end{array}$$

where ψ_ρ is defined like φ, except that η is replaced by $\rho \leq \eta$. Note that this map is well defined for $\rho = 0$ too, though t is then not unique.

Next we consider the one parameter family of homeomorphisms

$$g_t' := G_t' \circ \varphi^{-1} : \bar{Y}_\eta \to \bar{Y}_{\eta e^{2\pi i t}}.$$

We claim that the homeomorphism $g_1' : \bar{Y}_\eta \to \bar{Y}_\eta$ is homotopic to the geometric monodromy h_ω along ω. Let $g_t : \bar{Y}_\eta \to \bar{Y}_{\eta e^{-2\pi i t}}$ be the parallel transport along ω^{-1}. We define a homotopy

$$H : \bar{Y}_\eta \times I \longrightarrow \bar{Y}_\eta$$

by $H(z,t) = g_t^{-1} \circ g_{1-t}'(z)$. Then

$$\begin{aligned} H(z,0) &= g_0^{-1} \circ g_1'(z) = g_1'(z), \\ H(z,1) &= g_1^{-1} \circ g_0'(z) = g_1^{-1}(z) = h_{\omega^{-1}}^{-1} = h_\omega, \end{aligned}$$

for all $z \in \bar{Y}_\eta$. Now this homotopy is stationary on $\partial\bar{Y}_\eta$, i.e., $H(z,t) = H(z,0)$ for all $z \in \partial\bar{Y}_\eta$ and $t \in I$.

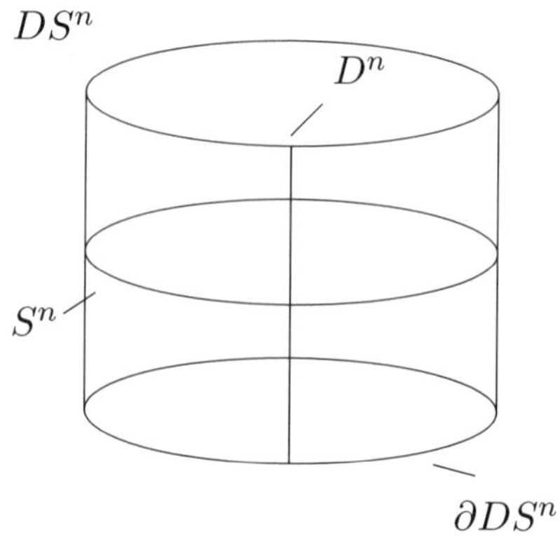

Figure 5.5. The disc bundle DS^n

If we now identify $\bar{Y}_\eta$ with DS^n via φ, then g_1' corresponds to the homeomorphism

$$\varphi^{-1} \circ G_1' = \varphi^{-1}(-\varphi \circ \psi_1) = -\psi_1,$$

with $-\psi_1 : DS^n \to DS^n$.

We now study the variation corresponding to $-\psi_1$. The diffeomorphism φ maps S_η to the sphere

$$S^n = \{u + iv \in DS^n \mid v = 0\}$$

and maps T_η to the fiber

$$D^n = \{u^0 + iv \in DS^n \mid |v| \leq 1\}$$

of DS^n over $u^0 = (1, 0, \ldots, 0)$ (see Figure 5.5). We orient S^n by taking $(e_2, \ldots, e_{n+1})$ as a positively oriented basis of $T_{e_1}S^n$, and the basis $(ie_2, \ldots, ie_{n+1})$ of $T_{e_1}D^n$ for D^n. Under $\varphi : DS^n \to \bar{Y}_\eta$ the orientation of $\bar{Y}_\eta$ is transferred via the complex structure on Y_η to the orientation

$$(e_2, ie_2, \ldots, e_{n+1}, ie_{n+1})$$

of DS^n. The intersection number $\langle D^n, S^n \rangle$ is then the signum of the permutation

$$(ie_2, \ldots, ie_{n+1}, e_2, \ldots, e_{n+1}) \longmapsto (e_2, ie_2, \ldots, e_{n+1}, ie_{n+1}).$$

Hence

$$\langle D^n, S^n \rangle = (-1)^{\frac{n(n+1)}{2}}.$$

We now want to compute $\mathrm{var}_\omega[D^n]$. Let $\pi : DS^n \to S^n$, $u + iv \mapsto u$, be the natural projection. Let $\sigma : S^n \to DS^n$ be the zero section. Then $\pi \circ \sigma = \mathrm{id}_{S^n}$ and $\sigma \circ \pi$ is homotopic to id_{S^n}. Thus it is enough to compute

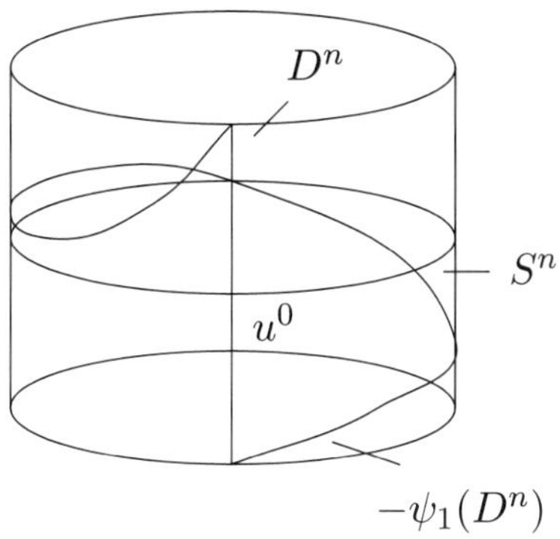

Figure 5.6. The image of Ψ

$\pi_* \operatorname{var}_\omega[D^n]$. The map $\pi_* \operatorname{var}_\omega[D^n]$ is represented by the real part of $-\psi_1|_{D^n}$. This is the map

$$\begin{array}{rcl} \Psi: \; (D^n, \partial D^n) & \longrightarrow & (S^n, \{u^0\}) \\ u^0 + iv & \longmapsto & -u^0 \cos(\pi|v|) - \frac{v}{|v|} \sin(\pi|v|). \end{array}$$

This mapping maps the interior $\overset{\circ}{D^n}$ of D^n homeomorphically onto $S^n - \{e_1\}$ and the boundary ∂D^n onto $\{e_1\}$ (see Figure 5.6). Thus $\pi_* \operatorname{var}_\omega[D^n] = \operatorname{grad}(\Psi) \cdot [S^n]$ with $\operatorname{grad}(\Psi) := \operatorname{sign} T_{u^0}(\Psi) = \pm 1$. It is now easy to calculate that $-\Psi(u^0) = u^0$ and that $-\Psi$ is differentiable at u^0 with derivative $T_{u^0}(-\Psi) = \mathrm{id}$. Hence $-\Psi$ is orientation preserving, and thus $\operatorname{grad}(-\Psi) = 1$, i.e., $\operatorname{grad}(\Psi) = (-1)^{n+1}$.

If now $\delta = \varphi_*[S^n] \in H_n(\bar{Y}_\eta)$ and if $\delta^* \in H_n(\bar{Y}_\eta, \partial \bar{Y}_\eta)$ is the uniquely determined element with $\langle \delta^*, \delta \rangle = 1$, then $\delta^* = (-1)^{n(n+1)/2} \varphi_*[D^n]$. It follows that

$$\begin{aligned} \operatorname{var}_\omega(\delta^*) &= (-1)^{n+1} \cdot (-1)^{\frac{n(n+1)}{2}} \delta \\ &= -(-1)^{\frac{n(n-1)}{2}} \delta, \end{aligned}$$

as was to be shown. □

Next we consider the global situation again. We consider the restriction of the function f to $\bar{X} \cap f^{-1}(\bar{\Delta}_\eta) \setminus B_\varepsilon$, where $\bar{\Delta}_\eta$ is the closed disc in $\mathbb{C}$ of radius η around 0. Then, by the Ehresmann fibration theorem,

$$f|_{\bar{X} \cap f^{-1}(\bar{\Delta}_\eta) \setminus B_\varepsilon} : \bar{X} \cap f^{-1}(\bar{\Delta}_\eta) \setminus B_\varepsilon \longrightarrow \bar{\Delta}_\eta$$

is the projection of a differentiable fiber bundle over the contractible disc $\bar{\Delta}_\eta$, so it must be trivial. We choose an Ehresmann connection and with it a product structure on $\bar{X} \cap f^{-1}(\bar{\Delta}_\eta) \setminus B_\varepsilon$. The geometric monodromy $h_\omega : \bar{X}_\eta \to \bar{X}_\eta$ belonging to the loop ω is then the identity on $\bar{X}_\eta \setminus Y_\eta$.

A relative cycle $c \in Z_q(\bar{X}_\eta, \partial\bar{X}_\eta)$ can be represented in the form

$$c = c_1 + c_2$$

with $c_1 \in Z_q(\bar{Y}_\eta, \partial\bar{Y}_\eta)$, $c_2 \in C_q(\bar{X}_\eta \setminus Y_\eta)$. Then

$$\begin{aligned} h_\omega(c) &= h_\omega(c_1) + c_2, \\ \mathrm{var}_\omega(c) &= \mathrm{var}_\omega(c_1). \end{aligned}$$

Lemma 5.3. *For $q \neq 0, n$*

$$\mathrm{var}_\omega : H_q(\bar{X}_\eta, \partial\bar{X}_\eta) \longrightarrow H_q(\bar{X}_\eta)$$

is the null map.

Proof. This follows from the above remarks and the fact that $H_q(\bar{Y}_\eta, \partial\bar{Y}_\eta) = 0$ for $q \neq 0, n$. □

We denote the image of δ in $H_n(\bar{X}_\eta)$ under $i_* : H_n(\bar{Y}_\eta) \to H_n(\bar{X}_\eta)$ by δ too. Then, from Theorem 5.1,

Corollary 5.2 (Picard-Lefschetz formulae). (i) *For $\alpha \in H_n(\bar{X}_\eta, \partial\bar{X}_\eta)$*

$$\mathrm{var}_\omega(\alpha) = -(-1)^{\frac{n(n-1)}{2}} \langle \alpha, \delta \rangle \delta.$$

(ii) *For $\alpha \in H_n(\bar{X}_\eta)$*

$$h_{\omega *}(\alpha) = \alpha - (-1)^{\frac{n(n-1)}{2}} \langle \alpha, \delta \rangle \delta.$$

Proof. For (i), let $\alpha = [c]$, $c \in Z_n(\bar{X}_\eta, \partial\bar{X}_\eta)$. We write

$$c = c_1 + c_2, \;\; c_1 \in Z_n(\bar{Y}_\eta, \partial\bar{Y}_\eta), \;\; c_2 \in C_n(\bar{X}_\eta \setminus Y_\eta).$$

Then $[c_1] = k \cdot \delta^*$ and since $\langle \delta^*, \delta \rangle = 1$, it follows that $k = \langle [c_1], \delta \rangle$. Since $\mathrm{var}_\omega(\alpha) = \mathrm{var}_\omega([c_1])$, formula (i) follows from Theorem 5.1.

For (ii), it follows from (i) that for $\alpha \in H_n(\bar{X}_\eta, \partial\bar{X}_\eta)$ and the relative monodromy $h_{\omega *} : H_n(\bar{X}_\eta, \partial\bar{X}_\eta) \to H_n(\bar{X}_\eta, \partial\bar{X}_\eta)$ we have

$$h_{\omega *}(\alpha) = \alpha - (-1)^{\frac{n(n-1)}{2}} \langle \alpha, \delta \rangle j_*(\delta).$$

Hence formula (ii) follows for $\alpha \in H_n(\bar{X}_\eta)$. □

The formulae in Corollary 5.2 are called the *Picard-Lefschetz formulae.* We list formula (ii) and the self-intersection number of δ as in Proposition 5.3 according to the residue classes of n modulo 4:

$$\begin{aligned} n \equiv 0 \pmod 4 &: \;\; h_{\omega *}(\alpha) = \alpha - \langle \alpha, \delta \rangle \delta, \;\; \langle \delta, \delta \rangle = 2, \\ n \equiv 1 \pmod 4 &: \;\; h_{\omega *}(\alpha) = \alpha - \langle \alpha, \delta \rangle \delta, \;\; \langle \delta, \delta \rangle = 0, \\ n \equiv 2 \pmod 4 &: \;\; h_{\omega *}(\alpha) = \alpha + \langle \alpha, \delta \rangle \delta, \;\; \langle \delta, \delta \rangle = -2, \\ n \equiv 3 \pmod 4 &: \;\; h_{\omega *}(\alpha) = \alpha + \langle \alpha, \delta \rangle \delta, \;\; \langle \delta, \delta \rangle = 0. \end{aligned}$$

When n is even, the intersection form $\langle\ ,\ \rangle$ is a symmetric bilinear form and we can combine both formulae for $n \equiv 0, 2 \pmod 4$ together as

$$h_{\omega*}(\alpha) = \alpha - \frac{2\langle\alpha, \delta\rangle}{\langle\delta, \delta\rangle}\delta.$$

This describes a reflection in the hyperplane of $H_n(\bar{X}_\eta)$ orthogonal to δ (with respect to $\langle\ ,\ \rangle$).

When n is odd, the intersection form $\langle\ ,\ \rangle$ is skew symmetric and $h_{\omega*}$ is what is called a symplectic transvection.

Now let

$$\bar{Y}_{[0,\eta]} = f^{-1}([0, \eta]) \cap \bar{Y}.$$

We show that one obtains $\bar{Y}_{[0,\eta]}$ from $\bar{Y}_\eta$ by attaching an $(n+1)$-cell to $\bar{Y}_\eta$.

Definition. The set

$$e = \{x \in \mathbb{R}^k \mid |x| \leq 1\}$$

is called a *k-cell*. Its boundary ∂e is the $(k-1)$-sphere S^{k-1}.

Now let Y be a topological space and $g : S^{k-1} \to Y$ a continuous map. Let us consider the topological space $Y \cup_g e$, defined as follows: In the disjoint union $Y \dot{\cup} e$ (with the topology as topological sum) we identify $x \in \partial e$ with $g(x) \in Y$, i.e., we consider the quotient space of $Y \dot{\cup} e$ under the equivalence relation

$$x \sim g(x) \text{ for all } x \in \partial e,$$
$$y \sim y \text{ for all } y \in Y \text{ or } y \in e,$$
$$x \sim y \text{ if } x, y \in \partial e \text{ and } g(x) = g(y)$$

with the quotient topology. The space $Y \cup_g e$ is called the space obtained from Y *by attaching a k-cell* via g.

The subset

$$e = \{x + iy \in \mathbb{C}^{n+1} \mid |x|^2 \leq \eta,\ y = 0\} \subset \bar{Y}$$

is homeomorphic to an $(n+1)$-cell. We have $\bar{Y}_\eta \cap e = \partial e = S_\eta = \varphi(S^n)$ (see Figure 5.7). Hence one obtains $\bar{Y}_\eta \cup e$ from $\bar{Y}_\eta$ by attaching an $(n+1)$-cell via $\varphi|_{S^n}$.

Lemma 5.4. *$\bar{Y}_\eta \cup e$ is a deformation retract of $\bar{Y}_{[0,\eta]}$.*

Proof. Consider the map

$$\begin{array}{rcl} \Phi : \ [0, \eta] \times DS^n & \longrightarrow & \bar{Y}_{[0,\eta]} \\ (\rho, u + iv) & \longmapsto & \sqrt{\frac{1}{2}(\varepsilon^2 - \rho)|v|^2 + \rho}\, u + i\sqrt{\frac{1}{2}(\varepsilon^2 - \rho)}\, v. \end{array}$$

This is the map $(\rho, u + iv) \mapsto \varphi_\rho(u + iv)$ in the notation of the proof of Proposition 5.3.

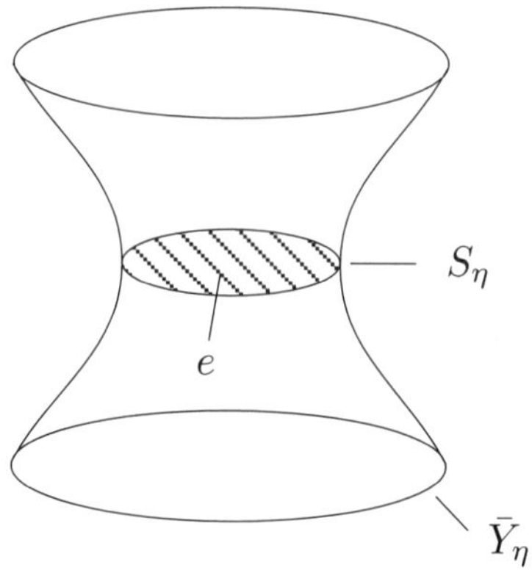

Figure 5.7. The $(n+1)$-cell e

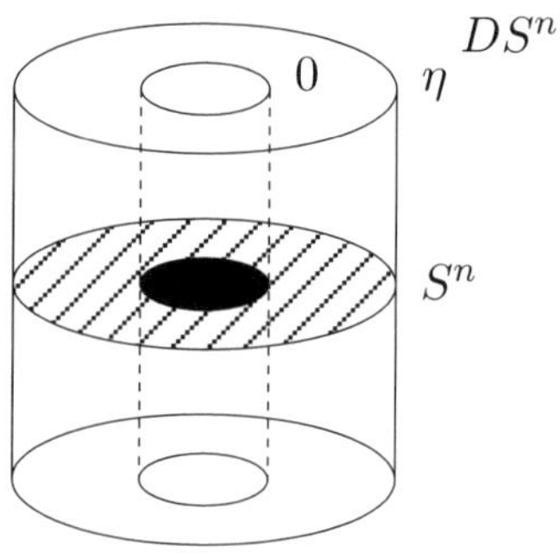

Figure 5.8. Proof of Lemma 5.4

Now $\Phi(\{0\} \times S^n) = \{0\}$ and Φ maps the complement of $\{0\} \times S^n$ in $[0,\eta] \times DS^n$ bijectively onto $\bar{Y}_{[0,\eta]} - \{0\}$. Hence Φ induces a homeomorphism from the space obtained from $[0,\eta] \times DS^n$ by the identification of $\{0\} \times S^n$ with a point to the space $\bar{Y}_{[0,\eta]}$:

$$\bar{\Phi} : ([0,\eta] \times DS^n)/\{0\} \times S^n \longrightarrow \bar{Y}_{[0,\eta]}.$$

Now $\bar{\Phi}$ maps $\{\eta\} \times DS^n \cup ([0,\eta] \times S^n)/\{0\} \times S^n$ onto $\bar{Y}_\eta \cup e$. However $\{\eta\} \times DS^n \cup ([0,\eta] \times S^n)/\{0\} \times S^n$ is a deformation retract of $([0,\eta] \times DS^n)/\{0\} \times S^n$ (see Figure 5.8). The lemma follows from this. □

5.4. The Milnor fibration

Let $f : (\mathbb{C}^{n+1}, 0) \to (\mathbb{C}, 0)$ be the germ of a holomorphic function with an isolated critical point at 0. As agreed in §3.6 we simply say that f is a singularity. We do not distinguish notationally between the germ f and a representative $f : M \to \mathbb{C}$, where M is an open neighborhood of $0 \in \mathbb{C}^{n+1}$.

By the implicit function theorem the level surface $f^{-1}(w)$ for $w \in \mathbb{C}$, $w \neq 0$, $|w|$ sufficiently small, is a complex submanifold of $\mathbb{C}^{n+1}$ in a neighborhood of $0 \in \mathbb{C}^{n+1}$. The zero set $f^{-1}(0)$ has a singularity at $0 \in \mathbb{C}^{n+1}$, but apart from 0 it is a complex submanifold of $\mathbb{C}^{n+1}$ in a neighborhood of 0. We wish to study the topology of this level surface and to apply the results of the preceding sections to it.

Let M be an open neighborhood of $0 \in \mathbb{C}^{n+1}$ and $f : M \to \mathbb{C}$ a representative of f. Let $\varepsilon > 0$ be as in Lemma 3.5 and let $X = B_\varepsilon \subset M$ be the open ball around $0 \in \mathbb{C}^{n+1}$ of radius ε. This is a complex submanifold of M. From Lemma 3.5 it follows that there is an $\eta_0 > 0$, $\eta_0 \ll \varepsilon$, such that $f^{-1}(w)$ intersects the sphere S_ε transversally for $w \in \mathbb{C}$, $|w| \leq \eta_0$. Let $\Delta := \{w \in \mathbb{C} \mid |w| < \eta_0\}$. Then $f : M \to \mathbb{C}$, with X, Δ, satisfy the hypotheses (1)–(3) of §5.1, where 0 is the only critical point of f on $\bar{X} \cap f^{-1}(\bar{\Delta})$.

Definition. The fibration

$$f|_{\bar{X}\cap f^{-1}(\bar{\Delta}\setminus\{0\})} : \bar{X} \cap f^{-1}(\bar{\Delta} \setminus \{0\}) \longrightarrow \bar{\Delta} \setminus \{0\},$$

which exists, by §5.1, is called the *Milnor fibration* of f. The fiber

$$\bar{X}_w = f^{-1}(w) \cap \bar{X}$$

over $w \in \bar{\Delta}\setminus\{0\}$ is called the *Milnor fiber* of f (over w). It is a $2n$-dimensional differentiable manifold with boundary and is uniquely determined up to a diffeomorphism.

Let $\omega_0 : [0,1] \to \bar{\Delta}$ be the loop with start and end point η_0 that goes round the critical value 0 in the positive direction (counterclockwise), i.e.,

$$\omega_0(t) = \eta_0 e^{2\pi i t} \text{ for } t \in [0,1].$$

The class $[\omega_0] \in \pi_1(\bar{\Delta} \setminus \{0\}, \eta_0)$ generates the fundamental group of the complement of the set of critical values in $\bar{\Delta}$.

Definition. The *classical geometric monodromy* of the singularity f is the geometric monodromy $h = h_{\omega_0} : \bar{X}_{\eta_0} \to \bar{X}_{\eta_0}$ of the loop ω_0. The *classical monodromy* (or the *classical monodromy operator*) of the singularity f is the monodromy $h_* = h_{\omega_0 *} : H_n(\bar{X}_{\eta_0}) \to H_n(\bar{X}_{\eta_0})$ corresponding to ω_0. The *variation* of the singularity f is the variation $\operatorname{Var}_f = \operatorname{var}_{\omega_0} : H_n(\bar{X}_{\eta_0}, \partial\bar{X}_{\eta_0}) \to H_n(\bar{X}_{\eta_0})$ corresponding to the loop ω_0.

We now want to compute the homology group $H_n(\bar{X}_{\eta_0})$. We put

$$\bar{\mathcal{X}} = \bar{B}_\varepsilon \cap f^{-1}(\bar{\Delta}_{\eta_0}).$$

We show first

Proposition 5.4. *The topological space $\bar{\mathcal{X}}$ is contractible.*

Proof. The proof proceeds in two steps.

a) We show first that the fiber $\bar{X}_0 = \bar{B}_\varepsilon \cap f^{-1}(0)$ is contractible. The punctured fiber $\bar{X}_0 \setminus \{0\}$ is a differentiable manifold. By Lemma 3.5, the spheres S_ρ of radius ρ with $0 < \rho \leq \varepsilon$ around $0 \in \mathbb{C}^{n+1}$ intersect the fiber $\bar{X}_0$ transversally. Hence the intersections $\bar{X}_0 \cap S_\rho$ are differentiable submanifolds of $\bar{X}_0 \setminus \{0\}$. We consider a vector field on $\bar{X}_0 \setminus \{0\}$ that is everywhere vertical to the differentiable submanifolds $\bar{X}_0 \cap S_\rho$ and that is directed inwards to the origin (see Figure 5.9). With the help of the corresponding one parameter group of diffeomorphisms we can define a contraction $g : \bar{X}_0 \to \{0\}$. In this way one can also show that $\bar{X}_0$ is homeomorphic to a cone over the differentiable manifold $f^{-1}(0) \cap S_\varepsilon$.

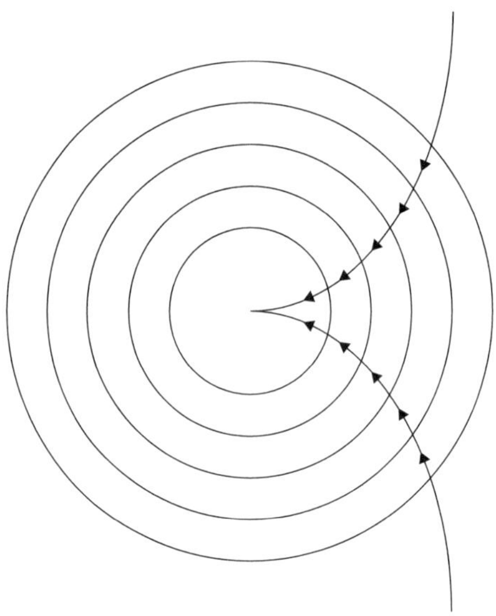

Figure 5.9. The vector field on $\bar{X}_0 \setminus \{0\}$

b) We next show that $\bar{X}_0$ is a deformation retract of $\bar{\mathcal{X}}$.

Let $\varepsilon = \varepsilon_0 > \varepsilon_1 > \varepsilon_2 > \ldots > 0$ be a strictly monotone decreasing null sequence. Correspondingly there exists a sequence $\eta_0 > \eta_1 > \eta_2 > \ldots > 0$ such that $f^{-1}(w)$ intersects transversally the sphere S_{ε_i} of radius ε_i around $0 \in \mathbb{C}^{n+1}$ for $|w| \leq \eta_i$. Then for each i

$$f|_{E_i} : E_i = f^{-1}(\bar{\Delta}_{\eta_i}) \cap (\bar{B}_\varepsilon \setminus B_{\varepsilon_i}) \longrightarrow \bar{\Delta}_{\eta_i}$$

is the projection of a trivial differentiable fiber bundle. Let the trivializations of this fiber bundle be chosen so that they agree on the intersections

$$E_i \cap E_{i-1} = f^{-1}(\bar{\Delta}_{\eta_i}) \cap (\bar{B}_\varepsilon \setminus B_{\varepsilon_{i-1}}).$$

Next we consider the homotopy

$$\begin{array}{rccl} g: & \bar{\Delta}_{\eta_0} \times I & \longrightarrow & \bar{\Delta}_{\eta_0} \\ & (w,t) & \longmapsto & \begin{cases} \eta_0 t \cdot \frac{w}{|w|} & \text{for } |w| \geq \eta_0 t, \\ w & \text{for } |w| \leq \eta_0 t. \end{cases} \end{array}$$

The map $g_t : \bar{\Delta}_{\eta_0} \to \bar{\Delta}_{\eta_0}$, $w \mapsto g(w,t)$, for $t \in (0,1]$, maps $\bar{\Delta}_{\eta_0}$ onto $\bar{\Delta}_{\eta_0 t}$ and leaves $\bar{\Delta}_{\eta_0 t}$ pointwise fixed. The map g is a homotopy between the identity g_1 on $\bar{\Delta}_{\eta_0}$ and the constant map $g_0 : \bar{\Delta}_{\eta_0} \to \{0\}$.

Let

$$\bar{\mathcal{X}}' = f^{-1}(\bar{\Delta}_{\eta_0} \setminus \{0\}) \cap \bar{B}_\varepsilon.$$

Then

$$f|_{\bar{\mathcal{X}}'} : \bar{\mathcal{X}}' \longrightarrow \bar{\Delta}_{\eta_0} \setminus \{0\}$$

is the projection of a differentiable fiber bundle. By the homotopy lifting theorem g can be lifted to a homotopy

$$\hat{g} : \bar{\mathcal{X}}' \times (0,1] \longrightarrow \bar{\mathcal{X}}'.$$

This homotopy can be chosen to agree with the product structure on E_i for $t \leq \eta_i/\eta_0$. Hence this homotopy can be extended to a homotopy

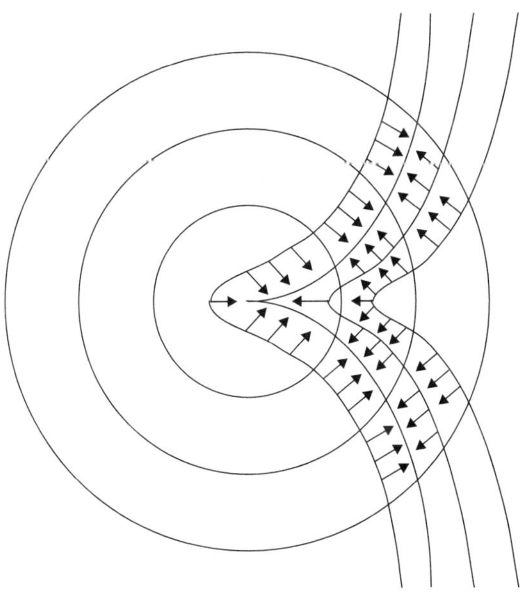

Figure 5.10. The homotopy $\hat{g}$

$$\hat{g} : \bar{\mathcal{X}} \times [0,1] \longrightarrow \bar{\mathcal{X}}$$

between $\hat{g}_1 = \mathrm{id}_{\bar{\mathcal{X}}}$ and $\hat{g}_0 : \bar{\mathcal{X}} \to \bar{\mathcal{X}}$ with $\hat{g}_0(\bar{\mathcal{X}}) = \bar{X}_0 \subset \bar{\mathcal{X}}$ (see Figure 5.10). □

Now let $F : M \times U \to \mathbb{C} \times U$ be a morsification of f. Then there is a $\lambda_0 > 0$ such that for all $\lambda \in \mathbb{C}$ with $|\lambda| \leq \lambda_0$ we have $\lambda \in U$, the level surface $f_\lambda^{-1}(w)$ for $|w| \leq \eta_0$ is transversal to the sphere S_ε, and the critical values of $f_\lambda|_{\bar{B}_\varepsilon}$ lie in Δ_{η_0}. We put

$$U = \{\lambda \in \mathbb{C} \mid |\lambda| \leq \lambda_0\}.$$

From the Ehresmann fibration theorem, applied to the restriction of $F : M \times U \to \mathbb{C} \times U$ to the complement of the critical set, it follows that $f_\lambda^{-1}(w) \cap \bar{X}$ is diffeomorphic to $f^{-1}(w) \cap \bar{X}$ so long as $w \neq 0$ and w is not a critical value of f_λ.

Now let $\lambda \in U$ so that f_λ is a Morse function. Next we consider $f_\lambda : M \to \mathbb{C}$ instead of $f : M \to \mathbb{C}$. The function satisfies the hypotheses (1)–(3) of §5.1 and is a Morse function so it also satisfies (4) of §5.2. We put

$$\bar{X}_w^{(\lambda)} = f_\lambda^{-1}(w) \cap \bar{X} \text{ for } w \in \bar{\Delta}_{\eta_0}.$$

Then $\bar{X}_{\eta_0}^{(\lambda)}$ is diffeomorphic to the Milnor fiber $\bar{X}_{\eta_0}$. Let the function f_λ have μ critical points $p_1, \ldots, p_\mu$ in $\bar{X}$ with the corresponding distinct critical values $s_1, \ldots, s_\mu \in \Delta_{\eta_0}$ (cf. §3.8).

Proposition 5.5. *A (strongly) distinguished system $(\delta_1, \ldots, \delta_\mu)$ of vanishing cycles forms a basis of $H_n(\bar{X}_{\eta_0}^{(\lambda)}) \cong H_n(\bar{X}_{\eta_0})$. We have $\tilde{H}_q(\bar{X}_{\eta_0}) = 0$ for $q \neq n$.*

Proof. We present a proof of this proposition which originates with E. Brieskorn (see [**Bri70**, Appendix]).

Let $\bar{B}_{\varepsilon_i}$ be a small closed ball around p_i as in §5.3 such that in $\bar{B}_{\varepsilon_i}$ the function f_λ can be written in the form

$$f_\lambda(z) = f_\lambda(p_i) + z_1^2 + \ldots + z_{n+1}^2$$

for local coordinates $(z_1, \ldots, z_{n+1})$ in $\bar{B}_{\varepsilon_i}$. Furthermore, let $\varepsilon_1, \ldots, \varepsilon_\mu$ be chosen so small that the balls $\bar{B}_{\varepsilon_i}$ are mutually disjoint.

Let $(\gamma_1, \ldots, \gamma_\mu)$ be a (strongly) distinguished system of paths that lead from the points $s_1, \ldots, s_\mu$ to η_0 such that δ_i is a vanishing cycle along γ_i. Since we can replace $\gamma_1, \ldots, \gamma_\mu$ by homotopic paths, if necessary, we can assume that for each s_i there is an $\eta_i > 0$ with $\eta_i \ll \eta_0$ such that $\gamma_i(t) = s_i + t$ for all $t \leq \eta_i$. We assume that η_i is chosen so small that $\bar{\Delta}_{\eta_i} \subset f_\lambda(\bar{B}_{\varepsilon_i})$ and that all the discs $\bar{\Delta}_{\eta_i}$ are mutually disjoint (see Figure 5.11).

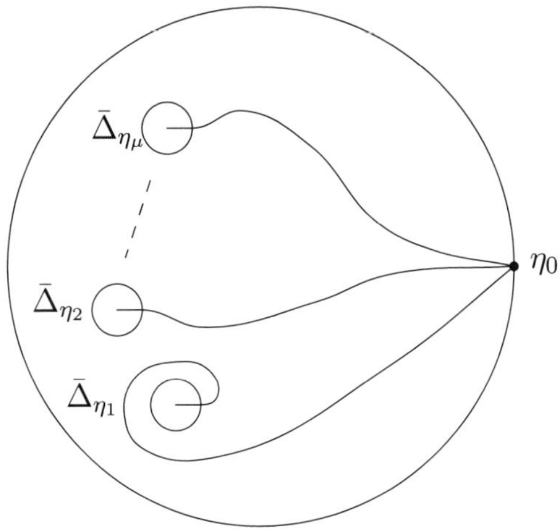

Figure 5.11. The discs $\bar{\Delta}_{\eta_i}$

We put

$$V := \bigcup_{i=1}^{\mu} \gamma_i([0,1]),$$
$$W := \bigcup_{i=1}^{\mu} \gamma_i([\eta_i,1]).$$

Then V is a deformation retract of $\bar{\Delta}_{\eta_0}$ and $\{\eta_0\}$ is a deformation retract of W (see Figure 5.12).

Let

$$\bar{\mathcal{X}}^{(\lambda)} = \bar{B}_\varepsilon \cap f_\lambda^{-1}(\bar{\Delta}_{\eta_0}).$$

The space $\bar{\mathcal{X}}^{(\lambda)}$ is (for sufficiently small λ) diffeomorphic to $\bar{\mathcal{X}}$, as a differentiable manifold with piecewise differentiable boundary. By Proposition 5.4, $\bar{\mathcal{X}}$ is contractible. Thus $\bar{\mathcal{X}}^{(\lambda)}$ is contractible too. To simplify the notation, we drop the index λ in what follows. We put

$$\bar{\mathcal{X}}_V := \bar{\mathcal{X}}^{(\lambda)} \cap f_\lambda^{-1}(V),$$
$$\bar{\mathcal{X}}_W := \bar{\mathcal{X}}^{(\lambda)} \cap f_\lambda^{-1}(W).$$

As in the proof of Proposition 5.4 we can show that the deformation retractions of $\bar{\Delta}_{\eta_0}$ on V and of W on $\{\eta_0\}$ can be lifted: $\bar{\mathcal{X}}_V$ is a deformation retract of $\bar{\mathcal{X}}$, and $\bar{X}_{\eta_0}^{(\lambda)}$ is a deformation retract of $\bar{\mathcal{X}}_W$.

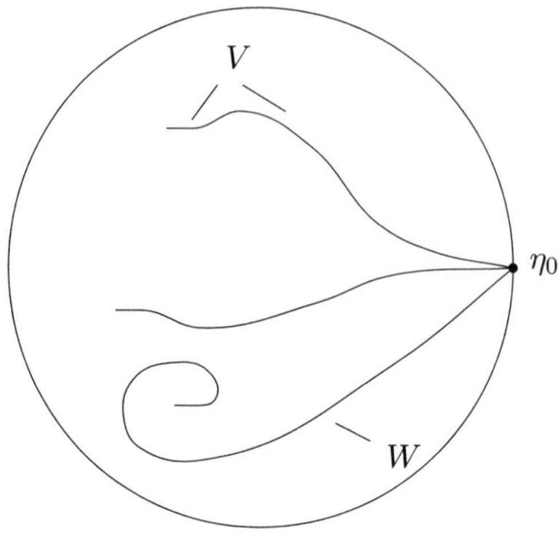

Figure 5.12. The sets V and W

Next we consider (see Figure 5.13)

$$\begin{aligned} Y_i &:= f_\lambda^{-1}([s_i, s_i+\eta_i]) \cap B_{\varepsilon_i}, \\ \bar{Y}_i &:= f_\lambda^{-1}([s_i, s_i+\eta_i]) \cap \bar{B}_{\varepsilon_i}, \\ \bar{X}_i &:= f_\lambda^{-1}([s_i, s_i+\eta_i]) \cap \bar{B}_{\varepsilon}, \\ \bar{Y}_{s_i+\eta_i} &:= f_\lambda^{-1}(s_i+\eta_i) \cap \bar{B}_{\varepsilon_i}. \end{aligned}$$

Since the map

$$f|_{\bar{X}_i - Y_i} : \bar{X}_i - Y_i \longrightarrow [s_i, s_i+\eta_i]$$

is the projection of a trivial differentiable fiber bundle, $\bar{X}^{(\lambda)}_{s_i+\eta_i} \cup \bar{Y}_i$ is a deformation retract of $\bar{X}_i$. Let e_i be the subset of $\bar{Y}_i$ that was considered in §5.3 ($e_i = e$ in the notation there) and that is homeomorphic to an $(n+1)$-cell. By Lemma 5.4, $\bar{Y}_{s_i+\eta_i} \cup e$ is a deformation retract of $\bar{Y}_i$. Hence $\bar{X}^{(\lambda)}_{s_i+\eta_i} \cup e_i$ is a deformation retract of $\bar{X}_i$.

All told, we see that $\bar{\mathcal{X}}$, as a deformation retract, has a topological space that arises from $\bar{X}^{(\lambda)}_{\eta_0}$ by attaching an $(n+1)$-cell to the vanishing cycle $S_i(1) \subset \bar{X}^{(\lambda)}_{\eta_0}$ with $[S_i(1)] = \delta_i$ for each $i = 1, \ldots, \mu$.

Next we consider the exact homology sequence of the pair $(\bar{\mathcal{X}}_V, \bar{\mathcal{X}}_W)$:

$$\ldots \to \tilde{H}_{q+1}(\bar{\mathcal{X}}_V) \to H_{q+1}(\bar{\mathcal{X}}_V, \bar{\mathcal{X}}_W) \to \tilde{H}_q(\bar{\mathcal{X}}_W) \to H_q(\bar{\mathcal{X}}_V) \to \ldots .$$

Since $\bar{\mathcal{X}}_V$ is a deformation retract of $\bar{\mathcal{X}}$ and since $\bar{\mathcal{X}}$ is contractible,

$$\tilde{H}_q(\bar{\mathcal{X}}_V) = 0 \text{ for all } q.$$

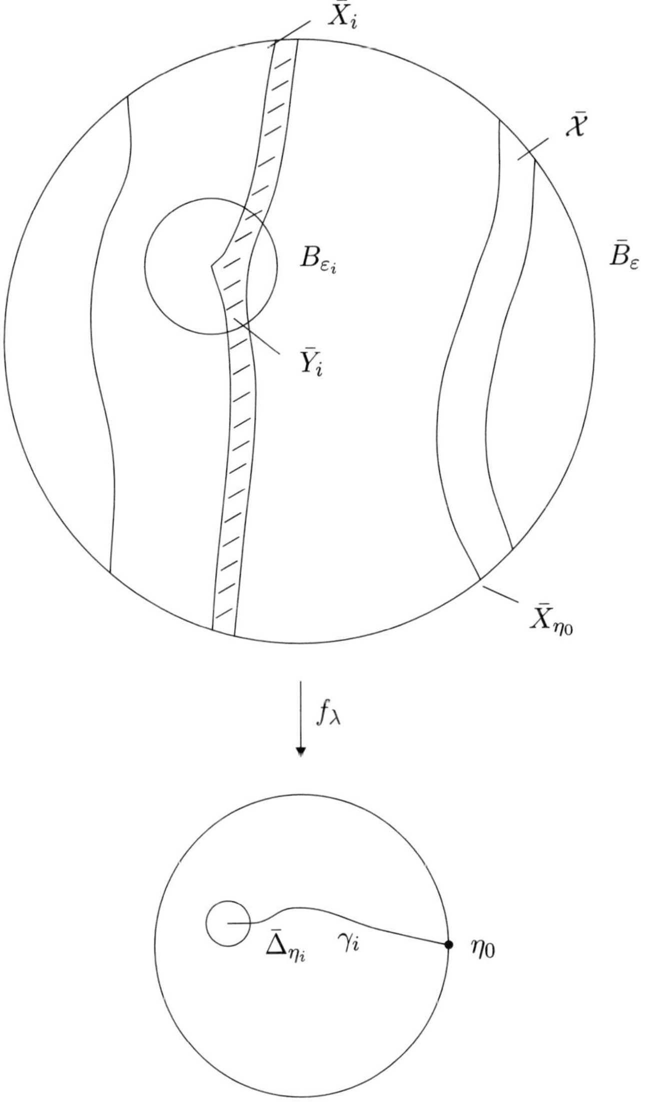

Figure 5.13. The sets $\bar{X}_i$ and $\bar{Y}_i$

Since $\bar{X}_{\eta_0}^{(\lambda)}$ is a deformation retract of $\bar{\mathcal{X}}_W$ it follows that

$$\tilde{H}_q(\bar{\mathcal{X}}_W) = \tilde{H}_q(\bar{X}_{\eta_0}^{(\lambda)}) \text{ for all } q.$$

Since $\bar{\mathcal{X}}^{(\lambda)}$ arises from $\bar{X}_{\eta 0}^{(\lambda)}$ by attaching μ $(n+1)$-cells,

$$\begin{aligned} H_{q+1}(\bar{\mathcal{X}}_V, \bar{\mathcal{X}}_W) &= H_{q+1}\left(\bar{X}_{\eta 0}^{(\lambda)} \cup \bigcup_{i=1}^{\mu} e_i, \bar{X}_{\eta 0}^{(\lambda)}\right) \\ &\overset{(*)}{=} H_{q+1}\left(\bigcup_{i=1}^{\mu} e_i, \bigcup_{i=1}^{\mu} \partial e_i\right) \\ &= \bigoplus_{i=1}^{\mu} H_{q+1}(e_i, \partial e_i). \end{aligned}$$

The equation $(*)$ follows from Theorem 4.4. Furthermore,

$$H_{q+1}(e_i, \partial e_i) = \begin{cases} \mathbb{Z} & \text{for } q = n, \\ 0 & \text{otherwise.} \end{cases}$$

From the exactness of the homology sequence of the pair $(\bar{\mathcal{X}}_V, \bar{\mathcal{X}}_W)$ we obtain

$$\begin{aligned} \tilde{H}_q(\bar{X}_{\eta 0}^{(\lambda)}) &\cong H_{q+1}(\bar{\mathcal{X}}_V, \bar{\mathcal{X}}_W) \\ &= \begin{cases} 0 & \text{for } q \neq n, \\ \mathbb{Z}^{\mu} & \text{for } q = n. \end{cases} \end{aligned}$$

Moreover

$$\partial_* : H_{n+1}(\bar{\mathcal{X}}_V, \bar{\mathcal{X}}_W) \longrightarrow H_n(\bar{\mathcal{X}}_W)$$

maps the generators $[e_i] \in H_{n+1}(\bar{\mathcal{X}}_V, \bar{\mathcal{X}}_W)$ to the vanishing cycles $\delta_i = [\partial e_i] \in H_n(\bar{\mathcal{X}}_W) \cong H_n(\bar{X}_{\eta 0}^{(\lambda)})$. The proposition follows from this. □

Remark 5.2. Given a topological space X, the higher homotopy groups $\pi_q(X)$ are defined as in §4.9, and each pair (X, A) also possesses an exact homotopy sequence. If in the proof above one considers the homotopy sequence of the pair $(\bar{\mathcal{X}}_V, \bar{\mathcal{X}}_W)$ instead of the homology sequence, one can show that the fiber $\bar{X}_{\eta 0}^{(\lambda)}$ is simply connected for $n \geq 2$.

The homotopy type of the Milnor fiber $\bar{X}_{\eta 0}$ was determined in the proof of Proposition 5.5.

Definition. Let X, Y be topological spaces. X and Y are called *homotopy equivalent* (symbolically, $X \sim Y$) if there are continuous maps $f : X \to Y$ and $g : Y \to X$ with $fg \sim \mathrm{id}_Y$ and $gf \sim \mathrm{id}_X$.

Definition. Let $S_1^n, \ldots, S_\mu^n$ be μ disjoint copies of the n-sphere, $p_i \in S_i^n$, $i = 1, \ldots, \mu$. The quotient space of the topological sum $S_1^n \cup \ldots \cup S_\mu^n$ by the equivalence relation

$$p_i \sim p_j \text{ for } 1 \leq i, j \leq \mu,$$
$$x \sim x \text{ for all } x \in S_i^n,\ i = 1, \ldots, \mu,$$

Figure 5.14. $S^1 \vee S^1 \vee S^1 \vee S^1$

is called a *bouquet of n-spheres* and will be denoted by

$$\underbrace{S^n \vee \ldots \vee S^n}_{\mu}.$$

Example 5.1. $n = 1$: A 1-bouquet of 1-spheres is represented in Figure 5.14.

The proof of Proposition 5.5 establishes also the following result.

Theorem 5.2 (Milnor). *The Milnor fiber $\bar{X}_{\eta_0}$ of f is homotopy equivalent to a bouquet of μ n-dimensional real spheres,*

$$\bar{X}_{\eta_0} \sim \underbrace{S^n \vee \ldots \vee S^n}_{\mu}.$$

We next show that even a weakly distinguished system of vanishing cycles forms a basis of $H_n(\bar{X}_{\eta_0}^{(\lambda)})$.

Proposition 5.6. *A weakly distinguished system $(\delta_1, \ldots, \delta_\mu)$ of vanishing cycles forms a basis of $H_n(\bar{X}_{\eta_0}^{(\lambda)}) \cong H_n(\bar{X}_{\eta_0})$.*

Proof. Let $(\delta_1, \ldots, \delta_\mu)$ be a weakly distinguished system of vanishing cycles in $H_n(\bar{X}_{\eta_0}^{(\lambda)})$. Since $H_n(\bar{X}_{\eta_0}^{(\lambda)})$ has a basis of μ vanishing cycles, by Proposition 5.5, it is enough to show that each vanishing cycle δ is an integral linear combination of the $\delta_1, \ldots, \delta_\mu$.

Let $(\gamma_1, \ldots, \gamma_\mu)$ be a weakly distinguished system of paths from the points $s_1, \ldots, s_\mu$ to η_0 such that δ_i is a vanishing cycle along γ_i. Let ω_i be the simple loop corresponding to γ_i, $i = 1, \ldots, \mu$. We again put $\Delta' = \bar{\Delta}_{\eta_0} \setminus \{s_1, \ldots, s_\mu\}$. Then the fundamental group $\pi_1(\Delta', \eta_0)$ is the free group on the generators $(\omega_1, \ldots, \omega_\mu)$.

Let δ be a vanishing cycle of $H_n(\bar{X}_{\eta_0}^{(\lambda)})$ that vanishes along a path γ from s_j to η_0. Since we can replace γ by a homotopic path, we can without loss of generality assume that there is a $\theta > 0$ with $\gamma|_{[0,\theta]} = \gamma_j|_{[0,\theta]}$. Next we consider the loop (see Figure 5.15)

$$\omega = \gamma_j^{-1}|_{[\theta,1]}\gamma|_{[\theta,1]}.$$

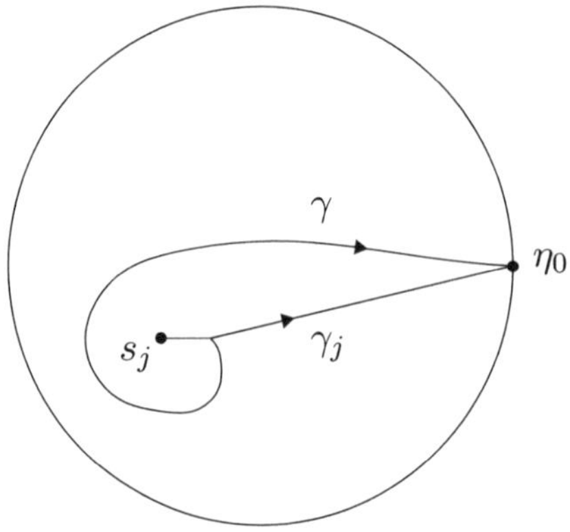

Figure 5.15. The loop ω

We interpret ω as an element of $\pi_1(\Delta', \eta_0)$. Now $\delta = \pm h_{\omega *}(\delta_j)$, where the sign depends on the orientation of δ and δ_j. Now by hypothesis ω can be expressed in terms of the $\omega_1, \omega_1^{-1}, \ldots, \omega_\mu, \omega_\mu^{-1}$, i.e.,

$$\omega = \omega_{i_1}^{\varepsilon_1} \omega_{i_2}^{\varepsilon_2} \cdots \omega_{i_r}^{\varepsilon_r}, \ \varepsilon_i \in \{\pm 1\}, \ i_j \in \{1, \ldots, \mu\}.$$

Then

$$h_{\omega *} = h_{\omega_{i_r} *}^{\varepsilon_r} h_{\omega_{i_{r-1}} *}^{\varepsilon_{r-1}} \cdots h_{\omega_{i_1} *}^{\varepsilon_1}.$$

From the Picard-Lefschetz formulae (Corollary 5.2) it follows that that $\delta = \pm h_{\omega *}(\delta_j)$ is an integral linear combination of the $\delta_1, \ldots, \delta_\mu$. □

Proposition 5.5 asserts that $H_n(\bar{X}_{\eta_0})$ is a free abelian group of rank μ. It is equipped with an integral bilinear form, the intersection form $\langle \ , \ \rangle : H_n(\bar{X}_{\eta_0}) \times H_n(\bar{X}_{\eta_0}) \to \mathbb{Z}$.

Definition. A pair $(L, \langle \ , \ \rangle)$ consisting of a free abelian group of finite rank and a bilinear form $\langle \ , \ \rangle : L \times L \to \mathbb{Z}$ with values in $\mathbb{Z}$ is called a *lattice*. If $\langle \ , \ \rangle$ is symmetric or skew symmetric, the lattice is called *symmetric* or *skew symmetric*. If $(L, \langle \ , \ \rangle)$ is a symmetric lattice and $\langle v, v \rangle \in 2\mathbb{Z}$ for all $v \in L$, the lattice $(L, \langle \ , \ \rangle)$ is called *even*.

Definition. Let $L = H_n(\bar{X}_{\eta_0})$ and let $\langle \ , \ \rangle$ be the intersection form on L. The pair $(L, \langle \ , \ \rangle)$ is called the *Milnor lattice* of the singularity f.

The Milnor lattice is symmetric: it is even when n is even and skew symmetric when n is odd.

5.5. Intersection matrix and Coxeter-Dynkin diagram

We retain the hypotheses and notation of the preceding section. We have seen that the Milnor fiber $\bar{X}_{\eta_0}$ is diffeomorphic to the corresponding fiber $\bar{X}_{\eta_0}^{(\lambda)}$ of a morsification f_λ of f. We therefore may define

Definition. A *vanishing cycle* $\delta \in H_n(\bar{X}_{\eta_0})$ is the image of a vanishing cycle in $H_n(\bar{X}_{\eta_0}^{(\lambda)})$ under the isomorphism $H_n(\bar{X}_{\eta_0}^{(\lambda)}) \cong H_n(\bar{X}_{\eta_0})$.

Definition. A basis $(\delta_1, \ldots, \delta_\mu)$ of $H_n(\bar{X}_{\eta_0})$ is called *strongly* (resp. *weakly*) *distinguished* if $(\delta_1, \ldots, \delta_\mu)$ is a strongly (resp. weakly) distinguished system of vanishing cycles.

By Proposition 5.5 and Proposition 5.6 every strongly or weakly distinguished system of vanishing cycles forms a basis.

Remark 5.3. The concepts "distinguished" and "weakly distinguished" are due to A. M. Gabrielov. In order to distinguish both concepts better, we shall, following a suggestion of E. Brieskorn, often say "strongly distinguished" instead of "distinguished". The term "geometric basis" is also used for a strongly distinguished basis.

Definition. The *monodromy group* Γ of the singularity f is the monodromy group of the Morse function f_λ of a morsification of f.

We shall show later that the set of vanishing cycles and the monodromy group of a singularity do not depend on the morsification chosen.

Definition. Let $(\delta_1, \ldots, \delta_\mu)$ be a weakly distinguished basis of $H_n(\bar{X}_{\eta_0})$. The matrix

$$(\langle \delta_i, \delta_j \rangle)_{j=1,\ldots,\mu}^{i=1,\ldots,\mu}$$

is called the *intersection matrix* of f with respect to $(\delta_1, \ldots, \delta_\mu)$.

By Proposition 5.3 the diagonal entries of the intersection matrix satisfy

$$\langle \delta_i, \delta_i \rangle = (-1)^{\frac{n(n-1)}{2}}(1 + (-1)^n) \text{ for all } i.$$

It is usual to represent the intersection matrix by a graph called the Coxeter-Dynkin diagram.

Definition. Let $(\delta_1, \ldots, \delta_\mu)$ be a weakly distinguished basis of $H_n(\bar{X}_{\eta_0})$. The *Coxeter-Dynkin diagram* of the singularity f with respect to $(\delta_1, \ldots, \delta_\mu)$ is the graph D defined as follows:

(i) The vertices of D are in one-to-one correspondence with the elements $\delta_1, \ldots, \delta_\mu$.

(ii) For $i < j$ with $\langle \delta_i, \delta_j \rangle \neq 0$ the i-th and the j-th vertex are connected by $|\langle \delta_i, \delta_j \rangle|$ edges, weighted with the sign $+1$ or -1 of $\langle \delta_i, \delta_j \rangle \in \mathbb{Z}$. We indicate the weight

$$\varepsilon = \begin{cases} (-1)^{\frac{n}{2}} & \text{for } n \text{ even,} \\ (-1)^{\frac{n+1}{2}} & \text{for } n \text{ odd} \end{cases}$$

by a dotted line, the weight $-\varepsilon$ by a solid line.

Example 5.2.

$\langle \delta_i, \delta_j \rangle = -\varepsilon \iff$ 

$\langle \delta_i, \delta_j \rangle = 2\varepsilon \iff$ 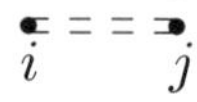

The intersection matrix, resp. the Coxeter-Dynkin diagram, determines the Milnor lattice of singularity f. It also determines the monodromy group of f: By §5.2 the monodromy group is generated by the Picard-Lefschetz transformations h_i corresponding to the paths γ_i of a weakly distinguished system of paths. By Corollary 5.2 the images under h_i of the basis elements of the corresponding weakly distinguished basis $(\delta_1, \ldots, \delta_\mu)$ are as follows:

$$h_i(\delta_j) = \delta_j - (-1)^{\frac{n(n-1)}{2}} \langle \delta_j, \delta_i \rangle \delta_i.$$

Thus the intersection matrix determines the operation h_i and hence the monodromy group.

The intersection matrix, resp. the Coxeter-Dynkin diagram, with respect to a strongly distinguished basis also determines the classical monodromy operator and with it also the variation of the singularity f. If indeed $(\delta_1, \ldots, \delta_\mu)$ is a strongly distinguished basis, then the classical monodromy operator h_* is equal to the product $h_1 \cdots h_\mu$ of the corresponding Picard-Lefschetz transformations. Indeed, the loop ω corresponding to h_* is homotopic to the composition $\omega_\mu \omega_{\mu-1} \cdots \omega_1$ of the simple loops corresponding to $h_\mu, h_{\mu-1}, \ldots, h_1$ (see Figure 5.16).

Example 5.3. We now want to study an example. We consider the function $f : \mathbb{C}^2 \to \mathbb{C}$ with $f(z, y) = z^3 + y^2$.

As a morsification we can choose

$$f_\lambda(z, y) = z^3 - 3\lambda z + y^2 \text{ for } \lambda > 0 \text{ small.}$$

Now

$$\frac{\partial f_\lambda}{\partial z} = 3z^2 - 3\lambda, \quad \frac{\partial f_\lambda}{\partial y} = 2y.$$

Hence f_λ has the critical points $p_1 = (\sqrt{\lambda}, 0)$ and $p_2 = (-\sqrt{\lambda}, 0)$ with the critical values $s_1 = -2\lambda\sqrt{\lambda}$ and $s_2 = 2\lambda\sqrt{\lambda}$. As a noncritical value we choose $s = 0$. We connect s_1 and s_2 to $s = 0$ by intervals on the real w-axis (see Figure 5.17).

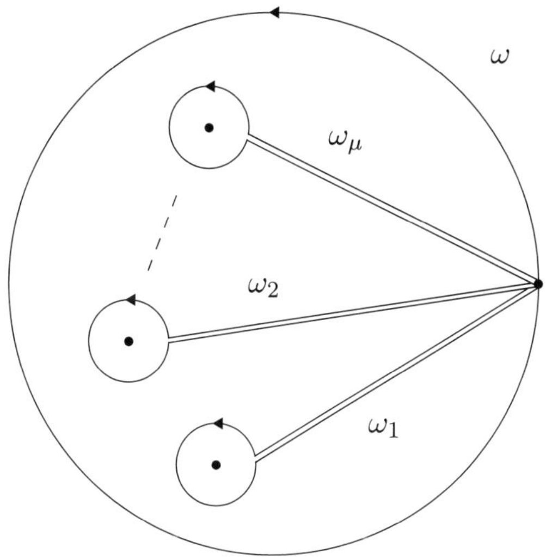

Figure 5.16. ω is homotopic to $\omega_\mu \omega_{\mu-1} \cdots \omega_1$

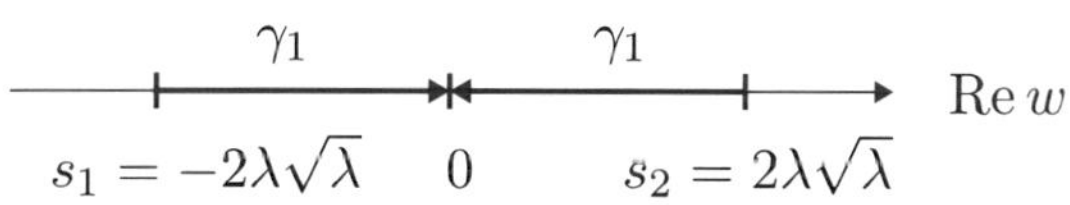

Figure 5.17. Critical values of f_λ

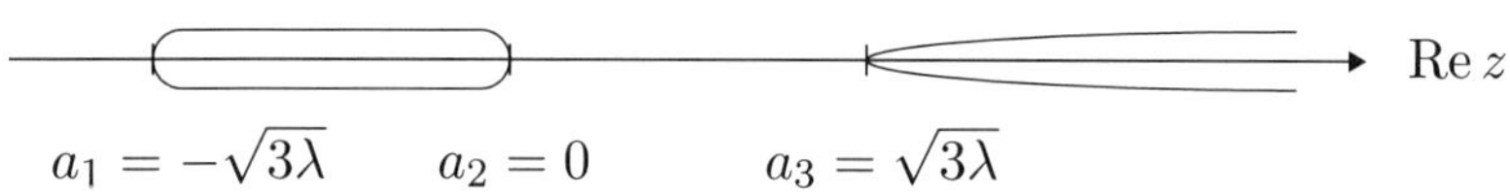

Figure 5.18. Riemann surface of $y = \pm(-z^3 + 3\lambda z)^{1/2}$

The Milnor fiber $\bar{X}_0^{(\lambda)}$ is a subset of the zero set of the function f_λ. This is the Riemann surface of the multivalued function

$$y = \pm\sqrt{-z^3 + 3\lambda z}$$

of a complex variable z. This Riemann surface is a twofold covering of the complex plane with the coordinate z, branched over the points $a_1 = -\sqrt{3\lambda}$, $a_2 = 0$, $a_3 = \sqrt{3\lambda}$. One obtains it from two copies of the complex plane with cuts from a_1 to a_2 and from a_3 to ∞ by pasting the edges crossways (see Figure 5.18). The level surface $f^{-1}(w)$ for $w \in \mathbb{R}$, $w \neq \pm 2\lambda\sqrt{\lambda}$, is the Riemann surface of the function

$$y = \pm\sqrt{-z^3 + 3\lambda z + w}.$$

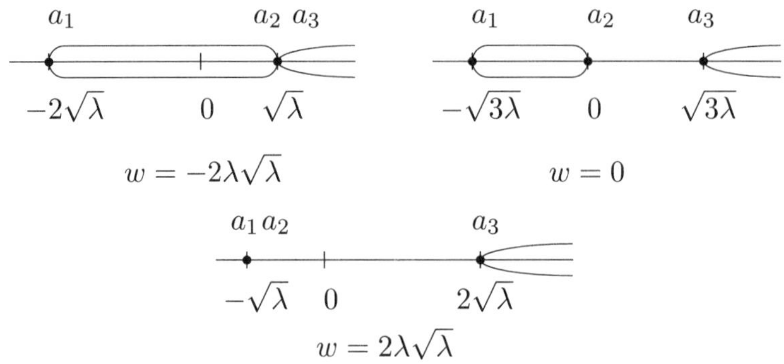

Figure 5.19. Riemann surface of $y = \pm(-z^3 + 3\lambda z + w)^{1/2}$ for $w = -2\lambda\lambda^{1/2}, 0, 2\lambda\lambda^{1/2}$

It is likewise a twofold covering of $\mathbb{C}$ branched over the three points $a_1(w), a_2(w), a_3(w) \in \mathbb{R}$. For $w = \pm 2\lambda\sqrt{\lambda}$ we have

$$y = \pm\sqrt{(z \pm \sqrt{\lambda})^2(z \mp 2\sqrt{\lambda})},$$

and the corresponding affine curve has a singular point.

If we now move w from $s_1 = -2\lambda\sqrt{\lambda}$ along the real axis to $s_2 = 2\lambda\sqrt{\lambda}$, the branch points $a_1(w), a_2(w), a_3(w)$ move as displayed in Figure 5.19. The corresponding fibers $\bar{X}_w^{(\lambda)}$ are pictured in Figure 5.20. Similar figures may already be found in the work of Felix Klein [**Kle73**].

It is now clear what the corresponding vanishing cycles are: They are the cycles δ_1, δ_2, as sketched, with a suitable orientation. If we go along γ_1^{-1} from 0 to s_1, then δ_1 shrinks to a point. If correspondingly we go along γ_2^{-1} from 0 to s_2, then δ_2 shrinks to a point.

We orient δ_1 and δ_2 so that $\langle \delta_1, \delta_2 \rangle = 1$. Then (δ_1, δ_2) is a strongly distinguished basis of the singularity f with the intersection matrix

$$\begin{pmatrix} 0 & 1 \\ -1 & 0 \end{pmatrix}$$

and Coxeter-Dynkin diagram as shown in Figure 5.21. This is a diagram of type A_2 (cf. 3.11).

Exercise 5.1. Compute the Picard-Lefschetz transformations h_1, h_2 for this example and verify the Picard-Lefschetz formulae.

5.6. Classical monodromy, variation, and the Seifert form

In this section we shall study the connection between the intersection matrix and the variation of a singularity.

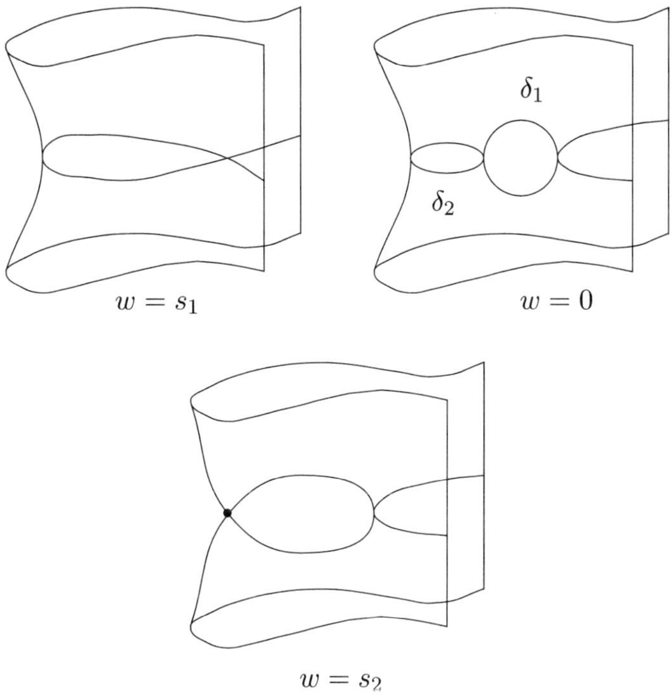

Figure 5.20. The fibers $\bar{X}_w^{(\lambda)}$ for $w = s_1, 0, s_2$

Figure 5.21. Coxeter-Dynkin diagram of type A_2

Let f_λ be a morsification of f and $(\delta_1, \ldots, \delta_\mu)$ a strongly distinguished basis of $H_n(\bar{X}_{\eta 0})$. Let $(\gamma_1, \ldots, \gamma_\mu)$ be a strongly distinguished system of paths such that δ_i is a vanishing cycle along γ_i. Furthermore, let ω_i be simple loop to γ_i, and h_i the corresponding Picard-Lefschetz transformation.

For the classical monodromy operator h_* we have, by §5.5,

$$h_* = h_1 \cdots h_\mu.$$

Correspondingly, for the variation Var = Var_f of f,

$$\mathrm{Var} = \mathrm{var}_{\omega_\mu \cdots \omega_1} .$$

If now τ is the composition of two loops τ_1 and τ_2, so that $\tau = \tau_1 \tau_2$, then

$$\mathrm{var}_\tau = \mathrm{var}_{\tau_1} + \mathrm{var}_{\tau_2} + \mathrm{var}_{\tau_2} \circ j_* \circ \mathrm{var}_{\tau_1} .$$

It follows that

$$
\begin{aligned}
\mathrm{Var} &= \mathrm{var}_{\omega_\mu \cdots \omega_1} \\
(*) \qquad &= \sum_{r=1}^{\mu} \sum_{i_1<i_2<\ldots<i_r} \mathrm{var}_{\omega_{i_1}} \circ j_* \circ \mathrm{var}_{\omega_{i_2}} \circ j_* \circ \cdots \circ j_* \circ \mathrm{var}_{\omega_{i_r}} .
\end{aligned}
$$

Proposition 5.7. *The variation*

$$\mathrm{Var} : H_n(\bar{X}_{\eta 0}, \partial \bar{X}_{\eta 0}) \longrightarrow H_n(\bar{X}_{\eta 0})$$

of the singularity f is an isomorphism.

Proof. By Theorem 4.5 there exists a basis $(\delta_1^*, \ldots, \delta_\mu^*)$ of $H_n(\bar{X}_{\eta 0}, \partial \bar{X}_{\eta 0})$ dual to the given strongly distinguished basis $(\delta_1, \ldots, \delta_\mu)$ of $H_n(\bar{X}_{\eta 0})$, i.e.,

$$\langle \delta_i^*, \delta_j \rangle = \delta_{ij} \text{ for } 1 \leq i, j \leq \mu.$$

From $\mathrm{var}_{\omega_i}(\alpha) = -(-1)^{n(n-1)/2} \langle \alpha, \delta_i \rangle \delta_i$ and the formula $(*)$ it follows that

$$\mathrm{Var}(\delta_i^*) = -(-1)^{\frac{n(n-1)}{2}} \delta_i + \sum_{j<i} c_{ji} \delta_j \text{ with } c_{ij} \in \mathbb{Z}.$$

Hence the matrix of Var with respect to the bases $(\delta_1^*, \ldots, \delta_\mu^*)$ and $(\delta_1, \ldots, \delta_\mu)$ is an upper triangular matrix with $-(-1)^{n(n-1)/2}$ on the diagonal. Such a matrix is invertible over $\mathbb{Z}$. This proves the proposition. □

By Proposition 5.7

$$\mathrm{Var}^{-1} : H_n(\bar{X}_{\eta 0}) \longrightarrow H_n(\bar{X}_{\eta 0}, \partial \bar{X}_{\eta 0})$$

is defined.

Definition. For $a, b \in H_n(\bar{X}_{\eta 0})$ we define $v(a, b) := \langle \mathrm{Var}^{-1} a, b \rangle$.

Proposition 5.8. *For $a, b \in H_n(\bar{X}_{\eta 0})$*

$$\langle a, b \rangle = -v(a, b) - (-1)^n v(b, a).$$

Proof. By Proposition 5.7 there are $a', b' \in H_n(\bar{X}_{\eta 0}, \partial \bar{X}_{\eta 0})$ with $a = \mathrm{Var}\, a'$, $b = \mathrm{Var}\, b'$. By Proposition 5.1

$$
\begin{aligned}
\langle a, b \rangle &= \langle \mathrm{Var}\, a', \mathrm{Var}\, b' \rangle \\
&= -\langle a', \mathrm{Var}\, b' \rangle - \langle \mathrm{Var}\, a', b' \rangle \\
&= -\langle \mathrm{Var}^{-1} a, b \rangle - \langle a, \mathrm{Var}^{-1} b \rangle \\
&= -\langle \mathrm{Var}^{-1} a, b \rangle - (-1)^n \langle \mathrm{Var}^{-1} b, a \rangle \\
&= -v(a, b) - (-1)^n v(b, a),
\end{aligned}
$$

as was to be shown. □

Let S be the intersection matrix of f with respect to the distinguished basis $(\delta_1, \ldots, \delta_\mu)$ and V the matrix of v with respect to $(\delta_1, \ldots, \delta_\mu)$, i.e.,

$$V = (v(\delta_i, \delta_j))_{j=1,\ldots,\mu}^{i=1,\ldots,\mu}.$$

Then we obtain the following corollary:

Corollary 5.3. (i) *The matrix* V *is an upper triangular matrix with* $-(-1)^{n(n-1)/2}$ *on the diagonal.*

(ii) *We have*

$$S = -V - (-1)^n V^t$$

(V^t denotes the transposed matrix.)

Proof. (i) follows from the proof of Proposition 5.7.

(ii) follows from Proposition 5.8. □

Remark 5.4. We have represented S as the sum of an upper and a lower triangular matrix and have given this matrix a geometric meaning.

Let H be the matrix of the classical monodromy h_* with respect to $(\delta_1, \ldots, \delta_\mu)$.

Proposition 5.9. *We have*

$$H = (-1)^{n+1} V^{-1} V^t.$$

Proof. Let

$$(\operatorname{Var}^{-1})^t : H_n(\bar{X}_{\eta 0}) \longrightarrow H_n(\bar{X}_{\eta 0}, \partial \bar{X}_{\eta 0})$$

be defined by

$$\langle (\operatorname{Var}^{-1})^t a, b \rangle = v(b, a) = \langle \operatorname{Var}^{-1} b, a \rangle$$

for $a, b \in H_n(\bar{X}_{\eta 0})$. By Proposition 5.7, $(\operatorname{Var}^{-1})^t a$ is well defined by this. Let again $j_* : H_n(\bar{X}_{\eta 0}) \to H_n(\bar{X}_{\eta 0}, \partial \bar{X}_{\eta 0})$ be the canonical homomorphism. Then $\langle a, b \rangle = \langle j_* a, b \rangle$ for all $a, b \in H_n(\bar{X}_{\eta 0})$. From Proposition 5.8 it follows that

$$j_* = -\operatorname{Var}^{-1} - (-1)^n (\operatorname{Var}^{-1})^t.$$

For the classical monodromy operator h_*, by §5.1,

$$\begin{aligned} h_* &= \operatorname{id} + \operatorname{Var} \circ j_* \\ &= \operatorname{id} + \operatorname{Var}(-\operatorname{Var}^{-1} - (-1)^n (\operatorname{Var}^{-1})^t) \\ &= (-1)^{n+1} \operatorname{Var} \circ (\operatorname{Var}^{-1})^t. \end{aligned}$$

Since V is the matrix of Var^{-1} with respect to the bases $(\delta_1, \ldots, \delta_\mu)$ and $(\delta_1^*, \ldots, \delta_\mu^*)$, the result is proved. □

From Corollary 5.3 and Proposition 5.9 it follows that the intersection matrix with respect to a distinguished basis determines the classical monodromy operator, resp. its matrix representation with respect to the same basis.

The converse is also true. This follows from a lemma in linear algebra.

Lemma 5.5. *Let A, B be upper triangular matrices with 1 on the diagonal, and let $C = AB^t$. Then A and B are uniquely determined by C.*

Proof. Let $A = (a_{ij})$, $B = (b_{ij})$ and $C = (c_{ij})$ be $m \times m$-matrices. It is easy to see that the system of equations

$$\begin{pmatrix} 1 & a_{12} & \cdots & a_{1m} \\ & 1 & \ddots & \vdots \\ & & \ddots & a_{m-1,m} \\ 0 & & & 1 \end{pmatrix} \begin{pmatrix} 1 & & & 0 \\ b_{12} & 1 & & \\ \vdots & \ddots & \ddots & \\ b_{1m} & \cdots & b_{m-1,m} & 1 \end{pmatrix} = \begin{pmatrix} c_{11} & c_{12} & \cdots & c_{1m} \\ c_{21} & c_{22} & \cdots & c_{2m} \\ \vdots & \vdots & \ddots & \vdots \\ c_{m1} & c_{m2} & \cdots & c_{mm} \end{pmatrix}$$

can be solved recursively for a_{ij} and b_{ij}. □

Proposition 5.10. *The matrix of the classical monodromy operator with respect to a strongly distinguished basis determines the intersection matrix.*

Proof. This follows from Proposition 5.9 and Corollary 5.3 with the help of Lemma 5.5. □

We now want to give yet another interpretation of the bilinear form $v(\cdot, \cdot)$.

Let S_ε^{2n+1} be the boundary of the ball $\bar{B}_\varepsilon$. We put

$$K = f^{-1}(0) \cap S_\varepsilon^{2n+1}.$$

Since, by the choice of ε, $f^{-1}(0)$ intersects the sphere S_ε^{2n+1} transversally, K is a $(2n-1)$-dimensional differentiable submanifold of S_ε^{2n+1}.

Definition. We call K the *link* of the singularity f.

Let NK be the normal bundle of the tangent bundle TK of K in TS_ε^{2n+1} with respect to the natural Riemannian metric which S_ε^{2n+1} has as a differentiable submanifold of $\mathbb{R}^{2n+2}$. Let

$$\mathring{N}_\rho K = \{v \in NK \mid \|v\| < \rho\}$$

be the corresponding open ρ-disc bundle with radius $\rho > 0$. By a result from differential topology (cf. [**BJ82**, (12.11)]) there are a $\rho > 0$ and an embedding $\tau : \mathring{N}_\rho K \to S_\varepsilon^{2n+1}$. The image of this embedding is called a(n open) *tubular neighborhood* of K. We choose $\rho > 0$ sufficiently small and a corresponding open tubular neighborhood T of K.

The following result is proved in Milnor's book [**Mil68**].

Proposition 5.11 (Milnor). *The map*

$$\begin{array}{rccc} \Phi : & S_\varepsilon^{2n+1} \setminus T & \longrightarrow & S^1 \subset \mathbb{C} \\ & z & \longmapsto & \frac{f(z)}{|f(z)|} \end{array}$$

is the projection of a differentiable fiber bundle.

Additional remark. The restriction

$$\Phi|_{\partial(S_\varepsilon^{2n+1}\setminus T)} : \partial(S_\varepsilon^{2n+1} \setminus T) = \partial\bar{T} \longrightarrow S^1$$

is the projection of a trivial differentiable fiber bundle.

The restriction of the Milnor fibration

$$f|_{\bar{B}_\varepsilon \cap f^{-1}(\bar{\Delta}\setminus\{0\})} : \bar{B}_\varepsilon \cap f^{-1}(\bar{\Delta} \setminus \{0\}) \longrightarrow \bar{\Delta} \setminus \{0\}$$

to the boundary $S^1_{\eta_0}$ of $\bar{\Delta}$ defines a fibration

$$\tilde{f} : f^{-1}(S^1_{\eta_0}) \cap \bar{B}_\varepsilon \longrightarrow S^1_{\eta_0}.$$

Indeed, the restriction of $\tilde{f}$ to the boundary $f^{-1}(S^1_{\eta_0}) \cap S_\varepsilon^{2n+1}$ defines a trivial differentiable fiber bundle over $S^1_{\eta_0}$.

Proposition 5.12. *The fibrations $\Phi : S_\varepsilon^{2n+1} \setminus T \to S^1$ and $\tilde{f} : f^{-1}(S^1_{\eta_0}) \cap \bar{B}_\varepsilon \to S^1_{\eta_0}$ are equivalent (relative to the diffeomorphism $S^1 \to S^1_{\eta_0}$ given by multiplication by η_0). In particular the fiber $\bar{F}_{w/|w|} = \Phi^{-1}(w/|w|)$ is diffeomorphic to $\bar{X}_w$ for $w \in S^1_{\eta_0}$.*

Proof. See [**Mil68**, §5]. □

Using Proposition 5.12, we can define the variation Var of f by means of the fibration Φ too. Let

$$g_t : \bar{F}_1 \longrightarrow \bar{F}_{e^{2\pi it}}$$

be the parallel transport along $\omega(t) = e^{2\pi it}$. Let a, b be piecewise differentiable n-cycles in the fiber $\bar{F}_1 \subset S_\varepsilon^{2n+1}$ that are null homologous in S_ε^{2n+1}. The cycle $g_{1/2}(b)$ lies in $\bar{F}_{-1}$ and so it does not intersect a. Hence, by §4.7, the linking number $l(a, g_{1/2}(b))$ of a and $g_{1/2}(b)$ in S_ε^{2n+1} is defined.

Definition. The *Seifert form* of f is the bilinear form $L : \tilde{H}_n(\bar{F}_1) \times \tilde{H}_n(\bar{F}_1) \to \mathbb{Z}$ defined by $L(a,b) = l(a, g_{1/2*}(b))$ for $a, b \in \tilde{H}_n(\bar{F}_1) \cong \tilde{H}_n(\bar{X}_{\eta_0})$.

Proposition 5.13. *For $a \in H_n(\bar{F}_1, \partial\bar{F}_1)$, $b \in \tilde{H}_n(\bar{F}_1)$*

$$L(\operatorname{Var} a, b) = \langle a, b \rangle.$$

Proof. We choose piecewise differentiable n-cycles that represent a and b, and we denote them by the same symbols. Consider the map

$$\begin{array}{rccc} \alpha : & [0,1] \times a & \longrightarrow & S_\varepsilon^{2n+1} \\ & (t, z) & \longmapsto & g_t(z)\,. \end{array}$$

Note that $\alpha(\{0\} \times a) = a$, $\alpha(\{1\} \times a) = g_1(a)$, $\alpha([0,1] \times \partial a) \subset \partial\bar{T}$, where T is the tubular neighborhood of K (see Figure 5.22).

The image A' of α is a piecewise differentiable $(n+1)$-chain in S_ε^{2n+1} with $\partial A' = g_1(a) - a + c$, where c is an n-cycle in $\partial\bar{T}$. Since the restriction

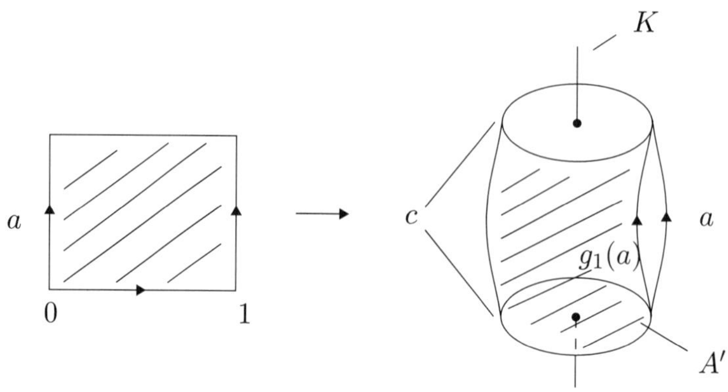

Figure 5.22. The map α

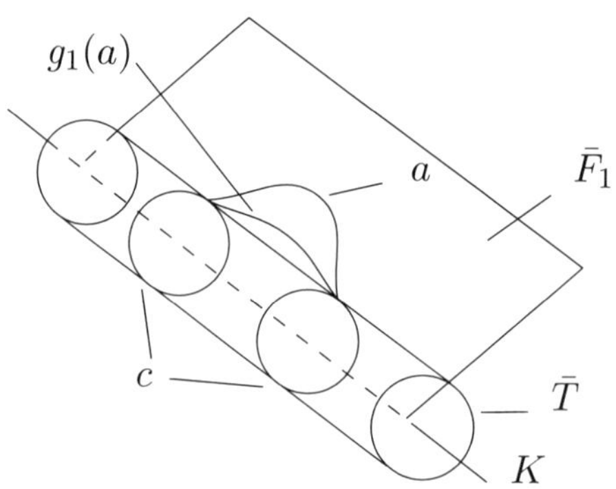

Figure 5.23. Retraction of c onto ∂a

of Φ to $\partial \bar{T}$ is a trivial fibration over S^1, we can retract c inside $\bar{T}$ onto $\partial a \subset \partial \bar{F}_1$ (see Figure 5.23).

The resulting piecewise differentiable $(n+1)$-chain in S^{2n+1} will be called A. Then $\partial A \subset \bar{F}_1$, $\partial A = \operatorname{Var} a$. Hence

$$\begin{aligned} L(\operatorname{Var} a, b) &= l(\operatorname{Var} a, g_{\frac{1}{2}*} b) \\ &= \langle A, g_{\frac{1}{2}*} b \rangle_{S_\varepsilon^{2n+1}} \\ &= \langle g_{\frac{1}{2}*} a, g_{\frac{1}{2}*} b \rangle_{\bar{F}_{-1}} \\ &= \langle a, b \rangle_{\bar{F}_1}, \end{aligned}$$

as was to be proved. □

Corollary 5.4. *For* $a, b \in \tilde{H}_n(\bar{X}_{\eta_0})$

$$v(a,b) = L(a,b).$$

Proof. We have

$$v(a,b) \underset{\text{Def.}}{=} \langle \operatorname{Var}^{-1} a, b\rangle \underset{\text{Proposition 5.13}}{=} L(a,b),$$

as was to be shown. □

We have thus identified the form v with the Seifert form L. We can now express Proposition 5.8 in this way:

Corollary 5.5. *For $a, b \in \tilde{H}_n(\bar{X}_{\eta_0})$*

$$\langle a, b\rangle = -L(a,b) - (-1)^n L(b,a).$$

5.7. The action of the braid group

A strongly or weakly distinguished system $(\gamma_1, \ldots, \gamma_\mu)$ of paths can be chosen in many various ways. Next we consider elementary operations on path systems with the property that they preserve strong or weak distinguishedness.

Let $(\gamma_1, \ldots, \gamma_\mu)$ be a strongly distinguished system of paths from $s_1, \ldots, s_\mu$ to s and let $(\delta_1, \ldots, \delta_\mu)$ be a corresponding strongly distinguished system of vanishing cycles. Furthermore, let $(\omega_1, \ldots, \omega_\mu)$ be a corresponding system of simple loops.

Definition. We define an operation α_j for $1 \leq j < \mu$. We define a new system of paths

$$(\tilde{\gamma}_1, \ldots, \tilde{\gamma}_\mu) = \alpha_j(\gamma_1, \ldots, \gamma_\mu)$$

as follows:

$$\begin{aligned} \tilde{\gamma}_i &= \gamma_i \text{ for } i \neq j, j+1, \\ \tilde{\gamma}_{j+1} &= \gamma_j, \\ \tilde{\gamma}_j &\sim \gamma_{j+1}\omega_j, \end{aligned}$$

where $\tilde{\gamma}_j$ is a small homotopic deformation of $\gamma_{j+1}\omega_j$ such that $\tilde{\gamma}_j$ has no self-intersection points and intersects the other paths only at s, for $t = 1$ (see Figure 5.24).

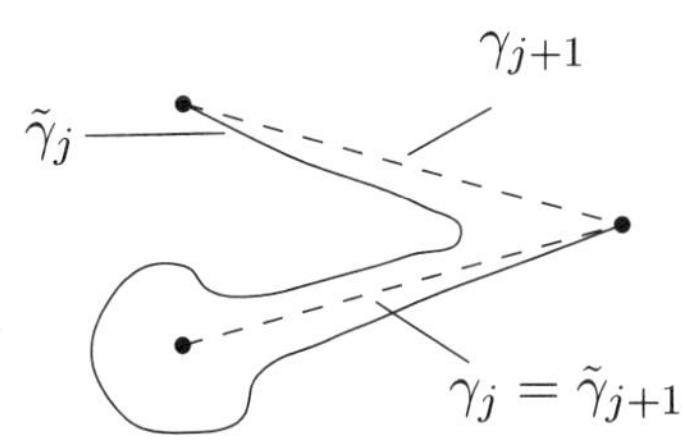

Figure 5.24. The operation α_j

Then $(\tilde{\gamma}_1, \ldots, \tilde{\gamma}_\mu)$ is itself a strongly distinguished system of paths.

For the corresponding system $(\tilde{\delta}_1, \ldots, \tilde{\delta}_\mu)$ of vanishing cycles we have

$$\begin{aligned}\tilde{\delta}_i &= \delta_i \text{ for } i \neq j, j+1, \\ \tilde{\delta}_{j+1} &= \delta_j, \\ \tilde{\delta}_j &= h_j(\delta_{j+1}) = \delta_{j+1} - (-1)^{n(n-1)/2}\langle \delta_{j+1}, \delta_j \rangle \delta_j.\end{aligned}$$

The operation $(\delta_1, \ldots, \delta_\mu) \mapsto (\tilde{\gamma}_1, \ldots, \tilde{\gamma}_\mu)$ will also be denoted by α_j.

Definition. We define an operation β_{j+1} for $1 \leq j < \mu$. We define a new system of paths

$$(\gamma'_1, \ldots, \gamma'_\mu) = \beta_{j+1}(\gamma_1, \ldots, \gamma_\mu)$$

as follows:

$$\begin{aligned}\gamma'_i &= \gamma_i \text{ for } i \neq j, j+1, \\ \gamma'_j &= \gamma_{j+1}, \\ \gamma'_{j+1} &\sim \gamma_j \omega_{j+1}^{-1},\end{aligned}$$

where the same holds for γ'_{j+1} as for $\tilde{\gamma}_j$ (see Figure 5.25). Then $(\gamma'_1, \ldots, \gamma'_\mu)$

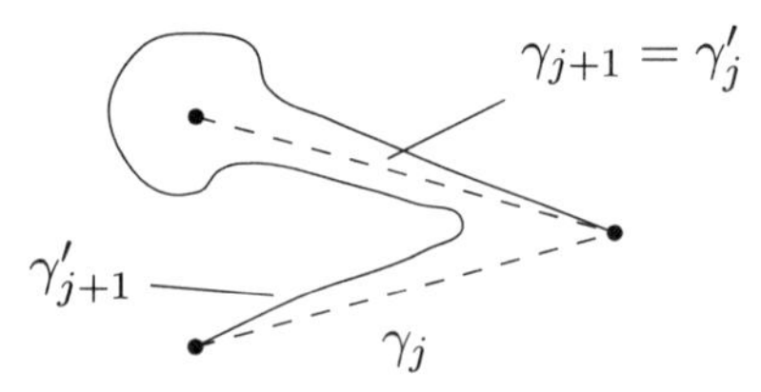

Figure 5.25. The operation β_{j+1}

is indeed a strongly distinguished system of paths. For the corresponding system $(\delta'_1, \ldots, \delta'_\mu)$ of vanishing cycles

$$\begin{aligned}\delta'_i &= \delta_i \text{ for } i \neq j, j+1, \\ \delta'_j &= \delta_{j+1}, \\ \delta'_{j+1} &= h_{j+1}^{-1}(\delta_j) = \delta_j - (-1)^{n(n-1)/2}\langle \delta_{j+1}, \delta_j \rangle \delta_{j+1}.\end{aligned}$$

The operation $(\delta_1, \ldots, \delta_\mu) \mapsto (\delta'_1, \ldots, \delta'_\mu)$ is again denoted by β_{j+1}.

Definition. Two strongly distinguished systems $(\gamma_1, \ldots, \gamma_\mu)$ and $(\tau_1, \ldots, \tau_\mu)$ of paths are called *homotopic*, symbolically, $(\gamma_1, \ldots, \gamma_\mu) \sim (\tau_1, \ldots, \tau_\mu)$, if there are homotopies $\varphi_i : I \times I \to \bar{\Delta}$ between γ_i and τ_i, $i = 1, \ldots, \mu$, such that for all $u \in I$ and paths $\varphi_i^u : I \to \bar{\Delta}$, $t \mapsto \varphi_i(u, t)$, $i = 1, \ldots, \mu$, the following properties are satisfied:

(i) $\varphi_i^u(0) = s_i$, $\varphi_i^u(1) = s$.

(ii) The paths φ_i^u are double point free.

(iii) Each two paths φ_i^u and φ_j^u have, for $i \neq j$, only the end point s in common.

Lemma 5.6. *The operations α_j and β_{j+1} are mutually inverse, i.e., the application of $\alpha_j \beta_{j+1}$ and $\beta_{j+1}\alpha_j$ to a strongly distinguished path system $(\gamma_1, \ldots, \gamma_\mu)$ yields a homotopic strongly distinguished path system.*

Proof. We consider the composition $\beta_{j+1}\alpha_j$. The proof for $\alpha_j\beta_{j+1}$ is similar (Exercise). We need to study the operation only on the paths γ_j and γ_{j+1}. Up to homotopy of strongly distinguished path systems this operation is as represented in Figure 5.26. □

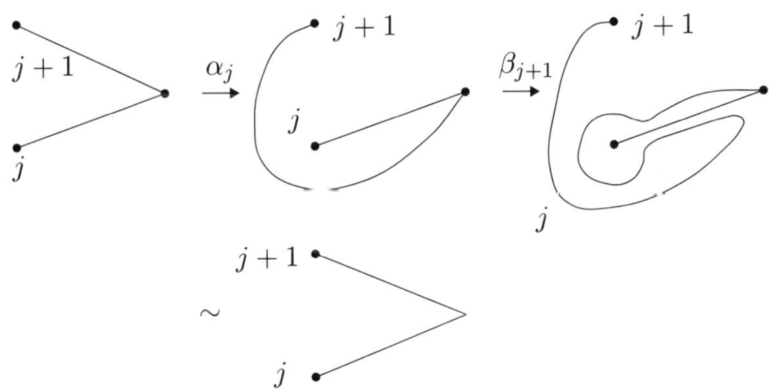

Figure 5.26. The operation $\beta_{j+1}\alpha_j$

Proposition 5.14. *Up to homotopy of strongly distinguished path systems we have*

(i) $\alpha_i\alpha_j = \alpha_j\alpha_i$ *for* i, j *with* $|i - j| \geq 2$,

(ii) $\alpha_j\alpha_{j+1}\alpha_j = \alpha_{j+1}\alpha_j\alpha_{j+1}$ *for* $1 \leq j < \mu - 1$.

Proof. (i) is clear from the definition.

For (ii), we need consider only the paths γ_j, γ_{j+1} and γ_{j+2}. The operation of the left side of (ii) is given in Figure 5.27, and the right side acts as represented in Figure 5.28. □

According to Proposition 5.14 we have an action of the braid group B_μ (cf. §4.8) on the set the homotopy classes of strongly distinguished path systems and so also on the set of all strongly distinguished systems of vanishing cycles.

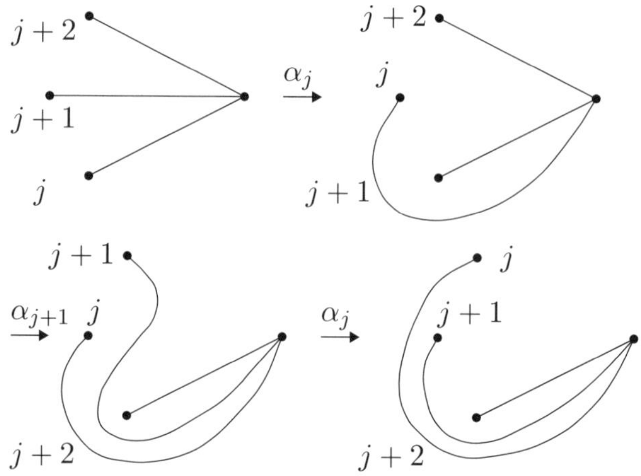

Figure 5.27. The operation $\alpha_j\alpha_{j+1}\alpha_j$

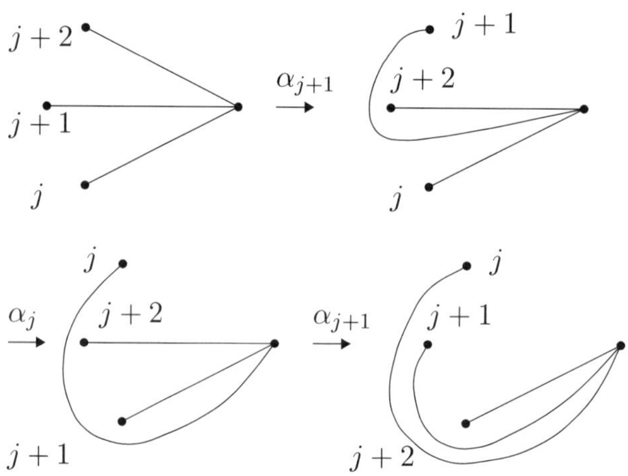

Figure 5.28. The operation $\alpha_{j+1}\alpha_j\alpha_{j+1}$

To a pair $(\gamma_1, \ldots, \gamma_\mu)$, $(\tau_1, \ldots, \tau_\mu)$ of strongly distinguished path systems one can now assign a braid b with μ strands as follows: After replacing $(\tau_1, \ldots, \tau_\mu)$ by a homotopic strongly distinguished path system, if necessary, we can assume that for the path system $(\tau_1, \ldots, \tau_\mu)$ we have: For each $i \in \{1, \ldots, \mu\}$ there is a $t_i \in (0, 1)$ such that $\tau_i(t) = \gamma_i(t)$ for all $t \in [t_i, 1]$. Then we consider the paths $\gamma_i|_{[0,t_i]}\tau_i|_{[0,t_i]}^{-1}$. These paths determine a braid b (see Figure 5.29). It is clear that this braid depends only on the homotopy classes of the two strongly distinguished path systems.

If $(\tau_1, \ldots, \tau_\mu)$ is a path system arising from $(\gamma_1, \ldots, \gamma_\mu)$ through the operation α_j, then the corresponding braid is just the braid a_j. In this way

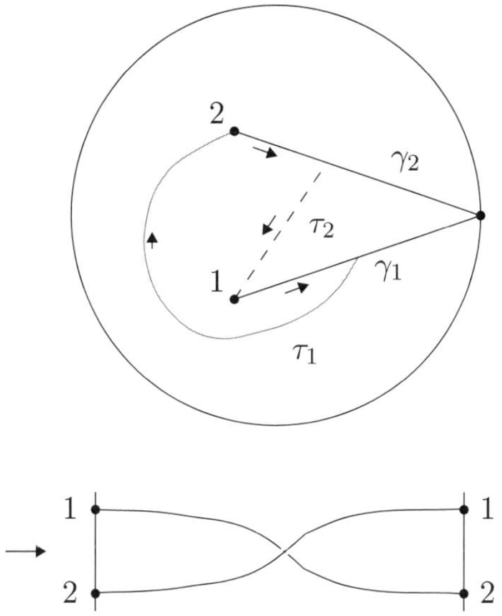

Figure 5.29. The braid b corresponding to a pair of strongly distinguished path systems $(\gamma_1, \gamma_2), (\tau_1, \tau_2)$

one sees that the assignment $\alpha_j \mapsto a_j$ induces a group isomorphism

$$B_\mu \cong \pi_1(Y_\mu, \bar{y})$$

(cf. §1.8).

Proposition 5.15. *The braid group B_μ acts transitively on the set of all homotopy classes of strongly distinguished path systems, i.e., any two strongly distinguished path systems can be transformed one to the other by iteration of the operations α_j and β_{j+1} and a succeeding homotopy.*

Proof. This follows from the fact that two distinguished path systems $\{\gamma_1, \dots, \gamma_\mu\}$ and $\{\tau_1, \dots, \tau_\mu\}$ determine a braid b. Under the isomorphism

$$B_\mu \cong \pi_1(Y_\mu, \bar{y}),$$

b corresponds to a word in α_j and $\beta_{j+1} = \alpha_j^{-1}$. This word prescribes the desired sequence of operations. □

Corollary 5.6. *Any two strongly distinguished systems of vanishing cycles can be transformed one to the other by iteration of the operations α_j and β_{j+1} and a succeeding change of orientation of some of the cycles.*

Next we consider operations that transform weakly distinguished path systems again into weakly distinguished path systems.

Let $(\gamma_1, \dots, \gamma_\mu)$ now be a weakly distinguished path system from the points $s_1, \dots, s_\mu$ to s, let $(\omega_1, \dots, \omega_\mu)$ be a corresponding system of simple loops and let $(\delta_1, \ldots, \delta_\mu)$ be a corresponding weakly distinguished system of vanishing cycles.

Definition. We define operations $\alpha_i(j)$ and $\beta_i(j)$ for $i, j \in \{1, \ldots, \mu\}$, $i \neq j$, as follows:

$$\begin{aligned} \alpha_i(j)(\gamma_1, \ldots, \gamma_\mu) &= (\gamma_1, \ldots, \gamma_{j-1}, \gamma_j \omega_i, \gamma_{j+1}, \ldots, \gamma_\mu), \\ \beta_i(j)(\gamma_1, \ldots, \gamma_\mu), &= (\gamma_1, \ldots, \gamma_{j-1}, \gamma_j \omega_i^{-1}, \gamma_{j+1}, \ldots, \gamma_\mu). \end{aligned}$$

These operations induce the following operations on the corresponding systems of simple loops, and we denote them by the same symbol:

$$\begin{aligned} \alpha_i(j)(\omega_1, \ldots, \omega_\mu) &= (\omega_1, \ldots, \omega_{j-1}, \omega_i^{-1} \omega_j \omega_i, \omega_{j+1}, \ldots, \omega_\mu), \\ \beta_i(j)(\omega_1, \ldots, \omega_\mu) &= (\omega_1, \ldots, \omega_{j-1}, \omega_i \omega_j \omega_i^{-1}, \omega_{j+1}, \ldots, \omega_\mu). \end{aligned}$$

If $\omega_1, \ldots, \omega_\mu$ forms a generating system for $\pi_1(\Delta', s)$, then $\pi_1(\Delta', s)$ is also generated by the new simple loops that arise from application of the operations $\alpha_i(j)$ and $\beta_i(j)$. Hence $\alpha_i(j)$ and $\beta_i(j)$ transfer weakly distinguished path systems again to weakly distinguished path systems.

These operations thus induce operations on the corresponding weakly distinguished systems of vanishing cycles too, which we denote by the same symbols, and they appear as follows:

$$\begin{aligned} \alpha_i(j)(\delta_1, \ldots, \delta_\mu) &= (\delta_1, \ldots, \delta_{j-1}, h_i(\delta_j), \delta_{j+1}, \ldots, \delta_\mu), \\ \beta_i(j)(\delta_1, \ldots, \delta_\mu) &= (\delta_1, \ldots, \delta_{j-1}, h_i^{-1}(\delta_j), \delta_{j+1}, \ldots, \delta_\mu). \end{aligned}$$

The operations $\alpha_i(j)$ and $\beta_i(j)$ are again mutually inverse in the sense above. For even n they even agree.

If $(\gamma_1, \ldots, \gamma_\mu)$ is a strongly distinguished path system and if $\tau_{j,j+1} \in S_\mu$ denotes the transposition of j and $j+1$, then, up to homotopy,

$$\begin{aligned} \alpha_j &= \tau_{j,j+1} \circ \alpha_j(j+1), \\ \beta_{j+1} &= \tau_{j,j+1} \circ \beta_{j+1}(j). \end{aligned}$$

We now also have the following proposition:

Proposition 5.16. *Let $(\omega_1, \ldots, \omega_\mu)$ and $(\omega'_1, \ldots, \omega'_\mu)$ be two free generating systems of the free group $\pi_1(\Delta', s)$ such that ω_i and ω'_i are conjugate to one another for $i = 1, \ldots, \mu$. Then one can obtain $(\omega'_1, \ldots, \omega'_\mu)$ from $(\omega_1, \ldots, \omega_\mu)$ by the application of a sequence of operations of type $\alpha_i(j)$ or $\beta_i(j)$.*

This proposition was conjectured by S. M. Gusein-Zade and proved by S. P. Humphries [**Hum85**] in 1985. It also follows, as remarked by R. Pellikaan, from an old result of J. H. C. Whitehead from the year 1936 (cf. [**LS77**, Proposition 4.20]). We refer to [**Hum85**].

Corollary 5.7. *Any two weakly distinguished systems of vanishing cycles can be transformed one to the other by iteration of the operations $\alpha_i(j)$ and $\beta_i(j)$ and a succeeding change of orientation of some of the cycles.*

To finish this section, we consider an example.

Example 5.4. Consider the function $f : \mathbb{C} \to \mathbb{C}$ with $f(z) = z^{k+1}$.

The Milnor fiber $\bar{X}_\eta$ consists of $k+1$ points, namely the $(k+1)$-th roots of η. We have

$$\tilde{H}_0(\bar{X}_\eta) \cong \mathbb{Z}^k$$

and $\mu = k$.

The function $f_\lambda(z) = z^{k+1} - \lambda z$ is, for $\lambda \neq 0$, a morsification of f. Now suppose that $\lambda \in \mathbb{R}$, $\lambda > 0$. The critical points of the function f_λ are given by the equation

$$f_\lambda'(z) = (k+1)z^k - \lambda = 0\,.$$

They are the points

$$p_i = \sqrt[k]{\frac{\lambda}{k+1}}\,\xi_i, \quad \xi_i = e^{-\frac{2\pi i\sqrt{-1}}{k}},$$

with the critical values

$$s_i = -\frac{\lambda k}{k+1}\sqrt[k]{\frac{\lambda}{k+1}}\,\xi_i, \quad i = 1, \ldots, k.$$

As a noncritical value we choose $-\eta_0$, where $\eta_0 \in \mathbb{R}$, $\eta_0 > 0$ and

$$\eta_0 \gg \frac{\lambda k}{k+1}\sqrt[k]{\frac{\lambda}{k+1}}.$$

Let $\gamma_i : [0,1] \to \bar{\Delta}$, $t \mapsto (1-t)s_i$, and let τ be a path from 0 to $-\eta_0$ that runs along the real axis and goes round the critical value

$$s_k = -\frac{\lambda k}{k+1}\sqrt[k]{\frac{\lambda}{k+1}}\,\xi_k \in \mathbb{R}$$

in the positive direction (see Figure 5.30).

Next we consider the path system $(\gamma_1\tau, \ldots, \gamma_k\tau)$. By a small deformation we can deform this path system into a strongly distinguished path system. Let $(\delta_1, \ldots, \delta_k)$ be a corresponding strongly distinguished system of vanishing cycles in $\tilde{H}_0(\bar{X}_{-\eta_0})$.

In order to compute the intersection numbers $\langle \delta_i, \delta_j \rangle$ of the vanishing cycles in $\tilde{H}_0(\bar{X}_{-\eta_0})$ we transport the system $(\delta_1, \ldots, \delta_k)$ by parallel transport along the path τ^{-1} to $\tilde{H}_0(\bar{X}_0)$. We thus consider a system of vanishing cycles in $\tilde{H}_0(\bar{X}_0)$, which we again denote by $(\delta_1, \ldots, \delta_k)$, and which is defined by the path system $(\gamma_1, \ldots, \gamma_k)$.

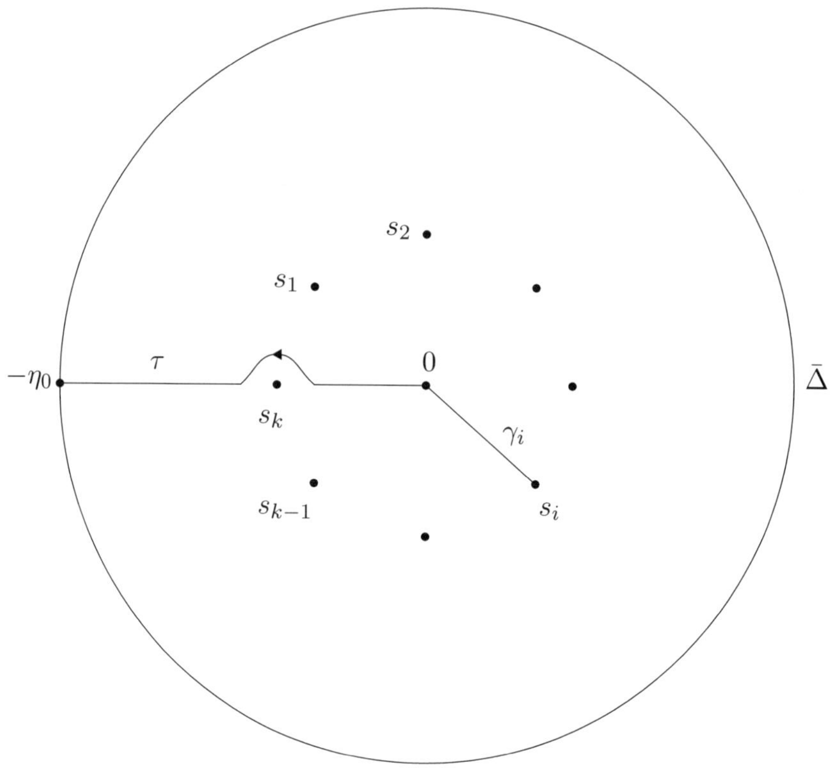

Figure 5.30. Critical values of the function f_λ and the paths γ_i and τ

The fiber $\bar{X}_0$ consists of the $k+1$ points

$$x_0 = 0, x_1 = \sqrt[k]{\lambda}\xi_1, \ldots, x_k = \sqrt[k]{\lambda}\xi_k.$$

Then up to orientation δ_i is represented by the cycle $x_i - x_0$ for it is easy to calculate that $x_i - x_0$ vanishes along γ_i, i.e., that the points x_i and x_0 fall together along γ_i. Let

$$\delta_i = [x_i - x_0], \ i = 1, \ldots, k.$$

Then

$$\langle \delta_i, \delta_j \rangle = \begin{cases} 2 & \text{for } i = j, \\ 1 & \text{for } i \neq j. \end{cases}$$

Thus the Coxeter-Dynkin diagram D with respect to $(\delta_1, \ldots, \delta_k)$ looks as follows: D is a complete graph with only dashed edges (i.e., each two vertices are joined by a dashed edge (see Figure 5.31 for $k = 5$)).

We now simplify this graph with the help of the action of the braid group

$$\begin{aligned} & (\delta_1, \ldots, \delta_k) \\ \stackrel{\alpha_{k-1}}{\longmapsto} \ & (\delta_1', \ldots, \delta_k') = (\delta_1, \ldots, \delta_{k-2}, \delta_k - \delta_{k-1}, \delta_{k-1}). \end{aligned}$$

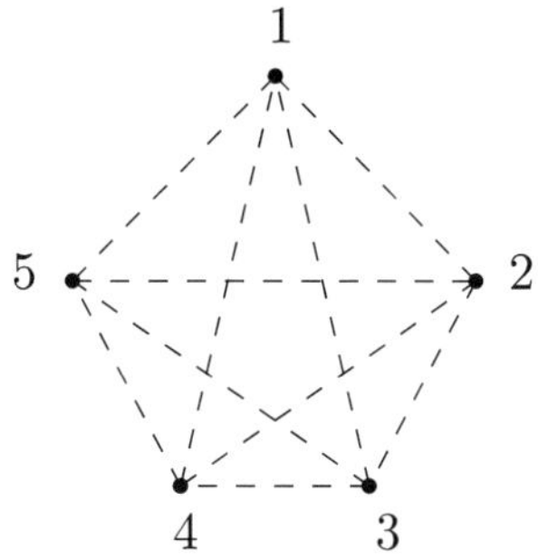

Figure 5.31. Coxeter-Dynkin diagram with respect to $(\delta_1, \ldots, \delta_k)$ for $k = 5$

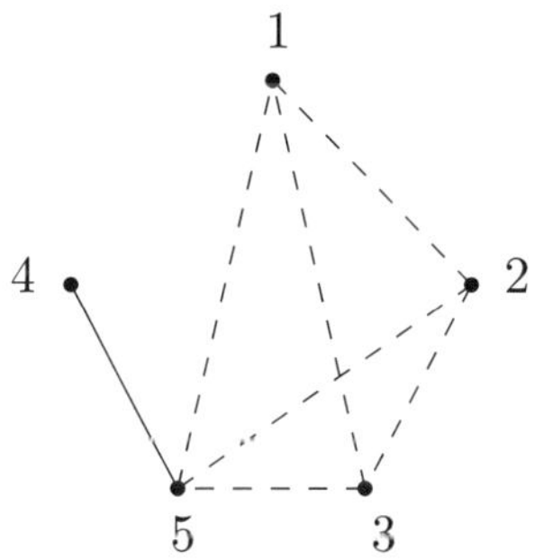

Figure 5.32. Coxeter-Dynkin diagram with respect to $(\delta_1', \ldots, \delta_k')$ for $k - 5$

We have

$$\begin{aligned} \langle \delta_i', \delta_{k-1}' \rangle &= 0 \text{ for } 1 \le i \le k-2, \\ \langle \delta_k', \delta_{k-1}' \rangle &= -1. \end{aligned}$$

The corresponding Coxeter-Dynkin diagram (for $k = 5$) is represented in Figure 5.32.

We now apply the operations $\alpha_{k-2}, \alpha_{k-3}, \ldots, \alpha_1$. Since the vertex δ_{k-1}' is not joined to the vertices $\delta_{k-2}', \ldots, \delta_1'$, these operations do not alter the form of the diagram, but only the numbering:

$$\begin{aligned} &\overset{\alpha_{k-2}}{\longmapsto} (\delta_1, \ldots, \delta_{k-3}, \delta_k - \delta_{k-1}, \delta_{k-2}, \delta_{k-1}) \\ &\overset{\alpha_{k-3}}{\longmapsto} \ldots \overset{\alpha_2}{\longmapsto} \\ &\overset{\alpha_1}{\longmapsto} (\delta_1^{(k-1)}, \ldots, \delta_k^{(k-1)}) = (\delta_k - \delta_{k-1}, \delta_1, \delta_2, \ldots, \delta_{k-1}) . \end{aligned}$$

The Coxeter-Dynkin diagram with respect to $(\delta_1^{(k-1)}, \ldots, \delta_k^{(k-1)})$ (for $k = 5$) is represented in Figure 5.33.

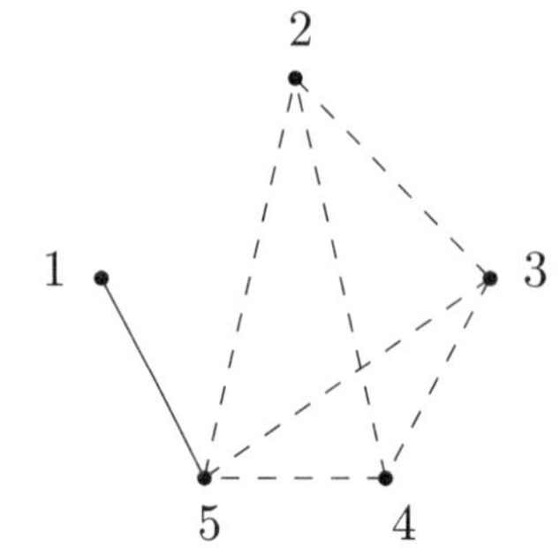

Figure 5.33. Coxeter-Dynkin diagram with respect to $(\delta_1^{(k-1)}, \ldots$ $\ldots, \delta_k^{(k-1)})$ for $k = 5$

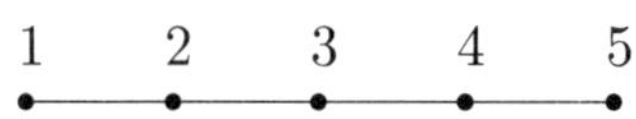

Figure 5.34. Coxeter-Dynkin diagram with respect to $(\tilde{\delta}_1, \ldots, \tilde{\delta}_k)$ for $k = 5$

Figure 5.35. Coxeter-Dynkin diagram of type A_k

The subgraph for $\delta_2^{(k-1)}, \ldots, \delta_k^{(k-1)}$ is again a complete graph and we can now apply the corresponding operations to this subgraph:

$$\begin{aligned}
&\stackrel{\alpha_{k-1}}{\longmapsto} \quad \cdots \quad \stackrel{\alpha_2}{\longmapsto} (\delta_k - \delta_{k-1}, \delta_{k-1} - \delta_{k-2}, \delta_1, \ldots, \delta_{k-2}) \\
&\stackrel{\alpha_{k-1}}{\longmapsto} \quad \cdots \quad \stackrel{\alpha_3}{\longmapsto} (\delta_k - \delta_{k-1}, \delta_{k-1} - \delta_{k-2}, \delta_{k-2} - \delta_{k-3}, \delta_1, \ldots, \delta_{k-3}) \\
&\qquad \vdots \\
&\stackrel{\alpha_{k-1}}{\longmapsto} \stackrel{\alpha_{k-2}}{\longmapsto} (\delta_k - \delta_{k-1}, \delta_{k-1} - \delta_{k-2}, \ldots, \delta_3 - \delta_2, \delta_1, \delta_2) \\
&\qquad \stackrel{\alpha_{k-1}}{\longmapsto} (\delta_k - \delta_{k-1}, \delta_{k-1} - \delta_{k-2}, \ldots, \delta_2 - \delta_1, \delta_1) =: (\tilde{\delta}_1, \tilde{\delta}_2, \ldots, \tilde{\delta}_k) \,.
\end{aligned}$$

The Coxeter-Dynkin diagram with respect to $(\tilde{\delta}_1, \ldots, \tilde{\delta}_k)$ (for $k = 5$) is the graph represented in Figure 5.34. We have

$$\tilde{\delta}_1 = [x_k - x_{k-1}],\ \tilde{\delta}_2 = [x_{k-1} - x_{k-2}],\ \ldots,\ \tilde{\delta}_{k-1} = [x_2 - x_1],\ \tilde{\delta}_k = [x_1 - x_0].$$

The basis $(\tilde{\delta}_1, \ldots, \tilde{\delta}_k)$ has the Coxeter-Dynkin diagram represented in Figure 5.35. This is a classical Coxeter-Dynkin diagram of type A_k (cf. §3.11).

5.8. Monodromy group and vanishing lattice

The Coxeter-Dynkin diagram depends a priori on the choice of morsification and on the choice of the strongly distinguished path system. We have discussed the dependence on the choice of the strongly distinguished path system in §5.7. We now want to examine the dependence on the choice of the morsification.

Let $f : (\mathbb{C}^{n+1}, 0) \to (\mathbb{C}, 0)$ be a holomorphic function germ with an isolated singularity at 0, $\operatorname{grad} f(0) = 0$. By Proposition 3.16 one obtains a universal unfolding F of f as follows: Let $g_0 = -1, g_1, \ldots, g_{\mu-1}$ be representatives of a basis of the $\mathbb{C}$-vector space

$$\mathcal{O}_{n+1} \Big/ \left(\frac{\partial f}{\partial z_1}, \ldots, \frac{\partial f}{\partial z_n} \right) \mathcal{O}_{n+1},$$

which has dimension μ; cf. Proposition 3.19. Then put

$$\begin{array}{rccc} F: & (\mathbb{C}^{n+1} \times \mathbb{C}^{\mu}, 0) & \longrightarrow & (\mathbb{C}, 0) \\ & (z, u) & \longmapsto & f(z) + \sum_{j=0}^{\mu-1} g_j(z) u_j. \end{array}$$

Let

$$F : M \times U \to \mathbb{C}$$

be a representative of the unfolding F, where M is an open neighborhood of 0 in $\mathbb{C}^{n+1}$ and U is an open neighborhood of 0 in $\mathbb{C}^{\mu}$. As in §3.8 we put

$$\begin{aligned} \mathcal{Y} &:= \{(z, u) \in M \times U \mid F(z, u) = 0\}, \\ \mathcal{Y}_u &:= \{z \in M \mid F(z, u) = 0\}. \end{aligned}$$

Since $F(z, 0) = f(z)$, by Lemma 3.5, there is an $\varepsilon > 0$ such that every sphere $S_\rho \subset M$ around 0 of radius $\rho \le \varepsilon$ intersects the set $\mathcal{Y}_0$ transversally. Let $\varepsilon > 0$ be so chosen. Then there is also an $\eta > 0$ such that for $|u| \le \eta$ the set $\{u \in \mathbb{C}^{\mu} \mid |u| \le \eta\}$ lies entirely in U and $\mathcal{Y}_u$ intersects the sphere S_ε transversally. Let η be so chosen. We put

$$\begin{aligned} \mathcal{X} &:= \{(z, u) \in \mathcal{Y} \mid |z| < \varepsilon, |u| < \eta\}, \\ \bar{\mathcal{X}} &:= \{(z, u) \in \mathcal{Y} \mid |z| \le \varepsilon, |u| < \eta\}, \\ \partial\bar{\mathcal{X}} &:= \{(z, u) \in \mathcal{Y} \mid |z| = \varepsilon, |u| < \eta\}, \\ S &:= \{u \in U \mid |u| < \eta\}, \end{aligned}$$

$$\begin{array}{rccc} p: & \bar{\mathcal{X}} & \longrightarrow & S \\ & (z, u) & \longmapsto & u. \end{array}$$

Let C be the set the critical points and $D = p(C) \subset S$ the discriminant of p. By the Ehresmann fibration theorem the map

$$p' := p|_{\bar{\mathcal{X}} - p^{-1}(D)} : \bar{\mathcal{X}} - p^{-1}(D) \longrightarrow S - D$$

is then the projection of a differentiable fiber bundle, and the fibers of p' are diffeomorphic to a Milnor fiber $\bar{X}_w$ of f.

Let $s \in S - D$ and $\bar{X}_s := p^{-1}(s) = (p')^{-1}(s)$. By §4.4, p' defines a representation

$$\rho : \pi_1(S - D, s) \longrightarrow \operatorname{Aut}(H_n(\bar{X}_s)).$$

Our aim is to prove the following proposition.

Proposition 5.17. *The image Γ of the homomorphism*

$$\rho : \pi_1(S - D, s) \longrightarrow \operatorname{Aut}(H_n(\bar{X}_s))$$

coincides with the monodromy group of the singularity.

Remark 5.5. Proposition 5.17 offers another definition of the monodromy group, in which there is no mention of a morsification. This shows the independence of the monodromy group of the choice of morsification. It is also easy to see that Γ is independent of the choice of the representative of the universal unfolding.

In order to prove Proposition 5.17, we wish to examine $\pi_1(S - D, s)$ more closely.

Lemma 5.7 (Zariski). *Let $\Delta \subset \mathbb{C}$ be a disc of radius η_0 around 0 and $T \subset \mathbb{C}^{\mu-1}$ an open and contractible neighborhood of $0 \in \mathbb{C}^{\mu-1}$. We put $S := \bar{\Delta} \times T$. Let D be an analytic hypersurface in $\Delta \times T$, which is closed in $\bar{\Delta} \times T$, and let $\pi : \bar{\Delta} \times T \to T$ be the projection. Then $\pi|_D : D \to T$ is a finite branched covering. Let $B \subset T$ be the image of the branch locus of $\pi|_D$. Furthermore, let $t \in T - B$, $S_t = \pi^{-1}(t)$, $D_t = D \cap S_t$, $s \in S_t - D_t$.*

Then the inclusion $i : S_t - D_t \to S - D$ induces a surjection $i_* : \pi_1(S_t - D_t, s) \to \pi_1(S - D, s)$ of the fundamental groups.

Proof. Let $(S - D)_{T-B} = S - D \cap \pi^{-1}(T - B)$. Since $\pi|_D : D \to T$ is a finite branched covering,

$$\pi : (S - D)_{T-B} \longrightarrow T - B$$

is, by the Ehresmann fibration theorem, the projection of a differentiable fiber bundle. Let $t \in T - B$ and $s = (\eta_0, t) \in \partial\bar{\Delta} \times T$. Then this fiber bundle has a section $\sigma : T - B \to (S - D)_{T-B}$ defined by $\sigma(t) = (\eta_0, t)$. By Proposition 4.20 we now have a split short exact sequence

$$0 \longrightarrow \pi_1(S_t - D_t, s) \xrightarrow{i_*} \pi_1((S - D)_{T-B}, s) \underset{\sigma_*}{\overset{\pi_*}{\rightleftarrows}} \pi_1(T - B, t) \longrightarrow 0$$

and an isomorphism

$$\begin{array}{rccc} (i_*, \sigma_*): & \pi_1(S_t - D_t, s) \times \pi_1(T - B, t) & \longrightarrow & \pi_1((S - D)_{T-B}, s) \\ & (\alpha, \beta) & \longmapsto & \alpha\beta. \end{array}$$

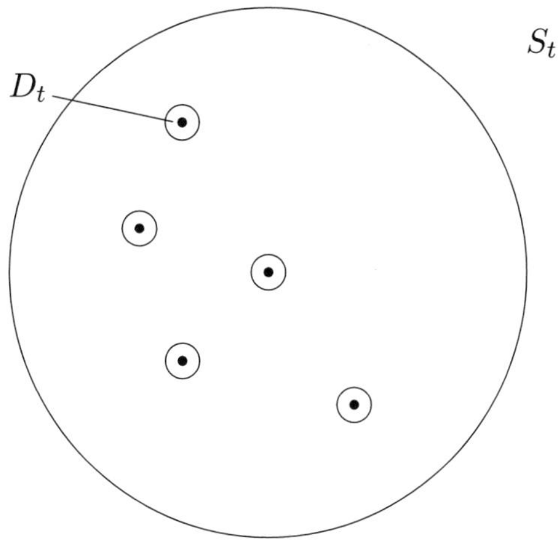

Figure 5.36. Small discs around the points of D_t in S_t

(The hypothesis that the fibers of the fiber bundle must be compact is not satisfied at first. However it can be satisfied by excising small open discs around the points of D_t from the fibers (see Figure 5.36). We pass over this small technical point.)

Now let ω be a closed path in $S - D$ with start and end point s, so that $[\omega] \in \pi_1(S - D, s)$. Since the real codimension of B in T is greater than or equal to 2, one can, by a small homotopic deformation of ω, arrange that $\pi \circ \omega$ is disjoint from B. We assume this done. Then ω lies in $(S - D)_{T-B}$. By the above isomorphism ω is then homotopic to $\omega_1 * \omega_2$ in $(S - D)_{T-B}$, where ω_1 is a closed path in $S_t - D_t$ and ω_2 is a closed path in $\{\eta_0\} \times T$. Since $\{\eta_0\} \times T$ is contractible, it follows that ω is homotopic to ω_1 in $S - D$. Thus $i_*[\omega_1] = [\omega]$, as was to be shown. □

Proof of Proposition 5.17. As in §3.8 let $T := \{t = (u_1, \ldots, u_{\mu-1}) \in \mathbb{C}^{\mu-1} \mid |t| < \eta_1\}$ be an open ball of radius $\eta_1 > 0$ around $0 \in \mathbb{C}^{\mu-1}$ and let $\Delta \subset \mathbb{C}$ be a disc of radius $\eta_0 > 0$ around 0 such that

$$\bar{\Delta} \times T = \{(u_0, u_1, \ldots, u_{\mu-1}) \in \mathbb{C}^\mu \mid u_0 \in \bar{\Delta}, (u_1, \ldots, u_{\mu-1}) \in T\} \subset S$$

(see Figure 3.8). We replace S by $\bar{\Delta} \times T$ and $\bar{\mathcal{X}}$ by $\bar{\mathcal{X}} \cap p^{-1}(\bar{\Delta} \times T)$, but we retain the same notation. For $t \in T$ we put

$$\begin{aligned} S_t &:= \bar{\Delta} \times \{t\} \\ &= \{(u_0, u_1, \ldots, u_{\mu-1}) \in \bar{\Delta} \times T \mid (u_1, \ldots, u_{\mu-1}) = t\}. \end{aligned}$$

We consider a suitable representative $\tilde{F} : \bar{\mathcal{X}} \to S = \bar{\Delta} \times T$ of

$$\begin{aligned} \tilde{F}: \quad (\mathbb{C}^{n+1} \times \mathbb{C}^\mu, 0) &\longrightarrow (\mathbb{C} \times \mathbb{C}^{\mu-1}, 0) \\ (z, u) &\longmapsto (F(z, u), u_1, \ldots, u_{\mu-1}) \end{aligned}$$

where F is of the form

$$F(z,u) = f(z) - u_0 + \sum_{j=1}^{\mu-1} g_j(z)u_j\,.$$

Let $t \in T$, $t \neq 0$. We define

$$f_{\lambda t} : M \longrightarrow \Delta \times \{\lambda t\}$$

by $f_{\lambda t}(z) = \tilde{F}(z, 0, \lambda t)$.

By the universal property of the universal unfolding F, and by Lemma 3.9, every morsification of f can be realized as an $f_{\lambda t}$ for some $\lambda t \neq 0$. Let λt be chosen so that $f_{\lambda t}$ is a Morse function. By Remark 3.18 then $\lambda t \notin B$, where B is the image of the branch locus of $\pi|_D : D \to T$ as in Lemma 5.7. Let $D_{\lambda t} = D \cap S_{\lambda t}$, $s \in S_{\lambda t} - D_{\lambda t}$. Let us choose a weakly distinguished path system $(\gamma_1, \ldots, \gamma_\mu)$ in $S_{\lambda t}$ such that the corresponding simple loops $\omega_1, \ldots, \omega_\mu$ freely generate the fundamental group $\pi_1(S_{\lambda t} - D_{\lambda t}, s)$. By Lemma 5.7, the $\omega_1, \ldots, \omega_\mu$ generate the fundamental group $\pi_1(S - D, s)$. It follows from this that the image of the representation

$$\rho : \pi_1(S - D, s) \longrightarrow \operatorname{Aut}(H_n(\bar{X}_s))$$

coincides with the monodromy group of the singularity f. □

We now want to draw further consequences from Proposition 3.21 and in particular to exploit the irreducibility of D.

Let $p : \bar{\mathcal{X}} \to S = \bar{\Delta} \times T$ be as above, and let D be the discriminant of p. We denote by $S(D)$ the set of singular points of D, and we put $D_{\mathrm{reg}} := D - S(D)$.

Lemma 5.8. *D_{reg} is path connected.*

Proof. For $\mu = 1$ the result is trivial. So assume $\mu > 1$. By Proposition 3.21(iii), D is an irreducible hypersurface in S. By Proposition 2.36, $S(D)$ is thus a proper analytic subset of D. Let

$$A := p^{-1}(S(D)) \cap C.$$

Then A is a proper analytic subset of C. By Proposition 3.21(ii), C is a nonsingular analytic subset of $\mathcal{X}$ of dimension $\mu - 1$. Let $a, b \in C - A$. Then a, b can be connected by a path $\gamma : [0,1] \to C$ with $\gamma(0) = a$ and $\gamma(1) = b$. By Proposition 2.24, A has local codimension ≥ 1 in C. Hence one can assume that $\gamma(I)$ intersects the set A in only finitely many points. Let $p \in A$ be such a point of intersection, and let $\varphi : U \to U' \subset \mathbb{C}^{\mu-1}$ be a chart of C around p, where U is an open connected neighborhood of p in C. Then $\varphi(U \cap A)$ is a proper analytic subset of $U' \subset \mathbb{C}^{\mu-1}$. Now U' is a region in $\mathbb{C}^{\mu-1}$. By Proposition 2.23, $U' - \varphi(U \cap A)$ is connected, and so also is $U - U \cap A$. Hence one can replace the path γ in U by a path

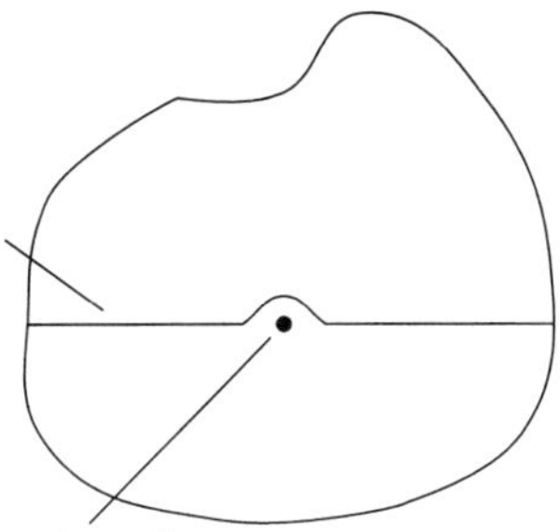

Figure 5.37. New path γ

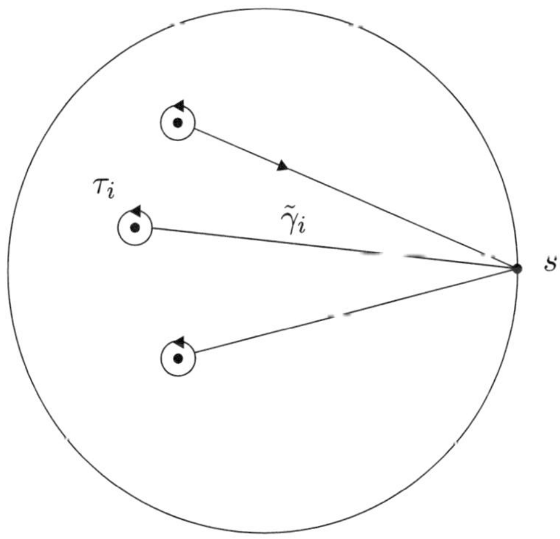

Figure 5.38. Definition of $\tilde{\gamma}_i$ and τ_i

that avoids A (see Figure 5.37). Hence $C - A$ is path connected. Therefore $p(C - A) = D - S(D)$ is path connected too. □

Now let $\pi : \bar{\Delta} \times T \to T$ be the projection and $B \subset T$ the image of the branch locus of $\pi|_D : D \to T$. Furthermore, let $t \in T - B$, $S_t := \pi^{-1}(t)$, $D_t := D \cap S_t =: \{s_1, \dots, s_\mu\}$, $s \in S_t - D_t$. Let $(\gamma_1, \dots, \gamma_\mu)$ be a weakly distinguished path system in S_t joining the points $s_1, \dots, s_\mu$ to s such that the corresponding simple loops $\omega_1, \dots, \omega_\mu$ freely generate the fundamental group $\pi_1(S_t - D_t, s)$. We define $\tilde{\gamma}_i$, τ_i again as in §5.2 (see Figure 5.38). Finally let $i : S_t - D_t \to S - D$ be the inclusion and $i_* : \pi_1(S_t - D_t, s) \to \pi_1(S - D, s)$ the induced map of the fundamental groups.

Lemma 5.9. *The homotopy classes $i_*[\omega_1], \dots, i_*[\omega_\mu]$ are conjugate to one another in $\pi_1(S - D, s)$.*

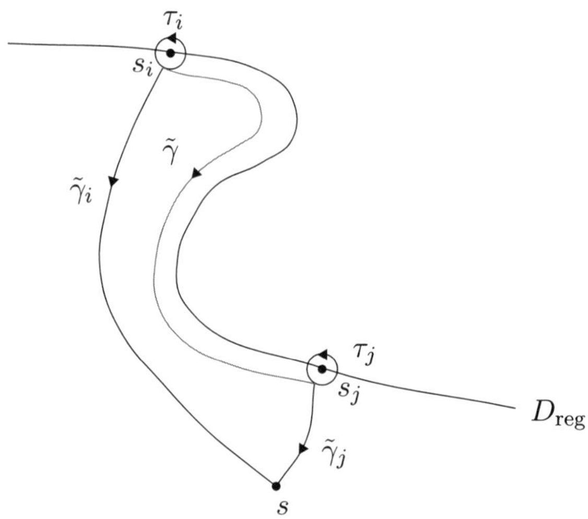

Figure 5.39. The path $\tilde{\gamma}$

Proof. By Lemma 5.8, D_{reg} is path connected. Let $1 \leq i, j \leq \mu$, $i \neq j$. Then the points s_i and s_j can be connected by a path $\gamma : [0,1] \to D_{\text{reg}}$ with $\gamma(0) = s_i$, $\gamma(1) = s_j$. Since D has codimension 1 in S, one can deform γ to a path $\tilde{\gamma}$ in $S - D$ so that $\tilde{\gamma}$ connects the start points of $\tilde{\gamma}_i$ and $\tilde{\gamma}_j$; see Figure 5.39.

We consider the path

$$\tilde{\gamma}_i^{-1}\tilde{\gamma}\tau_j\tilde{\gamma}^{-1}\tilde{\gamma}_i.$$

This path is homotopic to ω_i. Since $\omega_j = \tilde{\gamma}_j^{-1}\tau_j\tilde{\gamma}_j$, we have

$$\begin{aligned} i_*[\omega_i] &= [\tilde{\gamma}_i^{-1}\tilde{\gamma}\tilde{\gamma}_j\tilde{\gamma}_j^{-1}\tau_j\tilde{\gamma}_j\tilde{\gamma}_j^{-1}\tilde{\gamma}^{-1}\tilde{\gamma}_i] \\ &= [\tilde{\gamma}_i^{-1}\tilde{\gamma}\tilde{\gamma}_j]i_*[\omega_j][\tilde{\gamma}_i^{-1}\tilde{\gamma}\tilde{\gamma}_j]^{-1}. \end{aligned}$$

Thus $i_*[\omega_j]$ is conjugate to $i_*[\omega_i]$. □

Proposition 5.18. *Let $h_1, \ldots, h_\mu$ be the Picard-Lefschetz transformations of the weakly distinguished path system $(\gamma_1, \ldots, \gamma_\mu)$. Then $h_1, \ldots, h_\mu$ are conjugate to one another in the monodromy group Γ.*

Proof. This follows immediately from Lemma 5.9 and Proposition 5.17. □

Corollary 5.8 (Gabrielov, Lazzeri). *The Coxeter-Dynkin diagram with respect to a weakly distinguished system $(\delta_1, \ldots, \delta_\mu)$ of vanishing cycles is a connected graph.*

Proof. If the Coxeter-Dynkin diagram were not connected, we could divide the index set $\{1, \ldots, \mu\}$ into two nonempty disjoint subsets I, J such that $\langle \delta_i, \delta_j \rangle = 0$ for all $i \in I$, $j \in J$. But then $h_i h_j = h_j h_i$ for all $i \in I$, $j \in J$, i.e., h_i and h_j commute with one another. Thus Γ is isomorphic to $\Gamma_I \times \Gamma_J$, where Γ_I is the subgroup of Γ generated by the h_i, $i \in I$, and Γ_J is that generated by the h_j, $j \in J$. But then no h_i, $i \in I$, can be conjugate to any h_j, $j \in J$. □

We shall now employ the following proposition, which we will be able to prove only in the next section.

Proposition 5.19. *If* 0 *is neither a regular nor a nondegenerate critical point of* f*, then there are two vanishing cycles* δ, δ' *of* f *with* $\langle \delta, \delta' \rangle = 1$.

Definition. Let Δ be the set of the vanishing cycles of the singularity f. The orbit of an element $\delta \in \Delta$ under Γ will be denoted by $\Gamma \cdot \{\delta\}$.

Proposition 5.20. *If not both (i)* n *is odd and (ii)* 0 *is a nondegenerate critical point of* f*, then* Δ *is the only* Γ*-orbit, i.e., the monodromy group* Γ *acts transitively on* Δ.

Proof. Let $\delta_1, \delta_2 \in \Delta$. We have to show that under the hypotheses of Proposition 5.20 there is an $h \in \Gamma$ with $h(\delta_1) = \delta_2$.

We first employ the same construction as in the proof of Lemma 5.9. If δ_1 vanishes along a path γ_1 and δ_2 along a path γ_2 and if we put

$$\omega = \tilde{\gamma}_1^{-1} \tilde{\gamma} \tilde{\gamma}_2^{-1},$$

with $\tilde{\gamma}_1, \tilde{\gamma}, \tilde{\gamma}_2$ as in the proof of Lemma 5.9, then $h_{\omega *}^{-1}(\delta_1) = \pm \delta_2$.

If n is even, then $h_{\omega_2 *}(\delta_2) = -\delta_2$. It follows that in this case $\delta_2 \subset \Gamma \cdot \{\delta_1\}$.

If n is odd and 0 is neither a regular point (in which case $\Delta = \emptyset$ and the result is trivial) nor a nondegenerate critical point of f, then, by Proposition 5.19, there is a $\delta_3' \in \Delta$ with $\langle \delta_2, \delta_3' \rangle = 1$. For $\delta_3 := (-1)^{n(n-1)/2} \delta_3'$ we then have $\langle \delta_2, \delta_3 \rangle = (-1)^{n(n-1)/2}$ and

$$\begin{aligned} h_{\delta_3} h_{\delta_2} h_{\delta_2} h_{\delta_3}(\delta_2) &= h_{\delta_3} h_{\delta_2} h_{\delta_2}(\delta_2 - \delta_3) \\ &= h_{\delta_3} h_{\delta_2}(-\delta_3) \\ &= h_{\delta_3}(-\delta_3 - \delta_2) \\ &= -\delta_2. \end{aligned}$$

It follows that $\delta_2 \in \Gamma \cdot \{\delta_1\}$ in this case too. □

From Proposition 5.20 it follows that the pair $(\tilde{H}_n(\bar{X}_s), \Delta)$ is a vanishing lattice. We shall now define this concept.

Definition. Let $(L, \langle \, , \, \rangle)$ be a lattice. An *automorphism* of the lattice L is an isomorphism $h : L \to L$ that respects the bilinear form $\langle \, , \, \rangle$, i.e., for which $\langle h(v), h(w) \rangle = \langle v, w \rangle$ for all $v, w \in L$. The group of all automorphisms of the lattice L will be denoted by $\mathrm{Aut}(L)$.

Now let L be symmetric and even, or skew symmetric, $\varepsilon \in \{+1,-1\}$, $\delta \in L$. If L is symmetric, let $\langle\delta,\delta\rangle = 2\varepsilon$. We define an automorphism $s_\delta \in \mathrm{Aut}(L)$ by

$$s_\delta(v) := v - \varepsilon\langle v,\delta\rangle\delta$$

for all $v \in L$. Then s_δ is a reflection in the symmetric case and a symplectic transvection in the skew symmetric case.

Definition. Let $(L,\langle\,,\,\rangle)$ be an even symmetric or a skew symmetric lattice, let Λ be a subset of L, and let $\varepsilon \in \{\pm 1\}$. If $(L,\langle\,,\,\rangle)$ is symmetric, let $\langle\delta,\delta\rangle = 2\varepsilon$ for all $\delta \in \Lambda$. Let $\Gamma_\Lambda \subset \mathrm{Aut}(L)$ be the subgroup of $\mathrm{Aut}(L)$ generated by the transformations s_δ, $\delta \in \Lambda$. Suppose the pair (L,Λ) has the following properties:

(i) Λ generates L.

(ii) Λ is an orbit of Γ_Λ in L.

(iii) If the rank of L is greater than 1, then there are $\delta,\delta' \in \Lambda$ with $\langle\delta,\delta'\rangle = 1$.

Then the pair (L,Λ) is called a *vanishing lattice* and Γ_Λ the corresponding *monodromy group*.

Corollary 5.9. *Suppose it is not true that both n is odd and 0 is a nondegenerate critical point of f. Then the pair $(\tilde{H}_n(\bar{X}_s),\Delta)$ is a vanishing lattice with $\varepsilon = (-1)^{n(n-1)/2}$, and Γ is the corresponding monodromy group.*

Proof. This follows from Proposition 5.20 and Proposition 5.6. □

Example 5.5. We consider the case $n = 0$, so $f : (\mathbb{C},0) \to (\mathbb{C},0)$. Then the Milnor fiber $\bar{X}_s$ consists of $\mu+1$ points. The geometric monodromy group Γ_{geo} of the singularity f is a subgroup of the permutation group $\mathrm{Sym}(\mu+1)$ of these $\mu+1$ points.

Let f_λ be a morsification of f. Then f_λ has the form

$$z \longmapsto z^2$$

in a neighborhood of a nondegenerate critical point p_i (see Figure 5.40). If ω_i is a loop around the corresponding critical value s_i, then h_{ω_i} is the transposition of the two points of $\bar{X}_s$ that merge at s_i. Let $(\gamma_1,\ldots,\gamma_\mu)$ be a weakly distinguished path system for f_λ. By Proposition 5.18 we have

a) Γ_{geo} is generated by transpositions $h_1,\ldots,h_\mu$ that are all conjugate to one another in Γ.

Moreover

b) for each $x \in \bar{X}_s$ there is an i with $h_i(x) = h_{\omega_i}(x) \neq x$.

Proof of b). Suppose that there is an $x_0 \in \bar{X}_s$ with $h_i(x_0) = x_0$ for all $i = 1,\ldots,\mu$. Let $\bar{X}_s = \{x_0,x_1,\ldots,x_\mu\}$. Along each path γ_i a cycle $x_{j_i} - x_{k_i}$,

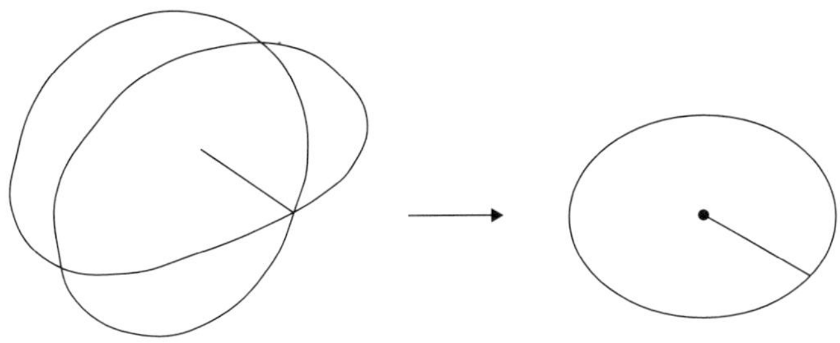

Figure 5.40. The map $z \mapsto z^2$

$j_i, k_i \in \{0, 1, \ldots, \mu\}$, vanishes. If $j_i = 0$ or $k_i = 0$, then $h_i(x_0) \neq x_0$. Hence $x_{j_i}, x_{k_i} \in \{x_1, \ldots, x_\mu\}$ for all $i = 1, \ldots, \mu$. Thus $[x_0 - x_1] \in \tilde{H}_0(\bar{X}_s)$ is not a linear combination of the elements $\delta_i := [x_{j_i} - x_{k_i}]$, $i = 1, \ldots, \mu$, of $\tilde{H}_0(\bar{X}_s)$. This is a contradiction to Proposition 5.6 which asserts that $(\delta_1, \ldots, \delta_\mu)$ forms a basis of $\tilde{H}_0(\bar{X}_s)$. □

From the properties a) and b) it now follows that $\Gamma = \mathrm{Sym}(\mu + 1)$ (Exercise in group theory).

The vanishing lattice (L, Δ) can in this case be described as follows: Let $e_1, \ldots, e_{\mu+1}$ be the canonical basis of $\mathbb{R}^{\mu+1}$ and $\langle\, , \rangle$ the Euclidean scalar product on $\mathbb{R}^{\mu+1}$. Then

$$\begin{aligned} L &= \tilde{H}_0(\bar{X}_s) = \{v = (v_1, \ldots, v_{\mu+1}) \in \mathbb{Z}^{\mu+1} \mid v_1 + \ldots + v_{\mu+1} = 0\}, \\ \Delta &= \{e_i - e_j \mid 1 \leq i, j \leq \mu + 1, i \neq j\} \\ &= \{v \in L \mid \langle v, v\rangle = 2\}, \\ \Gamma &= \Gamma_\Delta = \mathrm{Sym}(\mu + 1). \end{aligned}$$

This vanishing lattice is called a *root lattice of type* A_μ.

5.9. Deformation

We shall now append the proof of Proposition 5.19. We need a few preparations, which are of interest in themselves.

We retain the notation of §5.8 and consider again the map $p : \bar{\mathcal{X}} \to S = \bar{\Delta} \times T$ with discriminant D. Let $s_1 \in D$ and let $x_1, \ldots, x_k$ be the points of $C_{s_1} := p^{-1}(s_1) \cap C$. For each $i = 1, \ldots, k$ we choose a closed ball B_i in $\mathcal{X}$ around x_i and we further choose an open and connected neighborhood U of s_1 in S such that both the following conditions are satisfied:

(i) $B_i \cap B_j = \emptyset$ for all $i \neq j$.

(ii) For all $s \in U$ the fiber $p^{-1}(s)$ intersects the boundaries of the balls B_i transversally.

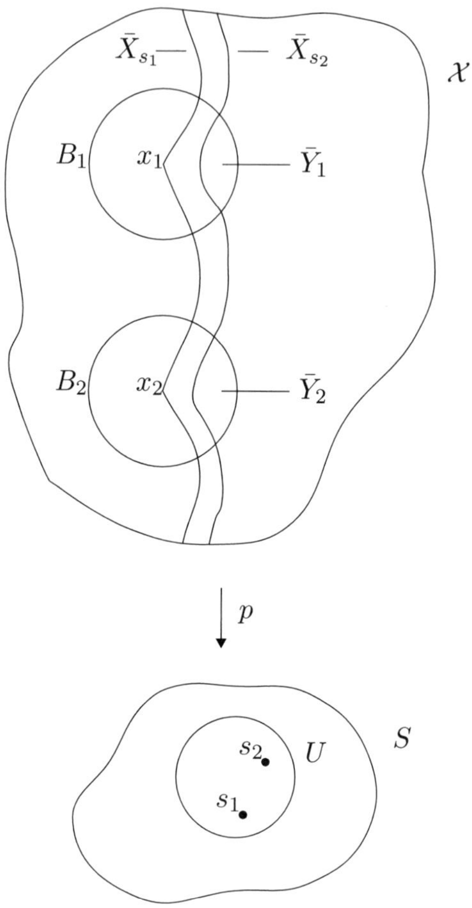

Figure 5.41. The local Milnor fibers $\bar{Y}_i$

Let $s_2 \in U - (U \cap D)$, $\bar{Y}_i = X_{s_2} \cap B_i$, $\mathcal{X}_U = \mathcal{X} \cap p^{-1}(U)$. Then

$$p|_{\mathcal{X}_U \cap B_i} : \mathcal{X}_U \cap B_i \longrightarrow U$$

is a representative of an unfolding of a singularity $f_i : (\mathbb{C}^{n+1}, x_i) \to (\mathbb{C}, s_1)$, and $\bar{Y}_i$ is a Milnor fiber of this singularity (see Figure 5.41).

Definition. Let M be a free $\mathbb{Z}$-module. A submodule $N \subset M$ is called *primitive* if M/N is torsion free.

Proposition 5.21. *The inclusion* $i : \bigcup_{j=1}^{k} \bar{Y}_j \to \bar{X}_{s_2}$ *induces an injective map of free* $\mathbb{Z}$*-modules*

$$i_* : \bigoplus_{j=1}^{k} \tilde{H}_n(\bar{Y}_j) \longrightarrow \tilde{H}_n(\bar{X}_{s_2}).$$

The image of i_* is a primitive submodule of $\tilde{H}_n(\bar{X}_{s_2})$.

The map i_* preserves the intersection form and maps vanishing cycles to vanishing cycles.

If

$$\delta_1', \ldots, \delta_m' \in \bigcup_{i=1}^{k} \tilde{H}_n(\bar{Y}_i)$$

is a strongly distinguished system of vanishing cycles for the singularities $f_1, \ldots, f_k$, one can extend $\delta_1 := i_*(\delta_1'), \ldots, \delta_m := i_*(\delta_m')$ to a strongly distinguished system of vanishing cycles $\delta_1, \ldots, \delta_m, \delta_{m+1}, \ldots, \delta_\mu$ of the singularity f.

Proof. Let

$$\begin{aligned} Z &:= \bar{\mathcal{X}}_u - \bigcup_{j=1}^{k}(B_j \cap \bar{\mathcal{X}}_u), \\ Z_{s_1} &:= \bar{X}_{s_1} - \bigcup_{j-1}^{k}(B_j \cap X_{s_1}). \end{aligned}$$

Then $(Z, p|_Z, U, Z_{s_1})$ is a trivial differentiable fiber bundle (by Proposition 4.11). Hence for each $q \subset \mathbb{Z}$ there is an isomorphism

$$\begin{aligned} \tilde{H}_q(\bar{X}_{s_2}, \bigcup_j \bar{Y}_j) &\cong \tilde{H}_q(\bar{\mathcal{X}}_U, \bigcup_j (\bar{\mathcal{X}}_U \cap B_j)) \\ &\cong \tilde{H}_q(\bar{X}_{s_1}, \bigcup_j (\bar{X}_{s_1} \cap B_j)). \end{aligned}$$

Since $\bar{X}_{s_1} \cap B_j$ is contractible for each $j = 1, \ldots, k$, by Proposition 5.4,

$$\tilde{H}_q(\bar{X}_{s_1}, \bigcup_j (\bar{X}_{s_1} \cap B_j)) \cong \tilde{H}_q(\bar{X}_{s_1}, C_{s_1}).$$

Since $\bar{X}_{s_2}$ has the homotopy type of a bouquet of n-spheres (Proposition 5.2) and since $\bar{X}_{s_1}$ arises from $\bar{X}_{s_2}$ by contraction of one of these spheres, $\bar{X}_{s_1}$ also has the homotopy type of a bouquet of n-spheres. From this follows

$$\tilde{H}_{n+1}(\bar{X}_{s_1}, C_{s_1}) = 0.$$

Thus we obtain for the exact reduced homology sequence of the pair $(\bar{X}_{s_1}, C_{s_1})$:

$$\begin{array}{ccccc}
\tilde{H}_{n+1}(\bar{X}_{s_1}, C_{s_1}) & \longrightarrow & \tilde{H}_n(C_{s_1}) & \longrightarrow & \tilde{H}_n(\bar{X}_{s_1}) \\
\| & \searrow & & \nearrow & \\
0 & & 0 & &
\end{array}$$

$$\longrightarrow \tilde{H}_n(\bar{X}_{s_1}, C_{s_1}) \xrightarrow{\partial_*} \tilde{H}_{n-1}(C_{s_1}) \longrightarrow \tilde{H}_{n-1}(\bar{X}_{s_1}) = 0$$

Since $\bar{Y}_i$ is actually homotopy equivalent to a bouquet of n-spheres, it follows that

$$\tilde{H}_{n-1}(\bigcup_i \bar{Y}_i) \cong \tilde{H}_{n-1}(C_{s_1}).$$

Hence the exact reduced homology sequence of the pair $(\bar{X}_{s_2}, \bigcup_i \bar{Y}_i)$ is

$$\begin{array}{ccccc}
\tilde{H}_{n+1}(\bar{X}_{s_2}, \bigcup_i \bar{Y}_i) & \longrightarrow & \tilde{H}_n(\bigcup_i \bar{Y}_i) & \xrightarrow{i_*} & \tilde{H}_n(\bar{X}_{s_2}) \\
\| & & \| & & \\
0 & & \bigoplus_i \tilde{H}_n(\bar{Y}_i) & &
\end{array}$$

$$\begin{array}{cccc}
\longrightarrow & \tilde{H}_n(\bar{X}_{s_2}, \bigcup_i \bar{Y}_i) & \xrightarrow{\partial_*} & \tilde{H}_{n-1}(\bigcup_i \bar{Y}_i) \\
& \| & & \| \\
& \tilde{H}_n(\bar{X}_{s_1}, C_{s_1}) & & \tilde{H}_{n-1}(C_{s_1})
\end{array}$$

Thus we obtain the following short exact sequence:

$$0 \longrightarrow \bigoplus_{i=1}^{k} \tilde{H}_n(\bar{Y}_i) \xrightarrow{i_*} \tilde{H}_n(\bar{X}_{s_2}) \longrightarrow \tilde{H}_n(\bar{X}_{s_1}) \longrightarrow 0.$$

Here every term is a free $\mathbb{Z}$-module since all the spaces are homotopy equivalent to bouquets of n-spheres. From this sequence it follows that i_* is injective. Since

$$\tilde{H}_n(\bar{X}_{s_2}) / \operatorname{Im} i_* \cong \tilde{H}_n(\bar{X}_{s_1}),$$

$\operatorname{Im} i_*$ is a primitive submodule of $\tilde{H}(\bar{X}_{s_2})$.

It is clear that i_* respects the intersection form and maps vanishing cycles to vanishing cycles.

We must finally prove the last assertion of the proposition. Let $S = \bar{\Delta} \times T$ and let $\pi : \bar{\Delta} \to T$ be the projection. Furthermore, let $t_1 = \pi(s_1)$. Then the line $\mathbb{C} \times \{t_1\}$ intersects the discriminant D in μ points (counting

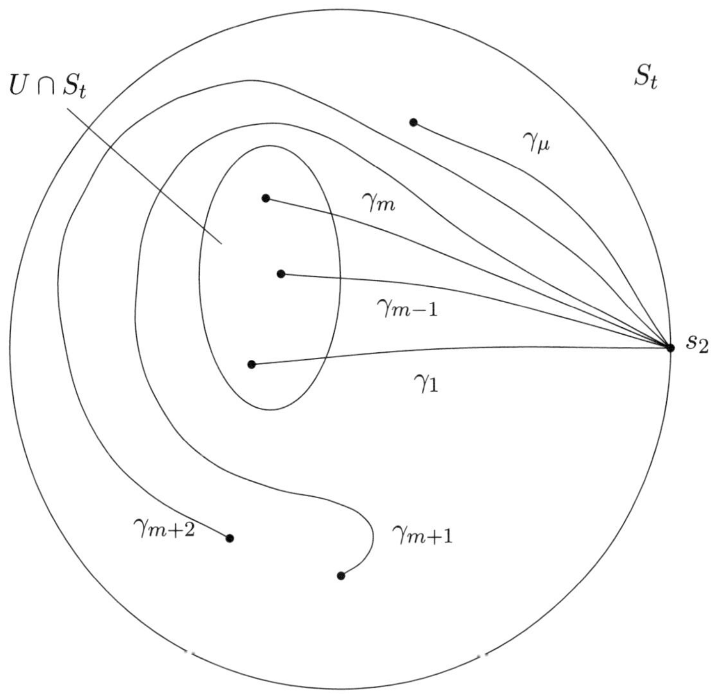

Figure 5.42. Extension of the strongly distinguished path system $(\gamma_1, \dots, \gamma_m)$ to $(\gamma_1, \dots, \gamma_\mu)$

multiplicity) that lie in $S_{t_1} = \bar{\Delta} \times \{t_1\}$. The point of intersection $s_1 \in D \cap S_{t_1}$ has multiplicity m. For general t near t_1 the line $\mathbb{C} \times \{t\}$ intersects the discriminant D in μ regular points that lie in S_t. Of these, m points lie in $U \cap S_t$. We assume that $s_2 \in \partial S_t$. We choose a strongly distinguished path system $(\gamma_1, \dots, \gamma_m)$ from the m points of $U \cap D \cap S_t$ to s such that the given cycles $\delta_1 = i_*(\delta_1'), \dots, \delta_m = i_*(\delta_m') \in \tilde{H}_n(\bar{X}_{s_2})$ vanish along $\gamma_1, \dots, \gamma_m$. We extend this path system to a strongly distinguished path system $(\gamma_1, \dots, \gamma_\mu)$ of the μ points of $D_t = S_t \cap D$ to s_2 (see Figure 5.42). The corresponding strongly distinguished system of vanishing cycles $(\delta_1, \dots, \delta_\mu)$ of $\tilde{H}_n(\bar{X}_{s_2})$ is then the extension sought. □

We now want to introduce a partial ordering on the set of all right equivalence classes of holomorphic function germs with isolated singularities. Let $f : (\mathbb{C}^{n+1}, 0) \to (\mathbb{C}, 0)$ be a holomorphic function germ with an isolated singularity at 0, $df(0) = 0$. Let F be a representative of the universal unfolding $F : (\mathbb{C}^{n+1} \times \mathbb{C}^\mu, 0) \to (\mathbb{C}, 0)$ of f as in §3.8.

Now let $g : (\mathbb{C}^{n+1}, 0) \to (\mathbb{C}, 0)$ be another holomorphic function germ with isolated singularity at 0.

Definition. We say that $[f]$ *deforms to* $[g]$, symbolically, $[f] \to [g]$ or $[f] \geq [g]$, if for every $\varepsilon > 0$ there are an $s_1 = (w_1, t_1) \in D$ with $|s_1| < \varepsilon$ and an $x \in C_{s_1}$ such that the germ at x of the holomorphic function F_{s_1} given by $F_{s_1}(z) = F(z, s_1)$ is right equivalent to $g : (\mathbb{C}^{n+1}, 0) \to (\mathbb{C}, 0)$.

Definition. Let (M, Λ), (L, Δ) be vanishing lattices. A *primitive embedding* of (M, Λ) into (L, Δ) is an injective $\mathbb{Z}$-module homomorphism $j : M \to L$ that respects the bilinear forms and maps Λ to Δ and for which $j(M)$ is a primitive submodule of L.

It now follows immediately from Proposition 5.21 that

Corollary 5.10. *If the singularity $[f]$ deforms to the singularity $[g]$, there exists a primitive embedding of the vanishing lattice (M, Λ) of g into the vanishing lattice (L, Δ) of f. In particular, $\mu(g) \leq \mu(f)$.*

Remark 5.6. It is clear that right equivalent singularities f_1 and f_2 have the same vanishing lattice.

Proposition 5.22. *Let $f : (\mathbb{C}^{n+1}, 0) \to (\mathbb{C}, 0)$ be a holomorphic function germ with an isolated singularity at 0. Let 0 be neither a regular nor a nondegenerate critical point of f. Then $[f]$ deforms to the singularity $A_2 = [g]$ with $g(z) = z_1^3 + z_2^2 + \ldots + z_{n+1}^2$.*

Proof. By the generalized Morse lemma (Proposition 3.20) f is right equivalent to a holomorphic function germ of the form

$$z \longmapsto f'(z_1, \ldots, z_r) + z_{r+1}^2 + \ldots + z_{n+1}^2,$$

where $r \geq 1$ and $f' \in \mathfrak{m}_r^3$. It is then enough to show that $[f'] \to [g']$ with $g'(z) = z_1^3 + z_2^2 + \ldots + z_r^2$, for a universal unfolding F' of f' determines a universal unfolding F of f by

$$F(z, u) = F'(z, u) + z_{r+1}^2 + \ldots + z_{n+1}^2.$$

Thus we can without loss of generality assume that $f \in \mathfrak{m}_{n+1}^3$.

Next we consider the unfolding $G : (\mathbb{C}^{n+1} \times \mathbb{C}^n, 0) \to (\mathbb{C}, 0)$ of f with

$$G(z, u) = f(z) + u_1 z_2^2 + \ldots + u_n z_{n+1}^2.$$

Let $G_u : (\mathbb{C}^{n+1}, 0) \to (\mathbb{C}, 0)$ be defined by $G_u(z) = G(z, u)$. For $u \neq 0$ arbitrarily small, G_u is right equivalent to $\tilde{g} : (\mathbb{C}^{n+1}, 0) \to (\mathbb{C}, 0)$ with

$$\tilde{g}(z) = h(z_1) + z_2^2 + \ldots + z_{n+1}^2,$$

$h \in \mathfrak{m}_1^3$. Next we consider the unfolding $H : (\mathbb{C} \times \mathbb{C}, 0) \to (\mathbb{C}, 0)$ of h with

$$H(z_1, v) = h(z_1) + v z_1^3.$$

For $v \neq 0$ arbitrarily small, H_v, with $H_v(z_1) = H(z_1, v)$, is of the form

$$H_v(z_1) = w^3$$

for some holomorphic function w of z_1. Thus H_v is right equivalent to $z_1 \to z_1^3$. The proposition follows from this. □

Proof of Proposition 5.19. Let $f : (\mathbb{C}^{n+1}, 0) \to (\mathbb{C}, 0)$ be a holomorphic function germ with an isolated singularity at 0 and let 0 be neither a regular nor a nondegenerate critical point of f. Then, by Proposition 5.22, $[f]$ deforms to the singularity A_2. As a generalization of Example 5.3 one sees easily that a singularity

$$g(z) = z_1^3 + z_2^2 + \ldots + z_{n+1}^2$$

of type A_2 has a strongly distinguished basis of vanishing cycles (δ_1, δ_2) with $\langle \delta_1, \delta_2 \rangle = \pm 1$. The proposition follows immediately from Corollary 5.10. □

5.10. Polar curves and Coxeter-Dynkin diagrams

We now want to present a method for the computation of Coxeter-Dynkin diagrams, due to A. M. Gabrielov, which is applicable to higher dimensional singularities. It allows one to compute Coxeter-Dynkin diagrams for some important singularities. In what follows we are concerned to present an overview of some newer results, so we shall omit some proofs.

Let $f : (\mathbb{C}^{n+1}, 0) \to (\mathbb{C}, 0)$ be a holomorphic function germ with an isolated singularity at 0. Let $\zeta : \mathbb{C}^{n+1} \to \mathbb{C}$ be a linear function which, for example, after a suitable change of coordinates can be chosen as the last coordinate function. We consider the map germ

$$\Phi = (\Phi_1, \Phi_2) = (f, \zeta) : (\mathbb{C}^{n+1}, 0) \longrightarrow (\mathbb{C}^2, 0).$$

We write $\Sigma_\zeta(f)$ for the critical set of this map; more precisely, the germ of this analytic set at 0.

Proposition 5.23. *We have* $\dim_{\mathbb{C}} \Sigma_\zeta(f) = 1$, *i.e.,* $\Sigma_\zeta(f)$ *is a (not necessarily reduced) curve.*

Proof. We consider a representative

$$\Phi : U \longrightarrow \mathbb{C}^2,$$

where $U \subset \mathbb{C}^{n+1}$ is a suitable open neighborhood of 0. Then

$$\begin{aligned} \Sigma_\zeta(f) &= \left\{ z \in U \;\middle|\; \operatorname{rank} \begin{pmatrix} \frac{\partial f}{\partial z_1} & \cdots & \frac{\partial f}{\partial z_{n+1}} \\ \frac{\partial \zeta}{\partial z_1} & \cdots & \frac{\partial \zeta}{\partial z_{n+1}} \end{pmatrix} < 2 \right\} \\ &= \{ z \in U \mid \operatorname{grad} f(z) = \varepsilon \operatorname{grad} \zeta(z) \text{ for an } \varepsilon \in \mathbb{C} \} \\ &= \{ z \in U \mid z \text{ is a critical point of } f - \varepsilon\zeta \text{ for an } \varepsilon \in \mathbb{C} \}. \end{aligned}$$

Since f has an isolated singularity at 0, the function $f - \varepsilon\zeta$ has only finitely many critical points in U for sufficiently small U and ε. The result follows from this. □

Definition. The curve $\Sigma_\zeta(f)$ is called the *polar curve* of the singularity f with respect to the linear function ζ.

Remark 5.7. From the proof of Proposition 5.23 we have another description of the polar curve $\Sigma_\zeta(f)$ of f: $\Sigma_\zeta(f)$ is the set of all points z from a neighborhood U of $0 \in \mathbb{C}^{n+1}$ in which the tangent space of the level surface of f through this point is parallel to the fixed hyperplane $\zeta = 0$: for this holds for a $z \in U$ if and only if the gradients $\operatorname{grad} f(z)$ and $\operatorname{grad} \zeta(z)$ are linearly independent.

Let $\Sigma_\zeta(f) = \bigcup\limits_i \Sigma_i$ be the decomposition of the polar curve into irreducible components. We have already seen that the critical points of the functions $f - \varepsilon\zeta$, $\varepsilon \in \mathbb{C}$, lie on $\Sigma_\zeta(f)$.

Definition. Let μ_i be the sum of the Milnor numbers of the critical points of the function $f - \varepsilon\zeta$ for small $\varepsilon \neq 0$ that lie on Σ_i and tend to 0 as $\varepsilon \to 0$.

We have $\mu(f) = \sum\limits_i \mu_i$.

Lemma 5.10. *The critical points of the function $f|_{\zeta=\varepsilon}$ on the hyperplane $\zeta = \varepsilon$ in $\mathbb{C}^{n+1}$ lie on the polar curve $\Sigma_\zeta(f)$.*

Proof. According to the Lagrange multiplier rule the critical points of $f|_{\zeta=\varepsilon}$ on $\zeta = \varepsilon$ are given by the equations

$$\begin{aligned} \zeta(z) &= \varepsilon, \\ \operatorname{grad} f(z) &= \lambda \operatorname{grad} \zeta(z). \end{aligned}$$

These equations describe the set

$$\Sigma_\zeta(f) \cap \{\zeta = \varepsilon\}.$$

This proves Lemma 5.10. □

Lemma 5.11. *The following assertions are equivalent:*

(i) *$f|_{\zeta=0}$ has an isolated singularity at 0.*

(ii) *$\Sigma_i \not\subset \{\zeta = 0\}$ for all i.*

(iii) *$f|_{\Sigma_i} \not\equiv 0$ for all i.*

Proof. (i)$\Rightarrow$(ii): From (i) it follows that 0 is an isolated point of $\Sigma_i \cap \{\zeta = 0\}$, for all i. Hence $\Sigma_i \not\subset \{\zeta = 0\}$ for all i.

(ii)$\Leftrightarrow$(iii):

$$\begin{aligned} f|_{\Sigma_i} \equiv 0 \quad &\Leftrightarrow \quad \operatorname{grad} f|_{\Sigma_i} = 0 \\ &\Leftrightarrow \quad \operatorname{grad} \zeta|_{\Sigma_i} = 0 \\ &\Leftrightarrow \quad \zeta|_{\Sigma_i} = 0 \\ &\Leftrightarrow \quad \Sigma_i \subset \{\zeta = 0\}. \end{aligned}$$

(ii)$\Rightarrow$(i): Suppose that $f|_{\zeta=0}$ has a nonisolated singularity at 0. Let $C \subset \{\zeta = 0\}$ be the set of singular points of $f|_{\zeta=0}$. Then C is an analytic subset of $\{\zeta = 0\}$ and has an irreducible component C_0 with $\dim C_0 \geq 1$. By Lemma 5.10

$$C_0 \subset C \subset \Sigma_\zeta(f).$$

Since C_0 is irreducible of dimension $\dim C_0 \geq 1$, we must have $C_0 = \Sigma_i$ for some i. For this i we have $\Sigma_i \subset \{\zeta = 0\}$. □

Definition. For i with $\Sigma_i \not\subset \{\zeta = 0\}$ let ν_i be the sum of the Milnor numbers of the critical points of the function $f|_{\zeta=\varepsilon}$, for a small $\varepsilon \neq 0$, that lie on Σ_i and tend to 0 as $\varepsilon \to 0$.

If the function $f|_{\zeta=0}$ has an isolated singularity at 0, then, by Lemma 5.11, $\Sigma_i \not\subset \{\zeta = 0\}$ for all i, and

$$\mu(f|_{\zeta=0}) = \sum_i \nu_i.$$

Suppose $\Sigma_i \not\subset \{\zeta = 0\}$. We consider the curve $\Phi(\Sigma_i)$. We choose coordinates (ζ, λ) of $\mathbb{C}^2$ such that the map $\Phi : U \to \mathbb{C}^2$ is given by $\Phi_2(z) = \zeta$ and $\Phi_1(z) = f(z) = \lambda$ for $z \in U \subset \mathbb{C}^{n+1}$. Since $\Sigma_i \not\subset \{\zeta = 0\}$, the set $\Phi(\Sigma_i)$ does not coincide with the coordinate axis $\zeta = 0$. Hence the open curve branch $\Phi(\Sigma_i)$ has, by Theorem 1.4, a Puiseux expansion

$$\lambda = a_i \zeta^{\alpha_i} + \text{ terms of higher order,}$$
$$a_i \neq 0, a_i \in \mathbb{C}, \alpha_i \in \mathbb{Q}.$$

By the definition of λ one can also interpret this as saying that $f|_{\Sigma_i}$ can be written as a Puiseux series, i.e., a power series in ζ with fractional exponents, beginning with the term $a_i \zeta^{\alpha_i}$.

Lemma 5.12. *If* $\Sigma_i \not\in \{\zeta = 0\}$, *then* $\alpha_i > 1$.

Proof. $\Sigma_\zeta(f)$ coincides, following the proof of Proposition 5.23, with the set of critical points of the function $f - \varepsilon\zeta$ in a neighborhood U of $0 \in \mathbb{C}^{n+1}$. For $\varepsilon \to 0$ the corresponding points must tend to 0. These points are solutions of the equations

$$\begin{aligned} \frac{\partial f}{\partial \zeta} &= a_i \alpha_i \zeta^{\alpha_i - 1} + \text{ terms of higher order} \\ &= \varepsilon. \end{aligned}$$

This proves Lemma 5.12. □

For i with $\Sigma_i \subset \{\zeta = 0\}$ we put $\alpha_i = 1$.

Lemma 5.13. *If* $\alpha_i > 1$, *then* $\mu_i = \nu_i(\alpha_i - 1)$.

Proof. Let Σ_i be an irreducible component of $\Sigma_\zeta(f)$ with $\alpha_i > 1$. Let n_i be the multiplicity of Σ_i. We write the Puiseux expansion of $\Phi(\Sigma_i)$ in the form

$$\zeta = t^{n_i k_i}, \ \lambda = a_i t^{n_i m_i} + \text{ terms of higher order},$$

where $k_i, m_i \in \mathbb{N}$. Then

$$\alpha_i = \frac{m_i n_i}{k_i n_i} = \frac{m_i}{k_i}.$$

The number μ_i is equal to the number of solutions distinct from 0 of the equation

$$\frac{\partial f}{\partial t} - \varepsilon \frac{\partial \zeta}{\partial t} = 0$$

that tend to 0 as $\varepsilon \to 0$. The number ν_i is by Lemma 5.10 the number of solutions of the equation $t^{n_i k_i} = \varepsilon$. It follows that $\mu_i = (m_i - k_i)m_i$, $\nu_i = n_i k_i$. Hence

$$\frac{\mu_i}{\nu_i} = \frac{(m_i - k_i)n_i}{k_i n_i} = \frac{m_i}{k_i} - 1 = \alpha_i - 1.$$

This proves Lemma 5.13. □

Gabrielov's method rests on establishing a connection between an intersection matrix of the singularity f and an intersection matrix of the singularity $f + \zeta^2$ (or $f|_{\zeta=0}$). Towards this end, we consider the behavior of the critical points and critical values for special unfoldings of these singularities for which the critical points lie on the polar curve of the singularity f. We consider first an unfolding

$$F_\varepsilon = f + (\zeta - \varepsilon)^2,$$

$\varepsilon \in \mathbb{C}$ small, of the function $f + \zeta^2$. It is easy to calculate that the critical points of the function F_ε lie on the polar curve $\Sigma_\zeta(f)$ of the function f. We are interested now in those critical points of F_ε that lie on the component Σ_i and tend to 0 as $\varepsilon \to 0$. They are solutions of the equation

$$\frac{\partial F_\varepsilon|_{\Sigma_i}}{\partial \zeta} = 0.$$

With the help of the Puiseux expansion of Σ_i we can write this equation as follows:

$$a_i \alpha_i \zeta^{\alpha_i - 1} + 2(\zeta - \varepsilon) + o(\zeta^{\alpha_i - 1}) = 0.$$

For $\alpha_i > 2$ it follows from this that

$$\zeta = \varepsilon + o(\varepsilon^{\alpha_i - 1}), \ F_\varepsilon|_{\Sigma_i}(\zeta) = a_i \varepsilon^{\alpha_i} + o(\varepsilon^{\alpha_i}).$$

The number of these critical points is ν_i.

For $\alpha_i = 2$, $a_i \neq -1$, we have

$$\zeta = \frac{\varepsilon}{a_i+1} + o(\varepsilon), \ F_\varepsilon|_{\Sigma_i}(\zeta) = \frac{a_i}{a_i+1}\varepsilon^2 + o(\varepsilon^2).$$

The number of these critical points is again ν_i.

For $\alpha_i < 2$ the equation reduces as follows:

$$a_i\alpha_i\zeta^{\alpha_i-1} - 2\varepsilon + o(\zeta) = 0.$$

To a first approximation this equation describes the critical points of the function $f - 2\varepsilon\zeta$. The number of these critical points is, by the definition of μ_i, equal to μ_i. Hence

$$\zeta = \left(\frac{2\varepsilon}{a_i\alpha_i}\right)^{\frac{1}{\alpha_i-1}} + o(\varepsilon) = o(\varepsilon), \ F_\varepsilon|_{\Sigma_i}(\zeta) = \varepsilon^2 + o(\varepsilon^2).$$

This proves both the following results.

Proposition 5.24. *The critical values of those critical points of $F_\varepsilon = f + (\zeta - \varepsilon)^2$ that lie on Σ_i and tend to 0 as $\varepsilon \to 0$ are*

$$a_i\varepsilon^{\alpha_i} + o(\varepsilon^{\alpha_i}), \ \text{if } \alpha_i > 2,$$
$$\frac{a_i}{a_i+1}\varepsilon^2 + o(\varepsilon^2), \ \text{if } \alpha_i = 2, a_i \neq -1,$$
$$\varepsilon^2 + o(\varepsilon^2), \ \text{if } \alpha_i < 2.$$

Proposition 5.25. *If $a_i \neq -1$ for $\alpha_i = 2$, then the Milnor number $\mu(f+\zeta^2)$ of the singularity $f + \zeta^2$ is*

$$\mu(f+\zeta^2) = \sum_{\substack{i \\ \alpha_i<2}} \mu_i + \sum_{\substack{j \\ \alpha_j\geq 2}} \nu_j.$$

Corollary 5.11.

$$\mu(f+\zeta^2) \leq \mu(f|_{\zeta=0})$$

with equality if and only if $\alpha_i \geq 2$ for all i.

Remark 5.8. It is easy to show that for $f \in \mathfrak{m}^k$ and sufficiently generally chosen ζ we have $\alpha_i \geq k$ for all i.

We now want to relate an intersection matrix of the singularity f to an intersection matrix of the singularity $f + \zeta^2$. To do this, we consider the unfolding $F_\varepsilon = f + (\zeta - \varepsilon)^2$ and construct a strongly distinguished path system for this unfolding. Let

$$A = \{\alpha_i\}.$$

By Proposition 5.24 the critical values of F_ε lie in annuli around the origin. Let $\varepsilon \neq 0$ be sufficiently small. For each $\alpha \in A$ with $\alpha > 2$ we can choose positive numbers r'_α and r''_α such that the critical values of all critical points of F_ε that belong to Σ_i with $\alpha_i = \alpha$, and tend to 0 as $\varepsilon \to 0$, are contained in the open annulus

$$\{u \in \mathbb{C} \mid r'_\alpha < |u| < r''_\alpha\}.$$

Equally, we can choose positive numbers r'_2 and r''_2 such that the critical values of all critical points belonging to Σ_i with $\alpha_i \leq 2$ that, furthermore, tend to 0 as $\varepsilon \to 0$, are contained in the open annulus

$$\{u \in \mathbb{C} \mid r'_2 < |u| < r''_2\}.$$

Furthermore, the numbers r'_α and r''_α ($\alpha \geq 2$) can be chosen so that $r''_\alpha < r'_\beta$ for $\alpha \geq \beta$.

In what follows we assume that the argument of the complex number u, $\arg u$, lies in the interval $[-\pi, \pi]$. Let $\sigma : \mathbb{R}_+ \to \mathbb{R}_+$ be a continuous monotone decreasing function with

$$\sigma(r) = \alpha - 1 \text{ for } r'_\alpha \leq r \leq r''_\alpha.$$

We define

$$V_m := \{u \in \mathbb{C} \mid \arg u + 2\pi\sigma(|u|) \geq (2m-1)\pi\},$$

where $m = 1, 2, \ldots$.

We assume that the following condition is satisfied:

$$(-a_i)^{q_i} \notin \mathbb{R}_+ \text{ for } \alpha_i = \frac{p_i}{q_i}, (p_i, q_i) = 1.$$

Then one can easily deduce from Proposition 5.24 that for sufficiently small $\varepsilon \in \mathbb{R}_+ \setminus \{0\}$ the critical values of F_ε are neither contained in $\mathbb{R}_-$ nor in the boundary of a set V_m.

Now we choose a strongly distinguished system $(\gamma_1, \ldots, \gamma_\nu)$ of paths ($\nu := \mu(f + \zeta^2)$) that connect the critical values of F_ε with the noncritical value 0. Strongly distinguished in this case means that the paths are double point free, crossing free ($\gamma_i \cap \gamma_j = \{0\}$), and are ordered in the following way: For $i < j$ let $\arg u_i > \arg u_j$, where u_i, resp. u_j, is the point of intersection of γ_i, resp. γ_j, with a small circle round the origin. We further assume that the path system has the following property:

(V) Each path intersects $\mathbb{R}_-$ only in the origin, and a path that begins at a critical value in V_m remains entirely in V_m.

This condition is satisfied e.g. for a system of line segments between the critical values of F_ε and 0 (with suitable alterations if two critical values lie on the same segment). Compare Figure 5.43.

Gabrielov proved the following in [**Gab79**].

Theorem 5.3 (Gabrielov). *Let $(\delta_1, \ldots, \delta_\nu)$ be a strongly distinguished basis of vanishing cycles of the singularity $f + \zeta^2$ defined by the strongly distinguished path system $(\gamma_1, \ldots, \gamma_\nu)$.*

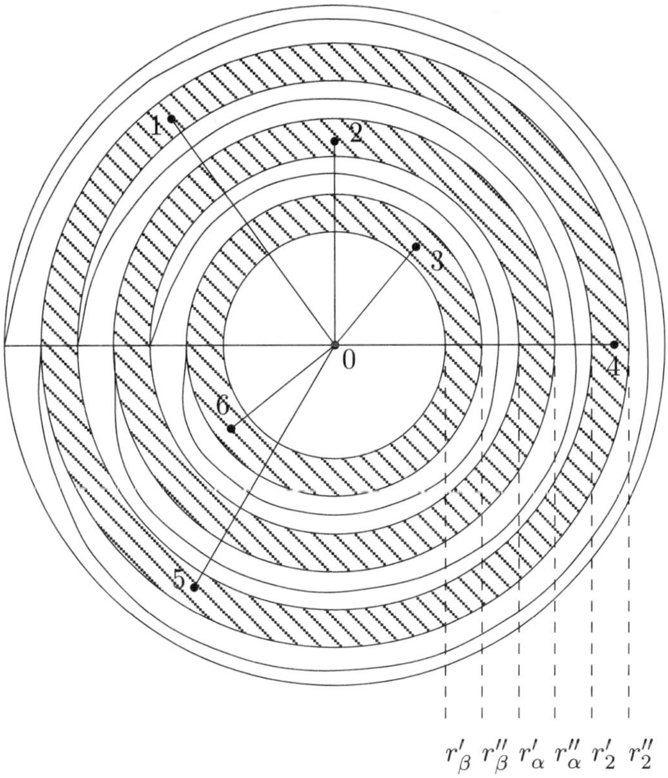

Figure 5.43. Choice of strongly distinguished path system $(\gamma_1, \ldots, \gamma_\nu)$ after Gabrielov

Then f has a strongly distinguished basis $(\delta_j^m \mid 1 \leq j \leq \nu, 1 \leq m \leq M_j)$ of vanishing cycles, ordered by the lexicographic order of the pairs (m, j), with the following properties:

(i) *For $1 \leq j, j' \leq \nu$, $1 \leq m \leq M_j - 1$,*

$$\langle \delta_j^1, \delta_{j'}^1 \rangle = \langle \delta_j, \delta_{j'} \rangle, \ \delta_j^{m+1} = h_*(\delta_j^m),$$

where h_* is the classical monodromy operator of the singularity f.

(ii) *A pair (m, j) is called* admissible *if $1 \leq j \leq \nu$ and $1 \leq m \leq M_j$. The first μ_i pairs of each set of pairs (m, j), where δ_j is a vanishing cycle that vanishes at a critical point on Σ_i, are admissible.*

(iii) *The basis (δ_j^m) has the following intersection matrix:*

$$\begin{aligned}
\langle \delta_j^m, \delta_{j'}^m \rangle &= \langle \delta_j, \delta_{j'} \rangle, \\
\langle \delta_j^m, \delta_{j'}^{m'} \rangle &= -(-1)^{n(n-1)/2}(m' - m)^n \text{ for } |m' - m| = 1, \\
\langle \delta_j^m, \delta_{j'}^{m'} \rangle &= -\langle \delta_j, \delta_{j'} \rangle \text{ for } |m' - m| = 1 \text{ and } (m' - m)(j' - j) < 0, \\
\langle \delta_j^m, \delta_{j'}^{m'} \rangle &= 0 \text{ for } |m' - m| > 1 \text{ or } (m' - m)(j' - j) > 0.
\end{aligned}$$

Now let $f|_{\zeta=0}$ have an isolated singularity at 0. Then there is an analogous connection between intersection matrices of the singularities f and $f|_{\zeta=0}$. Let us consider the unfolding

$$G_\varepsilon = f|_{\zeta=\varepsilon},\ \varepsilon \in \mathbb{C} \text{ small},$$

of the singularity $f|_{\zeta=0}$. By Lemma 5.10 the critical values of G_ε lie on the polar curve $\Sigma_\zeta(f)$. The critical values of those points of G_ε that lie on the component Σ_i and that tend to 0 as $\varepsilon \to 0$ are

$$a_i\varepsilon^{\alpha_i} + o(\varepsilon^{\alpha_i}).$$

Hence for sufficiently small $\varepsilon \neq 0$ and each $\alpha \in A$ we can choose positive numbers r'_α and r''_α such that

$$\begin{aligned} r'_\alpha &< r''_\alpha \text{ for all } a \in A,\\ r''_\alpha &< r'_\beta \text{ for } \alpha > \beta, \end{aligned}$$

and the critical values of all critical points of G_ε that belong to Σ_i with $\alpha_i = \alpha$ and that tend to 0 as $\varepsilon \to 0$ are contained in the open annulus

$$\{u \in \mathbb{C} \mid r'_\alpha < |u| < r''_\alpha\}.$$

We define the function $\sigma(r)$, the regions V_m and the path system $(\gamma_1, \ldots, \gamma_\nu)$ that joins the critical values of the function G_ε to the noncritical value 0 in just the same way as for the function F_ε, with the only difference being that we consider all $\alpha \in A$ and not only $\alpha \geq 2$.

We now cite a further proposition of Gabrielov's. Let us consider the concept of the stabilization of a singularity. Let $g : (\mathbb{C}^{n+1}, 0) \to (\mathbb{C}, 0)$ be a holomorphic function germ with an isolated singularity at 0. Let $(x_1, \ldots, x_{n+1})$ be the coordinates of $\mathbb{C}^{n+1}$ and $(x_1, \ldots, x_{n+1}, y_1, \ldots, y_m)$ the coordinates of $\mathbb{C}^{n+1+m}$.

Definition. The holomorphic function germ

$$\begin{aligned} g(x) + y_1^2 + \ldots + y_m^2 : \quad (\mathbb{C}^{n+1+m}, 0) &\longrightarrow (\mathbb{C}, 0)\\ (x, y) &\longmapsto g(x) + y_1^2 + \ldots + y_m^2 \end{aligned}$$

is called a *stabilization* of g.

Theorem 5.4 (Gabrielov). *Let g_λ be a morsification of the singularity g, let $(\gamma_1, \ldots, \gamma_\mu)$ be a strongly distinguished path system to g_λ, and let $(\delta_1, \ldots, \delta_\mu)$ be a corresponding strongly distinguished basis of vanishing cycles.*

Then $g_\lambda(x) + y_1^2 + \ldots + y_m^2$ is a morsification of the singularity $g(x) + y_1^2 + \ldots + y_m^2$, with the same critical values, $(\gamma_1, \ldots, \gamma_\mu)$ is also a strongly distinguished path system for this singularity, and for the corresponding strongly distinguished basis $(\tilde{\delta}_1, \ldots, \tilde{\delta}_\mu)$ of vanishing cycles we have

$$\langle \tilde{\delta}_i, \tilde{\delta}_j \rangle = [\operatorname{sign}(j - i)]^m (-1)^{(n+1)m + \frac{m(m-1)}{2}} \langle \delta_i, \delta_j \rangle$$

for $i \neq j$.

See [**BK91**] for the proof.

We continue our study of the function G_ε. Let $(\gamma_1, \ldots, \gamma_\nu)$ be a strongly distinguished path system as described. Let $(\tilde{\delta}'_1, \ldots, \tilde{\delta}'_\nu)$ be a strongly distinguished basis of vanishing cycles for $f|_{\zeta=0}$ defined by this. By Theorem 5.4, $(\gamma_1, \ldots, \gamma_\nu)$ also determines a strongly distinguished basis $(\tilde{\delta}_1, \ldots, \tilde{\delta}_\nu)$ of vanishing cycles for $f|_{\zeta=0} + \zeta^2$, and for the intersection matrix we have

$$\langle \tilde{\delta}_j, \tilde{\delta}_{j'} \rangle = (-1)^n \langle \tilde{\delta}'_j, \tilde{\delta}'_{j'} \rangle.$$

Theorem 5.5 (Gabrielov). *Suppose that $f|_{\zeta=0}$ has an isolated singularity at 0. Then in Theorem 5.3 we can replace the singularity $f + \zeta^2$ and the strongly distinguished basis $(\delta_1, \ldots, \delta_\nu)$ by $f|_{\zeta=0} + \zeta^2$ and $(\tilde{\delta}_1, \ldots, \tilde{\delta}_\nu)$.*

Theorem 5.5 reduces the computation of an intersection matrix of a singularity f of a function in $n+1$ variables to the computation of the intersection matrix of the singularity $f|_{\zeta=0}$ of a function in n variables with respect to a special strongly distinguished path systems and of the indices α_i for the components Σ_i of the polar curve $\Sigma_\zeta(f)$ of f.

Example 5.6. $f(x, \zeta) = x^3 + x^2\zeta^2 + \zeta^{6+l}$, $l \geq 0$. Then $f|_{\zeta=0} = x^3$ has an isolated singularity of type A_2 in 0. We have

$$\nu = \mu(f|_{\zeta=0}) = 2$$

and the intersection matrix of the strongly distinguished basis $(\tilde{\delta}_1, \tilde{\delta}_2)$ must, for a suitable orientation of the vanishing cycles $\tilde{\delta}_1, \tilde{\delta}_2$, have the form

$$\begin{pmatrix} 0 & 1 \\ -1 & 0 \end{pmatrix}.$$

The polar curve $\Sigma_\zeta(f)$ of f is given by the equation

$$0 = \frac{\partial f}{\partial x} = 3x^2 + 2x\zeta^2.$$

It has the irreducible components

$$\begin{aligned} \Sigma_1 &= \{x = 0\}, \\ \Sigma_2 &= \{3x + 2\zeta^2 = 0\}. \end{aligned}$$

We have $\nu_1 = \nu_2 = 1$,

$$f|_{\Sigma_1} = \zeta^{6+l},$$

whence $\alpha_1 = 6 + l$, $\mu_1 = 5 + l$, and

$$f|_{\Sigma_2} = -\frac{8}{27}\zeta^6 + \frac{4}{9}\zeta^6 + \zeta^{6+l},$$

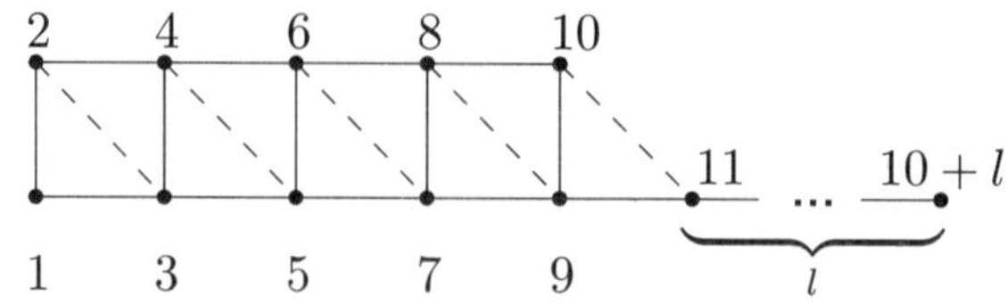

Figure 5.44. Coxeter-Dynkin diagram for the basis (δ_j^m) of the example

whence $\alpha_2 = 6$, $\mu_2 = 5$. By Theorem 5.5, f has a strongly distinguished basis (δ_j^m) of vanishing cycles with Coxeter-Dynkin diagram as shown in Figure 5.44.

5.11. Unimodal singularities

We classified the simple singularities in §3.10. We now want to present the classification of the unimodal singularities after Arnold.

The unimodal singularities are given, up to stabilization, by the function germs listed in Tables 5.1 and 5.2. In each case we list the name, the equation and a Coxeter-Dynkin diagram that one obtains by the method described in §5.10.

In most cases we have dropped the numbering of the Dynkin diagrams.

The Coxeter-Dynkin diagrams of the unimodular singularities can be brought into the forms given in Tables 5.3 and 5.4 by operations of the braid group.

For the singularities $\tilde{E}_6$, $\tilde{E}_7$, $\tilde{E}_8$ and $T_{p,q,r}$ Gabrielov has proved that the graphs shown are the Coxeter-Dynkin diagrams for strongly distinguished bases of each of the singularities. For the exceptional singularities Gabrielov has proved that the graphs shown are Coxeter-Dynkin diagrams for weakly distinguished bases of each of the singularities. The numbers p, q, r are therefore also called the *Gabrielov numbers*. The author has shown [**Ebe81**] that one can also find strongly distinguished bases with these Coxeter-Dynkin diagrams.

Let $f : (\mathbb{C}^3, 0) \to (\mathbb{C}, 0)$ be a holomorphic function germ with an isolated singularity at 0. Let L be the Milnor lattice of f and $(\delta_1, \ldots, \delta_\mu)$ a strongly distinguished basis of vanishing cycles of f. We consider the μ-dimensional real vector space

$$V = L \otimes \mathbb{R} = \{a_1\delta_1 + \ldots + a_\mu\delta_\mu \mid a_i \in \mathbb{R}\}.$$

The intersection form $\langle\ ,\ \rangle$ then induces a symmetric bilinear form

$$\langle\ ,\ \rangle : V \times V \longrightarrow \mathbb{R}$$

Name	Equation	Coxeter-Dynkin diagram
$\tilde{E}_6$	$x^3+y^3+z^3+\lambda xyz,$ $\lambda^3+27 \neq 0$	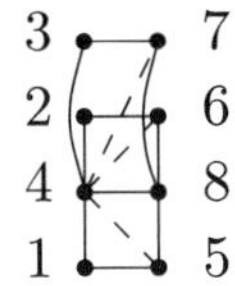
$\tilde{E}_7$	$x^4+y^4+z^2+\lambda x^2y^2,$ $\lambda^2 \neq 4$	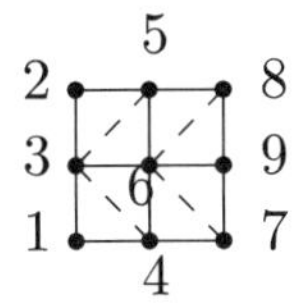
$\tilde{E}_8$	$x^3+y^6+z^2+\lambda x^2y^2,$ $4\lambda^3+27 \neq 4$	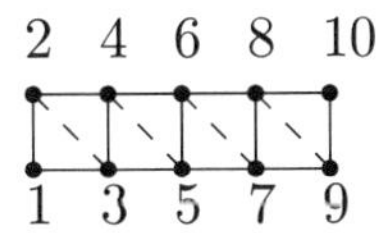
$T_{p,q,r}$	$x^p+y^q+z^r+\lambda xyz,$ $\frac{1}{p}+\frac{1}{q}+\frac{1}{r}<1, \lambda \neq 0$	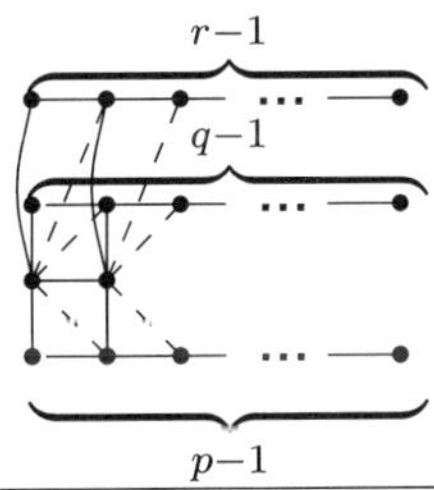

Table 5.1. The parabolic and hyperbolic unimodal singularities

on V. By the diagonalization theorem one can find a basis of V with respect to which the matrix of $\langle\ ,\ \rangle$ has a diagonal form. Let

$$\begin{aligned} \mu_+ &= \#\text{ positive diagonal entries}, \\ \mu_0 &= \#\ 0 \text{ on diagonal}, \\ \mu_- &= \#\text{ negative diagonal entries}. \end{aligned}$$

These numbers are independent of the basis and are also called the (Sylvester) *inertia indices*. We have

$$\mu = \mu_+ + \mu_0 + \mu_-.$$

We next want to determine the inertia indices for the unimodal singularities. Let $f : (\mathbb{C}^3, 0) \to (\mathbb{C}, 0)$ be a unimodal holomorphic function germ and $(\delta_1, \ldots, \delta_\mu)$ a strongly distinguished basis of vanishing cycles of f for each Coxeter-Dynkin diagram above.

First let f be from one of the classes $\tilde{E}_6$, $\tilde{E}_7$, $\tilde{E}_8$ or $T_{p,q,r}$. We consider the vector

$$\lambda_0 = \delta_{\mu-1} - \delta_\mu.$$

Name	Equation	Coxeter-Dynkin diagram
E_{12}	$x^3 + y^7 + z^2 + \lambda xy^5$	
E_{13}	$x^3 + xy^5 + z^2 + \lambda y^8$	
E_{14}	$x^3 + y^8 + z^2 + \lambda xy^6$	
Z_{11}	$x^3y + y^5 + z^2 + \lambda xy^4$	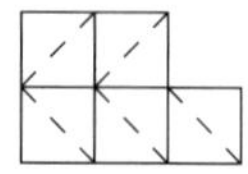
Z_{12}	$x^3y + xy^4 + z^2 + \lambda y^6$	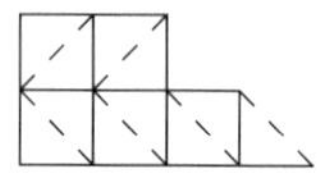
Z_{13}	$x^3y + y^6 + z^2 + \lambda xy^5$	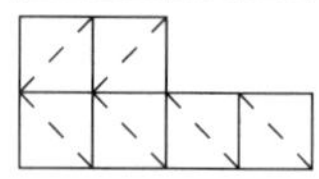
W_{12}	$x^4 + y^5 + z^2 + \lambda x^2y^3$	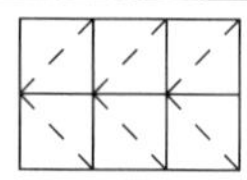
W_{13}	$x^4 + xy^4 + z^2 + \lambda y^6$	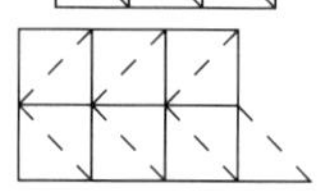
Q_{10}	$x^2z + y^3 + z^4 + \lambda yz^3$	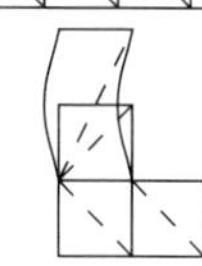
Q_{11}	$x^2z + y^3 + yz^3 + \lambda z^5$	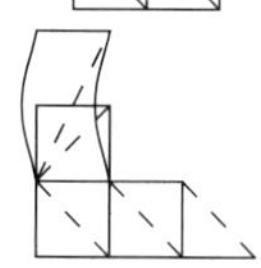
Q_{12}	$x^2z + y^3 + z^5 + \lambda yz^4$	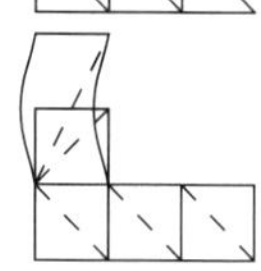
S_{11}	$x^2z + yz^2 + y^4 + \lambda y^3z$	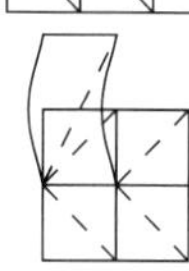
S_{12}	$x^2z + yz^2 + y^4 + \lambda y^5$	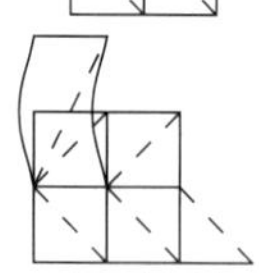
U_{12}	$x^3 + y^3 + z^4 + \lambda xyz^2$	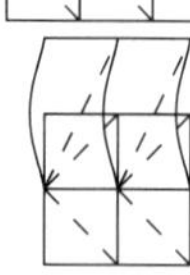

Table 5.2. The 14 exceptional unimodal singularities

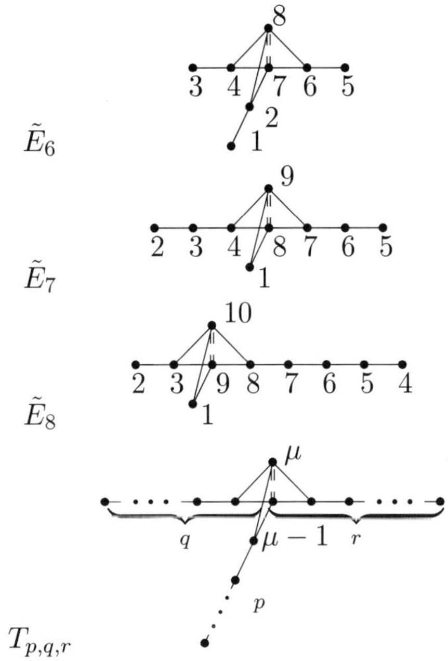

Table 5.3. Coxeter-Dynkin diagrams of the parabolic and hyperbolic unimodal singularities

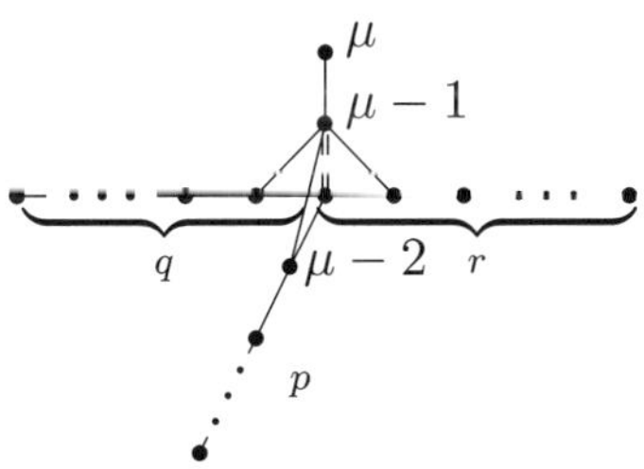

E_{12}	237	Z_{11}	245	Q_{10}	334
E_{13}	238	Z_{12}	246	Q_{11}	335
E_{14}	239	Z_{13}	247	Q_{12}	336
		W_{12}	255	S_{11}	344
		W_{13}	256	S_{12}	345
				U_{12}	444

Table 5.4. Coxeter-Dynkin diagrams of the exceptional unimodal singularities

Then

$$\begin{aligned}
\langle \lambda_0, \lambda_0 \rangle &= \langle \delta_{\mu-1} - \delta_\mu, \delta_{\mu-1} - \delta_\mu \rangle \\
&= \langle \delta_{\mu-1}, \delta_{\mu-1} \rangle - 2\langle \delta_{\mu-1}, \delta_\mu \rangle + \langle \delta_\mu, \delta_\mu \rangle \\
&= -2 + 4 - 2 \\
&= 0, \\
\langle \lambda_0, \delta_i \rangle &= 0 \text{ for } i = 1, \ldots, \mu - 1.
\end{aligned}$$

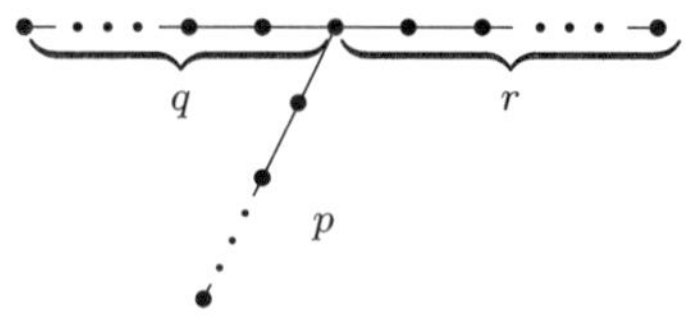

Figure 5.45. Coxeter-Dynkin diagram corresponding to S'

The matrix of the intersection form with respect to the new basis $(\lambda_0, \delta_1, \ldots, \delta_{\mu-1})$ therefore has the form

$$\left(\begin{array}{c|ccc} 0 & 0 & \cdots & 0 \\ \hline 0 & & & \\ \vdots & & S' & \\ 0 & & & \end{array}\right)$$

and S' corresponds to the Coxeter-Dynkin diagram of Figure 5.45, where for $\tilde{E}_6$, $\tilde{E}_7$, $\tilde{E}_8$ we have the following values of p, q, r:

$$\begin{array}{lcl} \tilde{E}_6 & : & 3\ 3\ 3, \\ \tilde{E}_7 & : & 2\ 4\ 4, \\ \tilde{E}_8 & : & 2\ 3\ 6. \end{array}$$

It is now easy to calculate

$$\begin{aligned} \det S' &= (-1)^{\mu-1}(qr + pr + pq - pqr) \\ &= (-1)^{\mu-1} pqr \left(\frac{1}{p} + \frac{1}{q} + \frac{1}{r} - 1\right). \end{aligned}$$

It follows that

$$(-1)^{\mu-1} \det S' \begin{cases} > 0 & \text{if } \frac{1}{p} + \frac{1}{q} + \frac{1}{r} > 1, \\ = 0 & \text{if } \frac{1}{p} + \frac{1}{q} + \frac{1}{r} = 1, \\ < 0 & \text{if } \frac{1}{p} + \frac{1}{q} + \frac{1}{r} < 1. \end{cases}$$

Hence we obtain

$$\begin{array}{rcl} \tilde{E}_6, \tilde{E}_7, \tilde{E}_8 & : & \mu_+ = 0,\ \mu_0 = 2, \\ T_{p,q,r} & : & \mu_+ = 1,\ \mu_0 = 1. \end{array}$$

The intersection forms of $\tilde{E}_6, \tilde{E}_7, \tilde{E}_8$ are hence (negative) semidefinite. Such forms are also called parabolic, so the singularities $\tilde{E}_6$, $\tilde{E}_7$ and $\tilde{E}_8$ are also called *parabolic singularities.*

A quadratic form with $\mu_+ = 1$ is called *hyperbolic*. The intersection forms of the singularities $T_{p,q,r}$ are therefore hyperbolic and so one also calls these singularities *hyperbolic* singularities.

Arnold has shown that the simple (hypersurface) singularities are the only singularities with a definite intersection form, the singularities $\tilde{E}_6, \tilde{E}_7, \tilde{E}_8$ are the only ones with parabolic intersection forms, and the $T_{p,q,r}$ are the only singularities with a hyperbolic intersection form.

Now let f be an exceptional unimodal singularity. We consider the vectors

$$\begin{aligned} \lambda_0 &= \delta_{\mu-2} - \delta_{\mu-1}, \\ \lambda_1 &= \delta_{\mu-2} - \delta_{\mu-1} - \delta_\mu. \end{aligned}$$

The matrix of the intersection form with respect to the basis $(\lambda_0, \lambda_1, \delta_1, \ldots, \delta_{\mu-2})$ then has the form

$$\left(\begin{array}{cc|ccc} 0 & 1 & 0 & \cdots & 0 \\ 1 & 0 & 0 & \cdots & 0 \\ \hline 0 & 0 & & & \\ \vdots & \vdots & & S' & \\ 0 & 0 & & & \end{array}\right).$$

The matrix $\begin{pmatrix} 0 & 1 \\ 1 & 0 \end{pmatrix}$ is equivalent to $\begin{pmatrix} 1 & 0 \\ 0 & -1 \end{pmatrix}$ over $\mathbb{R}$. It follows that

$$\mu_+ = 2, \ \mu_0 = 0$$

for the exceptional unimodal singularities.

The diagrams of Table 5.3 and Table 5.4 had the advantage that with their help one could easily determine the inertia indices of the intersection form. A disadvantage of these diagrams, however, is that they contain an edge of weight -2. One can, nevertheless, find Coxeter-Dynkin diagrams with respect to distinguished bases that are very similar to these diagrams and that contain edges of weight $+1$ or -1 only. By operations of the braid group the above diagrams can be transformed to the forms shown in Figure 5.46 and Figure 5.47 (cf. [**Ebe96**]).

G. G. Il'yuta [**Ily87**] has proved the following proposition, using a result of N. A'Campo.

Theorem 5.6 (Il'yuta). *A Coxeter-Dynkin diagram with respect to a strongly distinguished basis of a nonsimple singularity contains at least one edge of negative weight.*

The above diagrams also have the property of being minimal in the sense of Theorem 5.6. The diagrams have yet another property: A closed path in

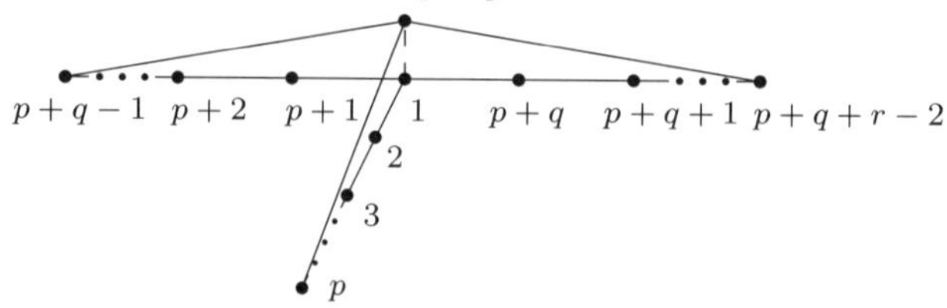

Figure 5.46. Normal form of the Coxeter-Dynkin diagrams for the parabolic and hyperbolic unimodal singularities

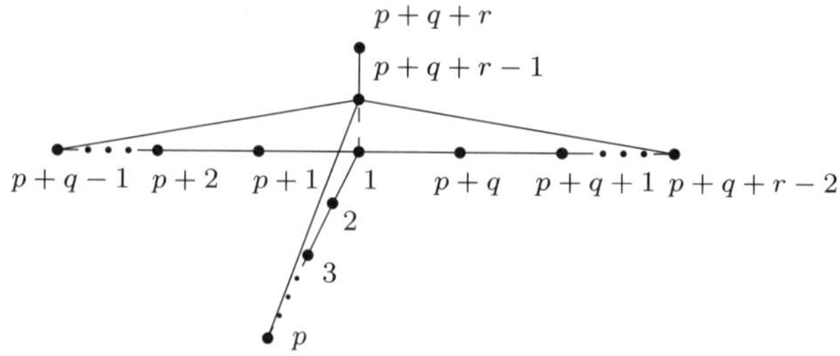

Figure 5.47. Normal form of the Coxeter-Dynkin diagrams for the exceptional unimodal singularities

a Coxeter-Dynkin diagram is called a *monotone cycle* if the numbering of the vertices is strictly monotone increasing as one traverses the cycle (up to the last step). The diagrams above contain exactly three monotone cycles and, after suppressing the edge from 1 to $p+q+r-1$, the monotone cycles are destroyed (cf. also [**Ebe96**]).

5.12. The monodromy groups of the isolated hypersurface singularities

In this section we shall present general results on the monodromy groups of vanishing lattices.

Let (L, Δ) be a vanishing lattice.

Definition. $L^{\#} = \mathrm{Hom}(L, \mathbb{Z})$ is called the *dual module.*

Let

$$\begin{array}{rrcl} j: & L & \longrightarrow & L^{\#} \\ & v & \longmapsto & l_v \text{ with } l_v(x) = \langle v, x\rangle, x \in L, \end{array}$$

be the canonical homomorphism. A homomorphism $h : L \to L$ induces a homomorphism $h^t : L^{\#} \to L^{\#}$ of the dual modules. If h leaves the bilinear form $\langle\, ,\, \rangle$ invariant, then $h^t(j(L)) \subset j(L)$. An automorphism $h \in \mathrm{Aut}(L)$ thus induces a homomorphism $h^t : L^{\#}/j(L) \to L^{\#}/j(L)$.

Definition. Let $\mathrm{Aut}^{\#}(L) \subset \mathrm{Aut}(L)$ be the subgroup of those automorphisms $h \in \mathrm{Aut}(L)$ with $h^t = \mathrm{id}_{L^{\#}/j(L)}$.

It is easy to see that $\Gamma_\Delta \subset \mathrm{Aut}^{\#}(L)$.

Now let L be symmetric and let $\varepsilon \in \{-1, +1\}$. We put

$$\begin{aligned} \bar{L} &:= L/\ker j, \\ \bar{L}_{\mathbb{R}} &:= \bar{L} \otimes \mathbb{R}. \end{aligned}$$

Then $\bar{L}_{\mathbb{R}}$ is a finite-dimensional real vector space with a nondegenerate symmetric bilinear form. We now define a homomorphism

$$\nu_\varepsilon : \mathrm{Aut}(L) \longrightarrow \{+1, -1\}$$

as follows. Let $h \in \mathrm{Aut}(L)$ and $\bar{h}$ the induced element in $O(\bar{L}_{\mathbb{R}})$. By a result from linear algebra one can write $\bar{h}$ as a product of reflections

$$\bar{h} = s_{v_1} \circ \ldots \circ s_{v_r}$$

with $v_i \in \bar{L}_{\mathbb{R}}$, $\langle v_i, v_i \rangle \neq 0$, $i = 1, \ldots, r$. We define

$$\nu_\varepsilon(h) := \begin{cases} +1 & \text{if } \varepsilon\langle v_i, v_i\rangle < 0 \text{ for an even number of indices,} \\ -1 & \text{otherwise.} \end{cases}$$

The homomorphism $\nu_\varepsilon : \mathrm{Aut}(L) \to \{-1, +1\}$ is called the *real ε-spinor norm.*

Definition. We define a subgroup $O^*_\varepsilon(L) \subset O(L)$ as follows:

$$O^*_\varepsilon(L) := \mathrm{Aut}^{\#}(L) \cap \ker \nu_\varepsilon.$$

It is easy to see that $\Gamma_\Delta \subset O^*_\varepsilon(L)$. If L is nondegenerate, then $L^{\#}/j(L)$ is a finite group and hence $O^{\#}(L) = \mathrm{Aut}^{\#}(L)$ is a subgroup of finite index in $O(L) = \mathrm{Aut}(L)$. The subgroup $\ker \nu_\varepsilon \subset O(L)$ is of index ≤ 2 in $O(L)$. Thus if L is nondegenerate, then $O^*_\varepsilon(L)$ is a subgroup of finite index in $O(L)$.

By using a result of M. Kneser the author has proved the following theorem (cf. [**Ebe87**]).

Theorem 5.7. *Let (L, Δ) be an even symmetric vanishing lattice. Suppose that Δ has six elements $\delta_1, \ldots, \delta_6$ with*

$$(\langle \delta_i, \delta_j \rangle) = \begin{pmatrix} 2\varepsilon & -\varepsilon & 0 & 0 & 0 & 0 \\ -\varepsilon & 2\varepsilon & -\varepsilon & -\varepsilon & 2\varepsilon & 0 \\ 0 & -\varepsilon & 2\varepsilon & -\varepsilon & -\varepsilon & 0 \\ 0 & -\varepsilon & -\varepsilon & 2\varepsilon & -\varepsilon & 0 \\ 0 & 2\varepsilon & -\varepsilon & -\varepsilon & 2\varepsilon & -\varepsilon \\ 0 & 0 & 0 & 0 & -\varepsilon & 2\varepsilon \end{pmatrix}$$

(see Figure 5.48).

Then

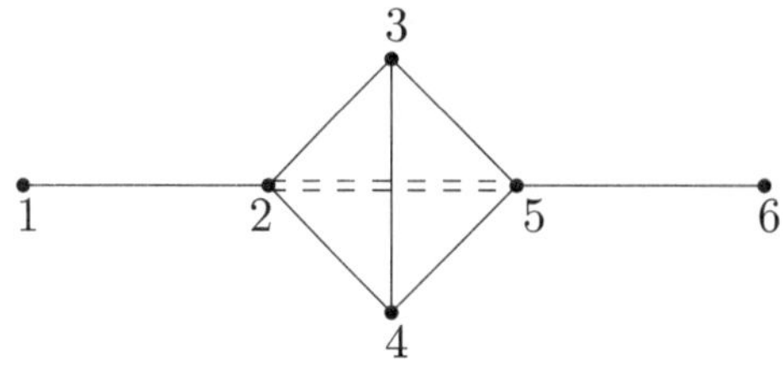

Figure 5.48. Coxeter-Dynkin diagram with respect to $(\delta_1, \ldots, \delta_6)$

(i) $\Delta = \{v \in L \mid \langle v, v \rangle = 2\varepsilon \text{ and } \langle v, L \rangle = \mathbb{Z}\}$,

(ii) $\Gamma = O_\varepsilon^*(L)$.

Here $\langle v, L \rangle = \mathbb{Z}$ means that there is a $y \in L$ with $\langle v, y \rangle = 1$.

Now let (L, Δ) be a skew symmetric vanishing lattice. A skew symmetric lattice L has a basis $(e_1, f_1, \ldots, e_m, f_m, g_1, \ldots, g_k)$ such that the matrix of the bilinear form $\langle \ , \ \rangle$ with respect to this basis has the following form:

$$\begin{pmatrix} \boxed{\begin{matrix} 0 & d_1 \\ -d_1 & 0 \end{matrix}} & & & & & & \\ & \boxed{\begin{matrix} 0 & d_2 \\ -d_2 & 0 \end{matrix}} & & & 0 & & \\ & & \ddots & & & & \\ & 0 & & \boxed{\begin{matrix} 0 & d_m \\ -d_m & 0 \end{matrix}} & & & \\ & & & & 0 & & \\ & & & & & \ddots & \\ & & & & & & 0 \end{pmatrix},$$

where $d_i \in \mathbb{N}$ with $d_i | d_{i+1}$ $(i = 1, \ldots, m-1)$. Such a basis is called a *symplectic* basis.

Let $(e_1, f_1, \ldots, e_m, f_m, g_1, \ldots, g_k)$ be a symplectic basis of L. Let $\eta_2(L)$ be the exponent of 2 in the prime factor decomposition of d_m. Let $\mu = 2m + k$. We identify L with $\mathbb{Z}^\mu$ through the symplectic basis $(e_1, f_1, \ldots, e_m, f_m, g_1, \ldots, g_k)$. A subgroup $G \subset \mathrm{Aut}^\#(L)$ then corresponds to a subgroup $\rho(G) \subset \mathrm{Sp}^\#(\mu, \mathbb{Z})$.

Definition. Let $r \in \mathbb{N} \setminus \{0\}$. A subgroup $G \subset \mathrm{Aut}^\#(L)$ is called a *congruence subgroup modulo r* if

$$\rho(G) = \{A \in \mathrm{Sp}^\#(\mu, \mathbb{Z}) \mid A \equiv E \bmod r\}.$$

Here E is the unit matrix and $A \equiv E \bmod r$ means that $a_{ij} \equiv \delta_{ij} \bmod r$ for all $1 \leq i, j \leq \mu$, where $A = (a_{ij})$.

A congruence subgroup is obviously of finite index in the group $\mathrm{Aut}^{\#}(L) = \mathrm{Sp}^{\#}(L)$.

Notation. Let $a \in L$. If there is an element $b \in L$ with $a - 2b \in \Delta$, then we write $a \in \Delta \bmod 2$.

W. A. M. Janssen [**Jan83**] has proved the following theorem:

Theorem 5.8 (Janssen). *For each skew symmetric vanishing lattice* (L, Δ)

(i) $\Delta = \{v \in L \mid \langle v, L\rangle = \mathbb{Z}$ *and* $v \in \Delta \bmod 2\}$,

(ii) Γ_Δ *contains the congruence subgroup modulo* $2^{\eta_2(L)+1}$ *of* $\mathrm{Sp}^{\#}(L)$.

Janssen also gives a complete classification of the skew symmetric vanishing lattices; cf. [**Jan85**].

To finish, we shall apply these results to singularities.

Theorem 5.9. *Let* $f : (\mathbb{C}^{n+1}, 0) \to (\mathbb{C}, 0)$ *be a holomorphic function germ with an isolated singularity at* 0 *and let* n *be even,* $\varepsilon = (-1)^{n(n-1)/2}$. *Suppose that* f *is not of type* $T_{p,q,r}$ *with* $(1/p) + (1/q) + (1/r) < 1$ *and* $(p, q, r) \neq (3, 3, 4), (2, 4, 5), (2, 3, 7)$. *Let* L *be the Milnor lattice,* Γ *the monodromy group, and* Δ *the set of vanishing cycles of* f. *Then*

(i) $\Gamma = O_\varepsilon^*(L)$,

(ii) $\Delta = \{v \in L \mid \langle v, v\rangle = 2\varepsilon$ *and* $\langle v, L\rangle = \mathbb{Z}\}$.

In (ii) *it is also assumed that* 0 *is not a regular point of* f.

Idea of Proof. For the simple, parabolic, and the three hyperbolic singularities $T_{3,3,4}, T_{2,4,5}, T_{2,3,7}$ one can show this directly.

For the exceptional unimodal singularities one can show that the hypothesis of Theorem 5.7 is satisfied. Besides, one can show that every not simple, not parabolic and not hyperbolic singularity deforms to an exceptional unimodal singularity. Hence, by Corollary 5.10, the hypothesis of Theorem 5.7 is satisfied for each such singularity. □

From Theorem 5.8 one can deduce directly:

Theorem 5.10. *Let* $f : (\mathbb{C}^{n+1}, 0) \to (\mathbb{C}, 0)$ *be a holomorphic function germ with an isolated singularity at* 0 *and let* n *be odd. Then*

(i) Γ *contains the congruence subgroup modulo* $2^{\eta_2(L)+1}$ *of* $\mathrm{Sp}^{\#}(L)$,

(ii) $\Delta = \{v \in L \mid \langle v, L\rangle = \mathbb{Z}$ *and* $v \in \Delta \bmod 2\}$.

This result was proved by S. V. Chmutov [**Chm82**] based on previous work of N. A'Campo [**A'C79**] and B. Wajnryb [**Waj80**] and generalized by W. A. M. Janssen [**Jan83**] to the case of isolated singularities of complete intersections. Theorem 5.9 also has an analog for complete intersections; cf. [**Ebe87**].

Bibliography

[A'C79] N. A'Campo: Tresses, monodromie et le groupe symplectique. Comment. Math. Helv. 54, 318–327 (1979).

[Ahl78] L. V. Ahlfors: Complex Analysis. Third Edition. McGraw-Hill Book Co., New York, 1978.

[Arn73] V. I. Arnold: Normal forms for functions near degenerate critical points, the Weyl groups of A_k, D_k, E_k and Lagrangian singularities. Funct. Anal. Appl. 6, 254–272 (1973).

[Arn06] V. I. Arnold: Ordinary Differential Equations. Springer-Verlag, Berlin, 2006.

[AGV85] V. I. Arnold, S. M. Gusein-Zade, A. N. Varchenko: Singularities of Differentiable Maps, I. Birkhäuser-Verlag, Basel, 1985.

[AGV88] V. I. Arnold, S. M. Gusein-Zade, A. N. Varchenko: Singularities of Differentiable Maps, II. Birkhäuser-Verlag, Basel, 1988.

[Arn93] V. I. Arnold (Ed.): Dynamical Systems VI – Singularity Theory I. (Encyclopaedia of Math. Sciences, Vol. 6) Springer-Verlag, New York, etc., 1993.

[BK91] D. Bättig, H. Knörrer: Singularitäten. Birkhäuser-Verlag, Basel, 1991.

[Bre93] G. Bredon: Geometry and Topology. (Graduate Texts in Math. 139) Springer-Verlag, New York, etc., 1993.

[Bri70] E. Brieskorn: Die Monodromie der isolierten Singularitäten von Hyperflächen. Manuscripta Math. 2, 103–160 (1970).

[Bri88] E. Brieskorn: Automorphic sets and braids and singularities. Contemporary Math. 78, 1988, pp. 45–115.

[BK86] E. Brieskorn, H. Knörrer: Plane Algebraic Curves. Birkhäuser-Verlag, Basel, 1986.

[BJ82] Th. Bröcker, K. Jänich: Introduction to Differential Topology. Cambridge University Press, Cambridge, 1982.

[Chm82] S. V. Chmutov: Monodromy groups of critical points of functions. Invent. Math. 67, 123–131 (1982).

[CJ89] R. Courant, F. John: Introduction to Calculus and Analysis. Volume I. Springer-Verlag, New York, etc., 1989.

[DFN85] B. A. Dubrovin, A. T. Fomenko, S. P. Novikov: Modern Geometry – Methods and Applications. Part II. The Geometry and Topology of Manifolds. (Graduate Texts in Math. 104) Springer-Verlag, New York, etc., 1985.

[Dur79] A. H. Durfee: Fifteen characterizations of rational double points and simple critical points. Enseign. Math. 25, 131–163 (1979).

[Ebe81] W. Ebeling: Quadratische Formen und Monodromiegruppen von Singularitäten. Math. Ann. 255, 463–498 (1981).

[Ebe87] W. Ebeling: The Monodromy Groups of Isolated Singularities of Complete Intersections. (Lecture Notes in Math., Vol. 1293) Springer-Verlag, Berlin, etc., 1987.

[Ebe96] W. Ebeling: On Coxeter-Dynkin diagrams of hypersurface singularities. J. of Math. Sciences 82, 3657–3664 (1996).

[Fis76] G. Fischer: Complex Analytic Geometry. (Lecture Notes in Math., Vol. 538) Springer-Verlag, Berlin, etc., 1976.

[FL88] W. Fischer, I. Lieb: Ausgewählte Kapitel aus der Funktionentheorie. Vieweg, Braunschweig Wiesbaden, 1988.

[FL92] W. Fischer, I. Lieb: Funktionentheorie, 6. Auflage. Vieweg, Braunschweig Wiesbaden, 1992.

[For81] O. Forster: Lectures on Riemann Surfaces. (Graduate Texts in Math., Vol. 81) Springer-Verlag, New York, etc., 1981.

[Gab79] A. M. Gabrielov: Polar curves and intersection matrices of singularities. Invent. Math. 54, 15–22 (1979).

[GG74] M. Golubitsky, V. Guillemin: Stable Mappings and their Singularities. (Graduate Texts in Math., Vol. 14) Springer-Verlag, Berlin, etc., 1974.

[GF76] H. Grauert, K. Fritzsche: Several Complex Variables. (Graduate Texts in Math. 38) Springer-Verlag, New York, 1976.

[GR71] H. Grauert, R. Remmert: Analytische Stellenalgebren. (Grundlehren der math. Wiss. 176) Springer-Verlag, Heidelberg, etc., 1971.

[GR04] H. Grauert, R. Remmert: Theory of Stein Spaces. Springer-Verlag, Berlin, 2004.

[GR84] H. Grauert, R. Remmert: Coherent Analytic Sheaves. (Grundlehren der math. Wiss. 265) Springer-Verlag, Heidelberg, etc., 1984.

[GH81] M. J. Greenberg, J. R. Harper: Algebraic Topology: A First Course. Benjamin/Cummings, Menlo Park, CA, 1981.

[GH78] P. Griffith, J. Harris: Principles of Algebraic Geometry. Wiley and Sons, New York, 1978.

[GR65] R. C. Gunning, H. Rossi: Analytic Functions of Several Complex Variables. Prentice-Hall, Englewood Cliffs, N. J., 1965.

[Hir76] M. W. Hirsch: Differential Topology. (Graduate Texts in Math. 33) Springer-Verlag, Berlin, etc., 1976.

[Hum85] S. P. Humphries: On weakly distinguished bases and free generating sets of free groups. Quart. J. Math. Oxford (2), 36, 215–219 (1985).

[Ily87] G. G. Il'yuta: On the Coxeter transformation of an isolated singularity. Russ. Math. Surveys 42:2, 279–280 (1987).

[Jan83] W. A. M. Janssen: Skew-symmetric vanishing lattices and their monodromy groups. Math. Ann. 266, 115–133 (1983).

[Jan85] W. A. M. Janssen: Skew-symmetric vanishing lattices and their monodromy groups II. Math. Ann. 272, 17–22 (1985).

[JS87] G. A. Jones, D. Singerman: Complex Functions: An Algebraic and Geometric Viewpoint. Cambridge University Press, Cambridge, 1987.

[dJP00] Th. de Jong, G. Pfister: Local Analytic Geometry. Vieweg, Braunschweig Wiesbaden, 2000.

[KK83] L. Kaup, B. Kaup: Holomorphic Functions of Several Variables. Walter de Gruyter, Berlin, New York, 1983.

[Kle73] F. Klein: Über Flächen dritter Ordnung. Math. Ann. 6, 551-581 (1873) (Gesammelte Math. Abhandlungen, Bd. II, J. Springer, Berlin, 1922, pp. 11–44, mit Zusätzen pp. 44–62).

[Kod86] K. Kodaira: Complex Manifolds and Deformation of Complex Structures. (Grundlehren der math. Wiss. 283) Springer-Verlag, New York, etc., 1986.

[Lam75] K. Lamotke: Die Homologie isolierter Singularitäten. Math. Z. 143, 27–44 (1975).

[Lan97] S. Lang: Undergraduate Analysis. 2nd Edition. Springer-Verlag, New York, 1997.

[Lan02] S. Lang: Algebra. 3rd Edition. (Graduate Texts in Math. 211) Springer-Verlag, New York, 2002.

[Lef75] S. Lefschetz: Applications of Algebraic Topology. (Applied Math. Sciences, Vol. 16) Springer-Verlag, Berlin, etc., 1975.

[Loj91] S. Lojasiewicz: Introduction to Complex Analytic Geometry. Birkhäuser, Basel, 1991.

[Loo84] E. J. N. Looijenga: Isolated Singular Points on Complete Intersections. (London Math. Soc. Lecture Note Series 77) Cambridge University Press, Cambridge, 1984.

[LS77] R. C. Lyndon, P. E. Schupp: Combinatorial Group Theory. (Ergebnisse der Mathematik und ihrer Grenzgebiete 89) Springer-Verlag, Berlin, Heidelberg, New York, 1977.

[Mal68] B. Malgrange: Analytic spaces. In: Topics of Several Variables. L'Enseignement Math. Monographie 14, Genève, 1968, pp. 1-28.

[Mar82] J. Martinet: Singularities of Smooth Functions and Maps. (London Math. Soc. Lecture Note Series 58) Cambridge University Press, Cambridge, 1982.

[Mil63] J. Milnor: Morse Theory. (Ann. of Math. Studies 51) Princeton University Press, Princeton, 1963.

[Mil65] J. Milnor: Topology from the Differentiable Viewpoint. The University Press of Virginia, Charlottesville, 1965.

[Mil68] J. Milnor: Singular Points of Complex Hypersurfaces. (Ann. of Math. Studies 61) Princeton University Press, Princeton, 1968.

[MK71] J. Morrow, K. Kodaira: Complex Manifolds. Holt, Rinehart, and Winston, New York, 1971.

[Mum76] D. Mumford: Algebraic Geometry I, Complex Projective Varieties. Springer-Verlag, Berlin, etc., 1976.

[Nar66] R. Narasimhan: Introduction to the Theory of Analytic Spaces. (Lecture Notes in Math., Vol. 25) Springer-Verlag, Berlin, etc., 1966.

[Nar85] R. Narasimhan: Analysis on Real and Complex Manifolds. Third printing. North-Holland, Amsterdam, New York, Oxford, 1985.

[Orl76] P. Orlik: The multiplicity of a holomorphic map at an isolated critical point. In: Real and Complex Singularities (P. Holm, ed.), Proc. Nordic Summer School, Oslo 1976, Sijthoff & Noordhoff, Alphen a/d Rijn 1976, pp. 405-474.

[RK82] W. Rothstein, K. Kopfermann: Funktionentheorie mehrerer komplexer Veränderlicher. B. I.-Wissenschaftsverlag, Mannheim, 1982.

[Ste51] N. Steenrod: The Topology of Fibre Bundles. Princeton University Press, Princeton, 1951.

[SZ88] R. Stöcker, H. Zieschang: Algebraische Topologie. Teubner Verlag, Stuttgart, 1988.

[Voi80] E. Voigt: Marokkanische Zöpfe: eine mathematische Bastelanleitung. Diplomarbeit, Bonn, 1980.

[War83] F. Warner: Foundations of Differentiable Manifolds and Lie Groups. (Graduate Texts in Math., Vol. 94) Springer-Verlag, New York, etc., 1983.

[Waj80] B. Wajnryb: On the monodromy group of plane curve singularities. Math. Ann. 246, 141–154 (1980).

[Wel80] R. O. Wells: Differential Analysis of Complex Manifolds. (Graduate Texts in Math. 65) Springer-Verlag, New York, etc., 1980.

[Whi72] H. Whitney: Complex Analytic Varieties. Addison-Wesley, Reading, Massachusetts, 1972.

[Wol64] J. Wolf: Differentiable fibre spaces and mappings compatible with Riemannian metrics. Michigan J. Math. 11, 65–70 (1964).

Index

Titles in This Series